EPSL FRONTIERS

COLLECTION 2004

EPSL FRONTIERS
Series Editor: A. Halliday (Oxford University)

ISSN: 1573-207X

Volume 1 (2004): Collection 2002–2003
Volume 2 (2005): Collection 2004

EPSL FRONTIERS

COLLECTION 2004

Edited by

A. HALLIDAY

Oxford University, Oxford, U.K.

2005

ELSEVIER

Amsterdam – Boston – Heidelberg – London – New York – Oxford
Paris – San Diego – San Francisco – Singapore – Sydney – Tokyo

ELSEVIER B.V.
Radarweg 29
P.O. Box 211, 1000 AE Amsterdam
The Netherlands

ELSEVIER Inc.
525 B Street, Suite 1900
San Diego, CA 92101-4495
USA

ELSEVIER Ltd
The Boulevard, Langford Lane
Kidlington, Oxford OX5 1GB
UK

ELSEVIER Ltd
84 Theobalds Road
London WC1X 8RR
UK

First edition 2005
The articles in this collection were earlier published in Earth and Planetary Science Letters (Volumes 219—226)

ISBN: 0-444-52051-1

Transferred to digital print 2008

Printed and bound in Great Britain by CPI Antony Rowe, Eastbourne

Preface

In this second volume of Frontiers articles we have, as usual, a broad array of fascinating topics. The first four papers focus on the Sun, the origin of the solar system and the early history of the Earth. The fifth, on komatiites, is also relevant to the early Earth but also forms a nice transition to three articles that focus on subduction and mantle processes. The final three articles focus on changes in our surface environment including biogeochemical cycles, the evolution of hominids and, finally, the future threat posed by near Earth asteroids.

A repeated issue in cosmochemistry and planetary science is whether the planetary objects of the solar system have similar or distinct isotopic compositions and relative chemical proportions to those found in the Sun. There are known explicable differences but the lack of knowledge of the isotopic compositions of oxygen and chromium, to name but two examples, severely limit solar system models. Our volume starts with an article by Roger Wiens, Peter Bochsler, Don Burnett and Robert Wimmer-Schweingruber who provide us with an update on the composition of the Sun and solar wind. This is a timely contribution given the eagerly anticipated results from the samples returned in the Genesis Mission.

A subject of widespread current interest is the origin of our solar system and the mechanisms by which the terrestrial planets accreted. John Chambers provides a very useful overview of recent developments in our understanding of the dynamics in particular. With powerful new computational codes and more realistic equations of state there have been major steps forward in this arena.

The most common but probably most controversial objects of primitive meteorites are *chondrules*, small solidified droplets that provide evidence of some kind of rapid heating process in the circumstellar disk from which the planets were made. As Brigitte Zanda explains, despite over a century of study there exists a considerable range of explanations for these objects. Recent progress is showing signs of building a consensus over certain kinds of models however.

The early differentiation of the Earth is shrouded in mystery, in part because of the lack of a geological record. As a result, our understanding is strongly dependent on theory, experimental simulations and isotopic data. The interpretations of the experimental work have been highly contentious. These latest breakthroughs in experimental petrology and how they interface with the results of other fields are explained by Mike Walter and Reidar Trønnes.

The origin of life is one of the great unsolved problems in science. Recently there has been a wide array of interesting new ideas presented on how life may have first started, fuelled in part by a better understanding of life in extreme environments such as the deep biosphere, as well as the development of astrobiology as a discipline. Jeff Bada provides us with a status report on some the latest theories.

Komatiites have long been of fascination because they provide insights into mechanisms of melting in the early Earth. For the past few years there has been a raging controversy over whether komatiite chemistry reflects higher temperatures of melting in the Precambrian or whether the high Mg concentration is a function of wet melting during Precambrian subduction. Tim Grove and Steve Parman attempt the difficult task of presenting us with a balanced viewpoint on the current status of this highly contentious debate.

Understanding the driving forces behind ancient and modern day plate tectonics has long been a challenge. The initiation of subduction is a particularly fascinating issue. Bob Stern provides us with a summary of current understanding of the different mechanisms that are thought to be involved.

It has long been argued that water and subduction are critical to the formation of continental crust. However, the lack of essential details on the chemistry of the fluids that are released during subduction has limited understanding of how this works. Craig Manning summarises the current constraints based on experiment and theory.

Tracing the products of subduction through the mantle to evaluate the origins and degree of mixing of mantle heterogeneities has been a primary goal of mantle geochemistry for decades. With the advent of new techniques like multiple collector inductively coupled plasma mass spectrometry, it has been possible to make major progress with the development of new stable isotope tracers of recycled surface materials. Tim Elliott, Alistair Jeffcoate and Claudia Bouman review for us the latest progress along these lines with one element in their paper on lithium isotopes.

David Archer, Pamela Martin, Bruce Buffett, Victor Brovkin, Stefan Rahmstorf and Andrey Ganopolski present us with a long-term view of climate change systems. The array of new data from ice cores can be used to predict what should happen over geological timescales. Archer and co. provide a very useful overview of current thinking on the likely long-term forcing and feedbacks between ocean temperature and the biogeochemistry of the Earth system.

The role of climate in biological evolution is a fascinating subject for study in Earth science. Peter DeMenocal provides us with a review of the data that are now available for mammalian evolution in Africa over the past few million years. There is evidence that speciation and emergence of new genera including *Homo*, are linked with changes in climate or increased climate instability.

This volume finishes with a paper by Clark Chapman describing and assessing the hazard of near-Earth asteroid impacts on Earth. He points out that the probabilities of impacts are not great but at least the technologies for detecting such objects and preventing impacts are tractable.

It has been a pleasure editing these papers and I am very grateful to the authors for taking the time to explain to their Earth and planetary science community, in simple terms, the wonders and interesting new developments in their own particular sub-fields. You can look forward to plenty more high quality *Frontiers* articles in the next volume. In the mean time if you think you would like to recommend a topic of broad interest and/or a good author (perhaps even yourself), please do not hesitate to get in touch.

Alex N. Halliday
27th March 2005

Contents

Reprinted from
Earth and Planetary Science Letters 226 (2004) 275–292

www.elsevier.com/locate/epsl

Subduction initiation: spontaneous and induced

Robert J. Stern[*]

Geosciences Department, University of Texas at Dallas, Richardson, TX 75083-0688, USA

Received 5 May 2004; received in revised form 13 July 2004; accepted 10 August 2004
Available online 11 September 2004
Editor: A.N. Halliday

Abstract

The sinking of lithosphere at subduction zones couples Earth's exterior with its interior, spawns continental crust and powers a tectonic regime that is unique to our planet. In spite of its importance, it is unclear how subduction is initiated. Two general mechanisms are recognized: induced and spontaneous nucleation of subduction zones. Induced nucleation (INSZ) responds to continuing plate convergence following jamming of a subduction zone by buoyant crust. This results in regional compression, uplift and underthrusting that may yield a new subduction zone. Two subclasses of INSZ, transference and polarity reversal, are distinguished. Transference INSZ moves the new subduction zone outboard of the failed one. The Mussau Trench and the continuing development of a plate boundary SW of India in response to Indo–Asian collision are the best Cenozoic examples of transference INSZ processes. Polarity reversal INSZ also follows collision, but continued convergence in this case results in a new subduction zone forming behind the magmatic arc; the response of the Solomon convergent margin following collision with the Ontong Java Plateau is the best example of this mode. Spontaneous nucleation (SNSZ) results from gravitational instability of oceanic lithosphere and is required to begin the modern regime of plate tectonics. Lithospheric collapse initiates SNSZ, either at a passive margin or at a transform/fracture zone, in a fashion similar to lithospheric delamination. The hypothesis of SNSZ predicts that seafloor spreading will occur in the location that becomes the forearc, as asthenosphere wells up to replace sunken lithosphere, and that seafloor spreading predates plate convergence. This is the origin of most boninites and ophiolites. Passive margin collapse is a corollary of the Wilson cycle but no Cenozoic examples are known; furthermore, the expected strength of the lithosphere makes this mode unlikely. Transform collapse SNSZ appears to have engendered new subduction zones along the western edge of the Pacific plate during the Eocene. Development of self-sustaining subduction in the case of SNSZ is signaled by the beginning of down-dip slab motion, causing chilling of the forearc mantle and retreat of the magmatic arc to a position that is 100–200 km from the trench. INSZ may affect only part of a plate margin, but SNSZ affects the entire margin in the new direction of convergence. INSZ and SNSZ can be distinguished by the record left on the upper plates: INSZ begins with strong compression and uplift, whereas SNSZ begins with rifting and seafloor spreading. Understanding conditions leading to SNSZ and how hinged subsidence of lithosphere changes to true subduction promise to be exciting and fruitful areas of future research.
© 2004 Elsevier B.V. All rights reserved.

Keywords: subduction; ophiolites; collision; tectonics

---

* Tel.: +1 9728832401; fax: +1 9728832537.
E-mail address: rjstern@utdallas.edu.

0012-821X/$ - see front matter © 2004 Elsevier B.V. All rights reserved.
doi:10.1016/j.epsl.2004.08.007

Jargon Box

Asthenosphere: weak, convecting upper mantle

ARM: amagmatic rifted margin, formed in the absence of extensive volcanism during continental rifting

DSDP: Deep Sea Drilling Project.

IBM: Izu-Bonin-Mariana arc system.

INSZ: induced nucleation of subduction zones.

Lithosphere: crust and uppermost mantle, cooled by conduction. Comprises the plate of plate tectonics.

MORB: mid-ocean ridge basalt.

MRC: Macquarie Ridge Complex.

NUVEL-1: model for estimating plate velocities over 1–5 million-year timescales.

ODP: Ocean Drilling Project.

OJP: Ontong Java Plateau.

Ophiolite: a fragment of oceanic (sensu lato) lithosphere.

SNSZ: spontaneous nucleation of subduction zones.

SSZ: supra-subduction zone (broadly, a convergent margin tectonic setting).

VRM: volcanic rifted margin, formed by volcanism during continental rifting.

Wadati-Benioff Zone: inclined zone of seismicity that marks the descending slab, named after the Japanese and US scientists who first recognized these features.

---

1. Earth: the subduction planet

Earth is a spectacularly unusual planet and one of its most remarkable features is the plate tectonic system. Missions to other planets reveal that ours is the only planet in the solar system with subduction zones and plate tectonics [1]. The unique nature of plate tectonics on Earth is equivalent to saying that only Earth has subduction zones [2]. In spite of this singularity, there are fundamental misconceptions that concern aspects of plate tectonics and mantle convection. Not only are these wrong, they are deeply embedded prejudices of many earth scientists that continue to be taught to students. The most important misconception is that mantle convection moves the lithosphere (see 'Jargon Box'), dragging the plates as it moves. This is repeatedly shown in introductory textbooks. In fact, Earth's mantle convects mostly because cold lithosphere sinks at subduction zones [3] with mantle plumes representing a '...clearly resolved but secondary mode of mantle convection' ([4], p. 159). The base of the continents may be dragged by circulating mantle (continental undertow of [5]), but the pioneering conclusion of Forsyth and Uyeda [6] that the excess density of the lithosphere in subduction zones drives the plates continues to be

supported by geodynamicists [7–9]. This negative buoyancy results from the small increase in density that silicates experience as temperature decreases, coupled with the fact that the thermal lithosphere thickens as it ages. Lithosphere becomes denser than the underlying asthenosphere within ~20–50 million years after it forms [7,10]. Density and mass excess continue to increase after this time, and subduction is Earth's way of returning to equilibrium by allowing great slabs of old, dense lithosphere to sink beneath underlying asthenosphere.

There is a consensus among geodynamicists that the sinking of cold, gravitationally unstable lithosphere drives the plates and indirectly causes mantle to well up beneath mid-ocean ridges. Some estimate that 90% of the force needed to drive the plates comes from the sinking of lithosphere in subduction zones, with another 10% coming from ridge push [11]. Cenozoic plate motions are well predicted from the distribution of age—and thus the mass excess—in subduction zones [9,12]. Mantle tomography shows that subducted lithosphere may sink through the 660-km discontinuity and into the deep mantle [13,14]—striking demonstration that the lithosphere as it ages and cools progressively develops a density excess that takes as long to dissipate as it does to develop. In recognition of the fact that plate motions are passive responses to sinking of the lithosphere at subduction zones, it is more accurate to describe Earth's geo-dynamic regime as one of 'subduction tectonics' rather than 'plate tectonics'.

In spite—or, perhaps, because—of the seity of Earth's regime of subduction tectonics, we have much to learn about the physics that allow this remarkable mode of planetary convection [1] and when this began. The intention here is to approach the problem by examining Cenozoic examples of subduction initiation. The hope is that a focus on simple models and the clearest examples will encourage teams of geologists, geophysicists and geodynamicists to attack the problem from new perspectives.

2. When did subduction begin?

To understand how new subduction zones form today, we must also consider when and why this tectonic style was established on Earth. It is widely acknowledged that the early Earth was hotter and that there was more mixing of the mantle [1], but this does not require plate tectonics. If the plates ultimately move because they are dense enough to sink in subduction zones and they are sufficiently dense because they are cold, then the much higher heat flow in the Archean required correspondingly more time for the lithosphere to cool, thicken and become gravitationally unstable. The relatively buoyant nature expected for Archean oceanic lithosphere is amplified by the fact that higher heatflow at this time should have resulted in increased melting and thicker oceanic crust [15]. Pre-1.0 Ga oceanic crust is inferred to have been much thicker than modern oceanic crust [16]. Crust is much less dense than asthenosphere, so thicker crust strongly counter-acts density excesses produced by lithospheric mantle. Both of these conclusions mitigate against the oper-ation of plate tectonics on the early Earth [17]. Such logic is supported by the absence of plate tectonics on Venus, which may be a good analogue for tectonics on the Earth during the Archean.

When plate tectonics began can also be examined using the ophiolite record, which is an unequivocal index of plate tectonic activity. Ophiolites are frag-ments of oceanic lithosphere tectonically emplaced on continental crust. These assemblages testify to sub-

duction and plate tectonics in two ways: their formation requires seafloor spreading and their emplacement requires plate convergence. The age distribution of ophiolites has been interpreted to suggest that seafloor spreading began in the late Archean [18], but evidence for Archean ophiolites is sparse and often controversial. Unequivocal ophiolites of this age are rare and the abundance of well-preserved Archean supracrustal sequences does not support contentions that the paucity of such ancient ophiolites is a preservation problem. There is evidence for generation and emplacement of ophiolites at about 1.95–2.0 Ga [19,20], but there was a long period after that for which little evidence for ophiolite formation and emplacement is preserved. It was not until Neoproterozoic time, ~900 Ma, that unequivocal ophiolites were produced, emplaced and abundantly preserved [21], and since that time ophiolites are ubiquitous in the geologic record. Thus, the ophiolite record suggests that seafloor spreading and creation of oceanic lithosphere followed by convergence of oceanic lithosphere may have occurred for brief periods at the end of the Archean and in the Paleoproterozoic, but most of the pre-Neoproterozoic geologic record lacks evidence for plate tectonics and subduction. Davies [7] argued from simple physics that the Earth did not cool sufficiently to allow the lithosphere to subduct until about 1 Ga. It is not clear what Earth's tectonic style was before the modern episode of plate tectonics, but Sleep [15] recognized three modes for convection in silicate planets: magma ocean, plate tectonics and stagnant lid. He noted that plate tectonics requires that a planet be in a delicate balance of thermal states. Plate/subduction tectonics can be shut down either by ridge lock, when the Earth's mantle is too cool to produce melt by adiabatic decompression, or by trench lock, which happens when Earth's interior is too hot and oceanic crust is too thick to subduct. The ophiolite record is thus consistent with geodynamic arguments that our planet wasn't cold enough for subduction to be securely established as the dominant tectonic mode until relatively recently.

3. Induced nucleation of subduction zones

Once we recognize that plate tectonics is the surface expression of a planet with subduction zones, and that these two linked phenomena are unusual among planetary bodies as well as through Earth history, it is easier to appreciate the difficulty of starting subduction zones. This essay focuses on the two ways that subduction must be able to start: spontaneous and induced (Fig. 2). In the case of spontaneous nucleation of a subduction zone, gravitationally unstable lithosphere collapses into the asthenosphere, whereas in the case of induced nucleation of a subduction zone, existing plate motions cause compression and lithospheric rupture. If it is true that the modern episode of plate tectonics began sometime in the Precambrian, and that the excess density of lithosphere drives plate motions, then spontaneous nucleation of at least one subduction zone is required to first set the plates in motion.

There is no question but that realistic numerical models are needed to understand subduction initiation. Still, much can be learned by focusing on what has and has not happened in the recent past, particularly since about half of the active subduction zones began in Cenozoic times [22]. The following discussion returns to these themes, emphasizing Cenozoic examples of subduction zone formation (locations of examples shown in Fig. 1). These examples are used because the geometric relationships of lithospheres, relevant plate motions and sequences of events can be reconstructed with relative confidence. Pre-Cenozoic examples are avoided because the evidence for subduction initiation is more obscure and reconstructions more conjectural the farther back in the geologic record we look.

Mode of initiation cannot be inferred from a subduction zone once it becomes 'self-sustaining' [22], but the two modes of subduction initiation should have distinctly different beginnings. For induced nucleation of a subduction zone (INSZ), the plates are already converging before the subduction zone forms. In one of its simplest forms, INSZ results from continued plate convergence after a subduction zone fails due to collision, the attempted subduction of buoyant crust [23]. A variation on this theme is when a component of convergence begins along a transform plate boundary as a result of a change in location of the associated plates' Euler pole. Fig. 2 shows two ways that continued plate convergence may yield a new subduction zone. In the case labelled 'Transference', a buoyant crustal block enters the

subduction zone and causes it to fail, and in the process is sutured to the original hanging wall of the subduction zone. Evidence of this event is preserved as a tectonic suture and an accreted terrane. Plate convergence may continue as a result of lithosphere sinking elsewhere along strike of the convergent plate margin, with the result that the oceanic lithosphere outboard of the collision zone ultimately ruptures to become a new subduction zone. The new site of subduction is transferred away from the collision site. A good example of this process can be seen in the formation of the Mussau Trench in the Western Pacific (Fig. 1), which may be forming in response to collision of thick crust of the Caroline Ridge with the Yap Trench farther west [24]. The Pacific plate continues to move west relative to the Philippine Sea plate and this convergence is partially accommodated by thrusting of the Pacific plate over the Caroline plate. Hegarty et al. [25] infer that about 10 km of Caroline lithosphere has been thrust beneath the Pacific plate. There is seismicity and structural evidence for thrust faulting but no arc-related igneous activity. Subduction-related igneous activity may commence if shortening continues and the Caroline plate descends to asthenospheric depths (magmatic arcs are associated with subduction zones beneath them at a depth of 65–130 km [26]).

A more problematic example of transference comes from the collision of India with Eurasia. India first collided with Eurasia in Eocene time but continues to converge. This caused widespread deformation, including thrusting in the Himalayas, crustal thickening beneath Tibet and 'tectonic escape' of portions of Eurasia to the southeast of the collision zone. A new subduction zone outboard of the colliding blocks has not formed but appears to be in the making. There has been a broad zone of deformation and seismicity in the Indian Ocean between India and Australia since late Miocene time (8.0–7.5 Ma) [27] (stippled zone in Fig. 1). In contrast to most intraplate regions, many earthquakes of magnitude 6 and 7 occur there, and that region may someday evolve into a new subduction zone.

The fact that a new subduction zone has not yet formed outboard of the Indo–Asian collision emphasizes the difficulty of nucleating subduction zones in old, cold oceanic lithosphere. It takes less work to deform large parts of continental Asia than it does to

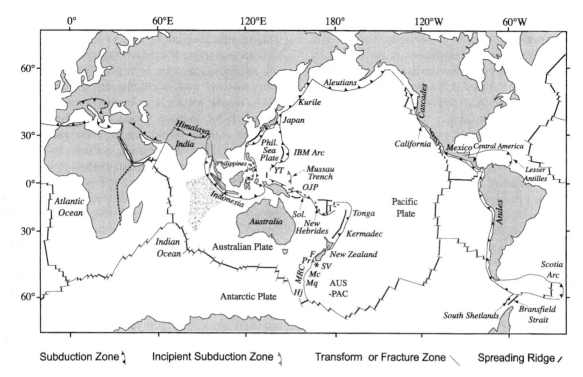

Fig. 1. Map showing location of subduction zones and where new subduction zones described in text have formed or are forming. YT=Yap Trench, Sol.=Solomon Arc, OJP=Ontong-Java Plateau. Stippled area in Indian Ocean between India and Australia is diffuse plate boundary of [88]. Area south of New Zealand is part of the MRC and Fiordland (F), including associated deeps: Puysgur (Pr), McDougall (Mc), Macquarie (Mq) and Hjort (Hj). General location of the pole of rotation for the Australian and Pacific plates is also shown (AUS–PAC).

start a new subduction zone in the Indian Plate. This is consistent with geodynamic models [28,29], which indicate the difficulty of rupturing intact oceanic lithosphere by compression. Successful nucleation of a subduction zone seems to require a pre-existing fault or other zone of lithospheric weakness. Note

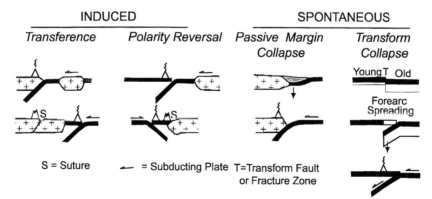

Fig. 2. General classes, subclasses and examples of how subduction zones form. See text for discussion.

that these models are solved in two dimensions, where orientation of the pre-existing fault is not important; in nature, the orientation of the fault with respect to the principal stress axis is very important. This may be the reason that a new subduction zone has not yet formed SE of India: no pre-existing faults have the proper orientation.

Toth and Gurnis [30] re-examined the issue of how subduction zones are born. Their numerical model imposed a convergence velocity of 2 cm/year on oceanic lithosphere with a fault that dipped 30°. Unless one plate is forced down at a rate of 1 cm/year or more, subduction is unlikely to become self-sustaining, because slower rates allow thermally induced density effects in the downgoing slab to dissipate. They also found that INSZ was accompanied by large amounts of uplift in the forearc, with the surface farther to the rear being strongly depressed. These predictions—early compression and uplift—are characteristics of INSZ. This response is also seen in physical models [31].

Another type of INSZ is polarity reversal (Fig. 2). Like transference, polarity reversal is triggered by buoyant crust entering the trench, but differs from transference, in that the new subduction zone nucleates in the originally overriding plate. The term 'polarity reversal' refers not only to the fact that the subduction zone dip changes direction, but also that the overriding and subducting plates exchange roles. The best Cenozoic example of INSZ polarity reversal is the Solomon arc, the result of the attempted subduction of the Ontong Java Plateau (OJP). OJP is the largest of the world's large igneous provinces, being the size of Alaska and with crust averaging 36 km thick [32]. This is much thicker than the ~20 km expected to result in subduction zone failure [32,33]. The OJP arrived at the Vitiaz Trench on the north side of the Solomon arc between 4 and 10 Ma ago [34,35], causing this south-dipping subduction zone to fail. Continued convergence between the Australian and Pacific plates forced a new subduction zone along the south flank of the Solomon Islands ~4 Ma ago [34,36].

The geodynamics of INSZ polarity reversal for arcs have not been quantitatively modeled, but upper plate responses are likely to be similar to those modelled for INSZ transference. Early stages should show upper plate uplift, thrusting and other expres-

sions of compression. The major difference may be that, because back-arc regions have elevated thermal regimes as a result of subduction-related advection of hot asthenosphere, this lithosphere should be relatively thin and weak. Consequently, polarity reversal INSZ should require less compression than transference INSZ because young, thin lithosphere of back-arc regions should rupture more readily than the cold, thick and strong oceanic lithosphere outboard of the trench. This suggestion is supported by the fact that subduction on the south side of the Solomons began almost immediately upon arrival of the OJP at the Vitiaz Trench, whereas the Mussau Trough and mid-Indian plate deformation zones are still in relatively early stages of subduction zone formation.

The continuing reorganization of the Pacific–Australian plate boundary south of New Zealand is a good example of INSZ. The tectonics of this boundary are similar to that discussed by Casey and Dewey [37], whereby relatively minor adjustments in relative plate motion convert a transform fault into a convergent plate boundary. The effect is magnified the closer the transform is to the original pole of rotation, which in the case of the Macquarie Ridge Complex (MRC; discussed below) is only 1500 km or so distant (Fig. 1). With this geometry, small changes in the location of the Euler pole have large effects on the tectonic expression of the plate boundary. The reasons for this reorganization are not known, but it may be a far-field response to collisions along the northern margin of the Australian plate.

The Australian and Pacific plates are now obliquely converging. In the Puysegur Ridge region, just south of New Zealand (Fig. 1), the NUVEL-1 plate motion model predicts ~3–3.5 cm/year relative plate motion trending N50–60°E. This is 30–45° oblique to the N15–20°E trending plate boundary [38]. The plate boundary is locally changing from a dextral strike-slip transform fault into a subduction zone [39]. A region of deformation, the MRC, parallels the plate boundary. The MRC consists of a shallow ridge and adjacent deeps (Puysegur, McDougall, Macquarie and Hjort deeps) and is continuous from Fiordland of southernmost New Zealand (Fig. 1) to the triple junction in the south. Earthquake focal mechanisms and epicenters give strike-slip solutions except under Fiordland and the adjacent Puysegur Ridge region, which have thrust-type earthquakes [40].

The Puysegur Ridge in particular preserves good evidence for a transpressional strike-slip environment that is evolving into a subduction zone. There is a trench in the west, where the Australian plate under-thrusts the Pacific plate [41]. The Puysegur Trench has an average depth of 5500 m, with a maximum of 6300 m. The trench shallows progressively south-wards, terminating at 49°50′ S. Fifty kilometers or so to the east, Puysegur Bank is cut by an elongate trough containing the strike-slip Puysegur Fault. Thrust and strike-slip systems partition boundary-parallel and boundary-normal components of oblique convergence. Beneath Fiordland, intermediate-depth earthquakes define a Wadati-Benioff zone that dips steeply east [42]. The deepest earthquakes at 150 km mark the northern end of the Puysegur–Fiordland subduction zone. Solander volcano on the northern Puysegur Ridge may be the magmatic expression of the subducted Australian plate descending below the ~125-km threshold depth for arc magmatism. At least 350 km of the Australian plate has been subducted at the Puysegur Trench [43]. Given this slab length and assuming that convergence was constant at 3–3.5 cm/year, subduction began at about 10–12 Ma; this is consistent with the region's history of uplift [44]. The Puysegur subduction system may be propagating southward from a mature subduction zone near New Zealand to an incipient stage farther south [45].

The sequence of events associated with INSZ along the Puysegur Ridge is well preserved and includes a record of uplift and wave-cut planation of the overriding plate followed by subsidence to present water depths of down to 1.5 km. This sequence qualitatively agrees with geodynamic models for INSZ, which predict the sequence compression-uplift-subsidence as lithospheric thrusting progresses to true subduction [22].

4. Spontaneous nucleation of subduction zones

A recent comprehensive overview of subduction initiation from the perspective of numerical modelling by Gurnis et al. [22] concludes that subduction initiation appears to have been forced in all known

Cenozoic cases. These arguments depend heavily on a variety of assumptions, most importantly for litho-spheric rigidity. The validity of this conclusion for pre-Cenozoic subduction zones is suspect from the perspective of the chicken and the egg: subduction must have begun spontaneously at least once without a priori plate convergence in order to begin the plate tectonic regime. The following discussion focuses on geologic evidence in support of the spontaneous nucleation of subduction zones (SNSZ). SNSZ begins when old, dense lithosphere spontaneously sinks into the underlying asthenosphere. Lithospheric collapse at first does not lead to a change in plate motion, but at some point sinking lithosphere develops a down-dip component of motion that pulls the plate towards the subduction zone; this is when true subduction begins and plate motion changes. In contrast to INSZ, SNSZ affects the entire plate margin. Because oceanic lithosphere gets stronger as well as denser with age, SNSZ requires a lithospheric weakness—such as a fracture zone—to overcome lithospheric strength and allow collapse.

SNSZ can best be distinguished from INSZ in the rock record by the sequence of early events preserved in the upper plate. Whereas early INSZ events are strongly compressional, early SNSZ events are strongly extensional. This dichotomy is dictated by the distinctly different ways in which subduction zones nucleate in the two general cases: INSZ by plate convergence leading to lithospheric failure and SNSZ by lithospheric failure leading to plate convergence.

Geodynamic models are not yet able to dynamically model SNSZ but provide important constraints and perspectives. The following paragraphs are intended to demonstrate the likelihood that SNSZ is an unusual yet critical process. This includes presenting the geologic and geodynamic evidence against passive margin collapse, emphasizing similarities between lithospheric delamination and SNSZ, and exploring the signifi-cance of ophiolites. Finally, we revisit the formation of the Izu-Bonin-Mariana (IBM) and Tonga-Kermadec convergent margins in mid-Eocene time as the best example of SNSZ in the Cenozoic record.

4.1. Passive margin collapse

Conversion of passive margins into subduction zones is a hypothesis that is more widely accepted by the geoscientific community than the evidence warrants. There are no Cenozoic examples of a passive continental margin transforming into a convergent margin, in spite of the fact that seafloor adjacent to SE Africa, NW Africa and eastern N. America is about 170 Ma old. The broad—and largely uncritical—acceptance that this hypothesis enjoys mostly reflects the fact that the conversion of a passive margin into an active margin is required for the closing phase of the 'Wilson cycle', the name given to the repeated opening and closing of ocean basins [46,47]. It will be shown later that, if the third dimension is considered, the location of the subduction zone responsible for closing the ocean can be quite different than suggested by 2D tectonic cartoons.

Because of the central role that passive margin collapse is thought to play in the Wilson cycle, how this could happen is of considerable interest to geodynamicists. Models for this entail both INSZ and SNSZ, although the reversal of plate motion implicit in the Wilson cycle concept would seem to require SNSZ. This is also difficult to reconcile with our understanding of lithospheric strength. Cloetingh et al. [48] concluded that transformation of passive continental margins into active plate margins was unlikely because by the time that the adjacent oceanic lithosphere becomes dense enough to founder, it is too strong to fail. Their models indicated that the overall strength of old oceanic lithosphere at passive margins is too great to be overcome, even with the additional stresses of sediment loading. This conclusion must be modified because water weakens lithosphere [49]. Situations in which the strength of old, dense passive margin lithosphere could be overcome were examined by Erickson [50]. Using elastic-plate bending theory, he concluded that sediment loading could cause the oceanic lithosphere to collapse if a previous zone of weakness such as a fault was reactivated, possibly as a result of a change in plate motion. Another way to overcome lithospheric strength was modeled by Kemp and Stevenson [51], who noted that lithosphere is significantly weaker under tension than compression. They argued that oceanic and continental lithospheres could be stretched even in an environment of moderate compression (due to ridge push) as a result of greater subsidence of oceanic lithosphere just outboard of the continent-ocean lithosphere boundary. Tensional stress could lead to rifting between con-

tinental and oceanic lithospheres to allow old, dense oceanic lithosphere to sink asymmetrically and asthenosphere to well up beneath the rift and flood over the sinking oceanic lithosphere. Note that both of these studies [50,51] emphasize early sinking, early extension or both to overcome plate strength.

These models need refinement because we now know that 'transitional crust' between continent and ocean can be quite variable. Transitional crust can range in composition and strength from 'volcanic rifted margins' (VRM) with unusually thick sections of strong basaltic crust [52], to amagmatic rifted margins (ARM) with abundant serpentinite dominating a weak crust [53]. It is likely that ARM lithosphere is weak due to abundant faults and serpentinites, whereas VRM lithosphere should strongly weld oceanic and continental lithospheres. The differing magma budgets of these two crustal types indicate that associated lithosphere will also vary from depleted mantle beneath VRMs and undepleted mantle beneath ARMs. It seems likely that the diminished lithospheric strength and denser nature of undepleted (Fe-rich) mantle beneath ARMs favor these over VRMs as sites of passive margin collapse. Similar comments apply to the "oceanic plateau model" for subduction initiation [54]. Regardless of these subtleties, extreme skepticism is warranted regarding the efficacy of passive margin collapse SNSZ, in light of geodynamic considerations and the absence of Cenozoic examples.

4.2. Similarities between lithospheric delamination and SNSZ

The SNSZ models of Erickson [50] and Kemp and Stevenson [51] have strong similarities to the hypothesis of lithospheric 'delamination' as first advanced to explain uplift of the Colorado Plateau [55]. Bird argued that subcontinental mantle lithosphere was gravitationally unstable relative to underlying asthenosphere. He concluded that delamination occurs when asthenospheric mantle can rise into and above the level of the lithosphere. Once this happens, the lithosphere will peel away and sink, resulting in '...wholesale replacement of cold mantle by hot in a geologically short time...' ([55], p. 7561). In spite of uncertainties, the overall concept is broadly accepted, mostly because it explains so many otherwise inexplicable observations. These include uplift of the Colorado and Tibetan plateaux [56,57] and the Altiplano [58], inclined seismic zones beneath continental interiors that are not associated with active subduction [59,60], and changes in melt source from lithospheric mantle to asthenosphere accompanying rapid uplift [61].

The concept of delamination continues to evolve, with one mechanism involving the 'peeling off' and sinking of lithospheric mantle as envisaged by Bird and another the sinking of gravitationally unstable mantle and lower crust by Rayleigh-Taylor instability [62]. Nevertheless, the general process whereby denser subcontinental lithosphere can sink into underlying asthenosphere provides a useful analogy for SNSZ because it supports the idea that gravitationally unstable lithosphere can collapse into underlying asthenosphere.

4.3. Ophiolites, the Izu-Bonin-Mariana forearc and SNSZ

Consensus exists that ophiolites are fragments of oceanic lithosphere obducted onto continental crust, but the tectonic setting in which they form is often controversial. In the late 1960s and early 1970s, ophiolites were thought to have originated at mid-ocean spreading ridges but geochemical evidence increasingly compels the conclusion that most ophiolites form above subduction zones [63,64]. Until recently, such ophiolites were generally ascribed to back-arc basin settings, principally in order to reconcile 'supra-subduction zone' (SSZ) chemical compositions with crustal structure indicating seafloor spreading. This combination is found among active systems solely in back-arc basins. However, the following discussion argues that most—if not all—well-preserved ophiolites formed in forearcs during the early stage of SNSZ, and explores the implications of this conclusion for understanding SNSZ.

The evolution of thinking about SSZ ophiolites is summarized by Pearce [64], who coined the term and who has long played a key role in understanding them. This somewhat non-specific term acknowledges that the chemical characteristics of these igneous rocks indicate subduction zone magmatism but does not specify the precise tectonic setting. The term is

defined [65] thus: "supra-subduction zone (SSZ) ophiolites have the geochemical characteristics of island arcs but the structure of oceanic crust and are thought to have formed by sea-floor spreading directly above subducted oceanic lithosphere. They differ from 'MORB' ophiolites in their mantle sequences, the more common presence of podiform chromite deposits, and the crystallization of clinopyroxene before plagioclase which is reflected in the high abundance of wehrlite relative to troctolite in the cumulate sequence. Most of the best-preserved ophiolites in orogenic belts are of this type". Subsequent studies confirm and expand on these observations, such as the fact that boninites—high-Mg andesites formed by hydrous melting of harzburgite—are common in some ophiolites, but are unknown from modern spreading ridges or older seafloor and do not erupt today.

The IBM arc system in the western Pacific is increasingly recognized as key for understanding how SSZ ophiolites form. This reflects the fact that the early record of this convergent margin is well preserved in IBM forearc crust, and that this crust is well exposed with limited sediment cover and no accretionary prism [66]. Forearc crust is readily studied on island exposures as well as in submerged realms by drilling, dredging and diving on the inner wall of the trench; serpentinite mud volcanoes bring up fragments of forearc and crust [67]. DSDP and ODP drilling in the 1970s and 1980s encouraged confidence that the limited exposures of pillowed tholeiites and boninites of Eocene age on islands in the Bonin and S. Mariana arc segments along with peridotite-gabbro-metavolcanic exposures in the IBM trench are representative of lithospheric structure for the entire IBM forearc [68] (although minor amounts of trapped older crust have been identified [69]). A consensus exists that the IBM forearc was created in a SSZ environment about the time that subduction began along the IBM trench [68]. Numerical models reveal the sequence of events leading to upper plate spreading as subduction begins [22,70]. A number of important controversies persist, including whether or not a transform fault was in the proper position and orientation to be converted into a subduction zone [70–72]. The bend in the Emperor–Hawaii seamount chain, which formed at about 43 Ma was an important if indirect argument for IBM SNSZ [12], but this may also record southward movement of the Hawaiian hotspot [73]. It was also suggested that boninites are abundant in the IBM forearc because subduction initiation occurred on the flanks of a mantle plume [74]. Geodynamic modelling results led to conclusions that IBM subduction was induced, not spontaneous [22,70].

In spite of these and other uncertainties, the IBM arc system is the best example of how subduction zones form by a process that entails abundant igneous activity in a broad region of the forearc. It is important to note that limited data indicate that the Tonga-Kermadec convergent margin began about the same time as the IBM arc system, about 45–50 Ma [75], satisfying a critical test of the SNSZ model: that subduction begins all along the 'downstream' margins of the affected plate (in this case, the Pacific plate) at about the same time. It should be noted that this does not preclude the possibility of INSZ for

these systems, which could also cause plate-scale initiation of subduction.

The point to be stressed is that the IBM and Tonga-Kermadec forearcs represent ophiolites waiting to be obducted. These 'all-but' ophiolites have the structure, crystallization sequence and chemical composition observed in well-preserved ophiolites, and are in a tectonic setting that favors obduction of coherent lithospheric slices once a continent or other tract of buoyant crust clogs the subduction zone (Fig. 3). I extrapolate from this that most well-preserved SSZ ophiolites represent forearc lithosphere that formed during SNSZ events and which are often emplaced during collision events. Shervais [76] explicitly identified the classic Mesozoic ophiolites of Cyprus, Oman and California as representing forearcs that formed during episodes of subduction

How to Emplace an Ophiolite

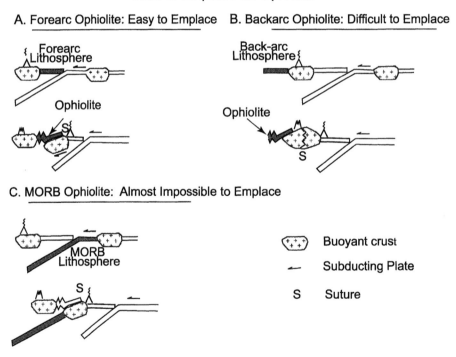

Fig. 3. Cartoon showing the relative feasibility of emplacing oceanic lithospheres created in forearc, back-arc basin and mid-ocean ridge settings. (A) It is relatively easy to emplace oceanic lithosphere of forearcs. Subduction of buoyant material leads to failure of subduction zone. Isostatic rebound of buoyant material follows, with ophiolite on top. (B) It is difficult to emplace back-arc basin oceanic lithosphere. Compression and shortening across the arc will lead to uplift of the arc. (C) It is almost impossible to emplace true MORB crust at a convergent plate boundary. Sediments and fragments of seamounts may be scraped off of the downgoing plate, but decollements do not cut deeply into the subducting lithosphere.

initiation. He argued that these and most—but not all—ophiolites form as outlined in the 'subduction zone infancy' model of [77]. This is shown in the right panel of Fig. 2 and details are outlined in Fig. 4. Some object to Tethyan ophiolites being infant arc products because they are not associated with a magmatic arc [78]. However, the portion of the Tethys being subducted was relatively narrow, so that collision occurred and subduction was arrested before the sinking slab reached the threshold depth (60–135 km) for mature arc magmagenesis.

4.4. Lithospheric collapse along transform faults and SNSZ

The model for SNSZ of the IBM arc system was first suggested by Natland and Tarney [79] and Hawkins et al. [80] and developed more fully by Stern and Bloomer [77]. Two lithospheres of differing density are juxtaposed across a lithospheric weakness, either a transform fault or fracture zone. Differences in density, elevation of the seafloor and depth to asthenosphere for the juxtaposed lithospheres are

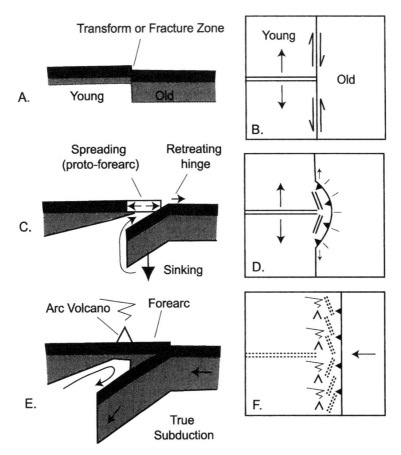

Fig. 4. Subduction infancy model of [77], modified to show the third dimension. Left panels are sections perpendicular to the plate boundary (parallel to spreading ridge) and right panels are map views. (A) and (B) show the initial configuration. Two lithospheres of differing density are juxtaposed across a transform fault or fracture zone. (C, D) Old, dense lithosphere sinks asymmetrically, with maximum subsidence nearest fault. Asthenosphere migrates over the sinking lithosphere and propagates in directions that are both orthogonal to the original trend of the transform/fracture zone as well as in both directions parallel to it. Strong extension in the region above the sinking lithosphere is accommodated by seafloor spreading, forming infant arc crust of the proto-forearc. (E, F) Beginning of down-dip component motion in sinking lithosphere marks the beginning of true subduction. Strong extension above the sunken lithosphere ends, which also stops the advection of asthenosphere into this region, allowing it to cool and become forearc lithophere. The locus of igneous activity retreats to the region where asthenospheric advection continues, forming a magmatic arc.

maximized where a spreading ridge intersects the transform, and this is where the process shown in Fig. 4C and D is most likely to begin. Sinking of old, dense lithosphere is accomplished by asymmetric or 'hinged' subsidence, accelerated by mass wasting of shallower crust adjacent to the transform as structural relief across this feature increases. Recent geodynamic modeling of this process requires a small amount of convergence (~2 cm/year) to overcome plate strength to cause subduction [70]. Propagation of the spreading ridge across the transform fault may accelerate depression of dense lithosphere adjacent to the fault. When the top of the sinking lithosphere is depressed beneath the level of the asthenosphere, the latter floods over the former, further accelerating the process. The pool of asthenosphere spreads laterally as it advances over the sinking lithosphere in directions that are both orthogonal to the original trend of the transform/fracture zone as well as parallel to it (Fig. 4C,D). The rate of propagation of the asthenospheric wedge is unknown but may be on the order of propagation rates of ~5 cm/year, as suggested for lithospheric delamination [55].

Geodynamic models show that asthenosphere wells up as the lithosphere continues to sink [70]. Upwelling asthenosphere can melt due to decompression alone, and this will be stimulated by water from the sinking lithosphere. Increasing pressure and temperature on the sinking plate will result in a massive expulsion of water, setting up conditions for unusually high extents of melting. Boninites (harzburgite melts) and unusually depleted arc tholeiites (lherzolite melts) and their fractionates are the hallmarks of this phase, although normal MORB-like tholeiites may also be produced in drier regions of upwelling mantle. Hinged subsidence of sinking lithosphere results in an environment of crust formation in the hanging wall that is indistinguishable from seafloor-spreading at mid-ocean ridges. Melting will be unusually extensive during this phase, and residual mantle will be unusually depleted harzburgite, characterized by Cr-rich spinel, such as that now found in peridotite exposures of inner trench walls [81] and recovered from serpentine mud volcanoes of the IBM forearc [82]. Similar lava and peridotite compositions are characteristic of the great, well-preserved SSZ ophiolites, such as Cyprus, Oman and California [75].

The phase of hinged subsidence of dense lithosphere beneath an expanding wedge of upwelling asthenosphere associated with seafloor spreading defines the infant arc phase. The distribution, orientation and length of spreading ridges associated with the 'proto-forearc' are speculatively shown in Fig. 4D aligned obliquely to the evolving trench and in an en-echelon pattern to account for along-strike expansion of the infant arc or proto-forearc. The infant arc phase may continue for 5 or 10 million years. Strong extension will exist in the region above the sinking lithosphere as long as the hinge continues to retreat rapidly. This will continue until the sinking lithosphere develops a significant component of down-dip motion, when true subduction begins (Fig. 4E,F). It is not clear what causes the change from hinged subsidence to true subduction; Stern and Bloomer [77] slab speculated that this may reflect the increasing difficulty of transferring asthenosphere from beneath and around the subsiding lithosphere, with down-dip slab motion possibly aided by transformation of basaltic crust into eclogite [83].

Once true subduction begins, the hingeline stops retreating (or 'rolls back' more slowly, typically at ~10% of the plate convergence rate [84]) and extensional forces in the forearc diminish (Fig. 4E,F). Asthenosphere is no longer advected beneath the proto-forearc and seafloor spreading ceases there. The subjacent mantle is conductively cooled by the subducting slab and becomes lithosphere. Flow of asthenosphere must reorganize during this transition. Migration of asthenosphere from beneath the sinking lithosphere is cut off and is supplied instead from convection induced beneath the overriding plate. The magmatic axis migrates away from the trench to form a fixed (relative to the trench) magmatic arc and vertical edifice building replaces spreading. The mantle beneath the forearc quickly cools following the establishment of a true subduction regime.

The general model outlined above explains the origin of the IBM subduction zone and reconciles the structure, composition and emplacement of most ophiolites. The model suffers from the fact that this style of subduction initiation is not now occurring and so cannot be studied directly (although the eastern part of the Mendocino Fracture Zone may be in the early stages of this evolution). The model also suffers because much of the geodynamic community thinks the oceanic lithosphere is too strong to flex as called for in the model, and the concept of the 'retreating hingeline' is especially difficult to reconcile with our understanding of lithospheric strength. Numerical models for subduction initiation require plate convergence in order to overcome the resistance of the plate to bending [22,70]. The objection based on lithospheric strength may not be fatal, because oceanic lithosphere may be weakened as a result of serpentinization accompanying reactivation of deep faults by down-flexing [85,86]. Normal faulting and concave-downward flexure of the lithosphere open cracks, which must be filled with seawater, promoting serpentinization of lithospheric peridotites and further weakening the lithosphere. Evidence for this process is seen in deep earthquakes in the outer rise associated with several circum-Pacific trenches and in associated double Benioff Zones [87]. Repetition and acceleration of the sequence flexure-fracture-serpentinization may so weaken the lithosphere that it is much easier to bend than present estimates of plate strength allow.

5. Implications for reconstructing subduction initiation in the geologic record

The forgoing discussion contrasts two fundamentally different ways that subduction zones form, which can be distinguished by whether or not the pertinent plates were converging prior to nucleation of the subduction zone. Alternatively, if the interior of a single plate ruptures (as is the case in transference INSZ), the question is whether or not strong compressional stresses preceded subduction initiation. In the case of INSZ, the plate convergence or intra-plate compression precedes subduction nucleation, so that subduction nucleation localizes shortening where lithospheric strength can be overcome. SNSZ contrasts with INSZ in that plate convergence may not precede the initiation of a subduction zone. Rather, the lithosphere descends many kilometers into the mantle before the pertinent plates begin to converge. In order to understand these fundamentally different modes of formation, fundamentally different geodynamic models must be created, and these models must be tested against our understanding of the early history of a convergent margin.

Unfortunately, it is difficult to determine relative plate motions before subduction nucleation from the geologic record. The continuing controversy about W. Pacific plate paleogeography and the significance of the bend in the Emperor–Hawaii Seamount chain are good examples of such uncertainties in the relatively robust Cenozoic record. These examples do not encourage confidence that we can be sure enough about past relative plate motion and plate margin orientations to allow us to distinguish ancient INSZ from SNSZ. We may need to rely instead on the expected consequences of the two subduction initiation modes preserved in the upper plate. Because INSZ results from compression, this mode results in early reverse faulting, folding, and uplift of the overriding plate as one plate is thrust over the other. This is predicted from geodynamic numerical experiments as well as physical modelling and is observed in late Cenozoic examples such as the MRC. As a starting point, all examples of subduction initiation where the early record in the overriding plate is one of compression, thrust faulting, folding and/or uplift should be interpreted as reflecting INSZ. In contrast, because SNSZ results from gravitational instability, its earliest expressions should reflect subsidence of the descending lithospheric plate. Again, as a starting point, all examples of subduction initiation where the early record in the overriding plate is one of extension, normal faulting, seafloor spreading and/or subsidence should be interpreted as manifesting SNSZ.

Two additional points need to be emphasized. First, most large, well-preserved ophiolites represent lithosphere produced during the infant arc phase of SNSZ. This is the simplest way to reconcile ophiolite chemistries with seafloor spreading structures in a setting that favors obduction. The classic ophiolites of Cyprus and Oman in particular should be regarded as infant arc lithosphere, which were emplaced when continental crust of Africa and Arabia slid underneath Eurasia [76]. Second, the identification of an ophiolite as manifesting SNSZ does not distinguish whether the lithosphere collapsed beneath a passive margin or along a transform fault. This requires an understanding of paleogeography that is often lacking. On a global scale, however, episodes of SNSZ due to passive margin collapse should result in Wilson cycles where the paths of the plates and continents during the closing phase parallel (in reverse) their paths during the opening phase. In contrast, SNSZ due to transform margin collapse should result in closing phases orthogonal to the opening phase. Examination of the first-order tectonic changes that the Earth has witnessed over the past 120 million years suggests that an overall N–S motion of the continents around an opening and closing Tethys has been supplanted by an overall E–W motion today (Fig. 5). This suggests that the most recent mode of major SNSZ was lithospheric collapse along transform margins.

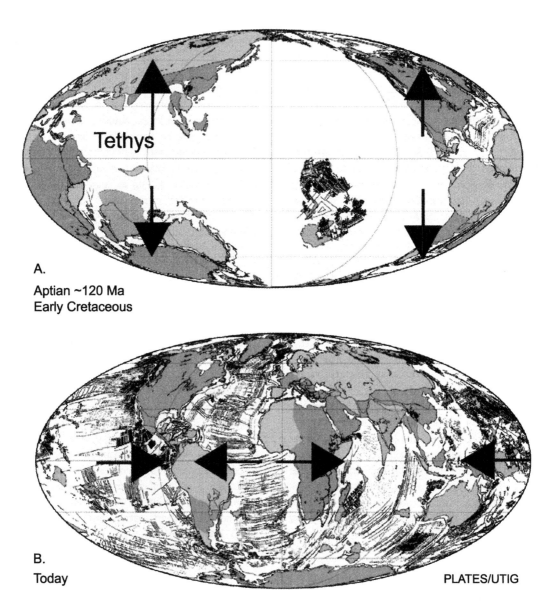

Fig. 5. Contrasting motions of the plates, early Cretaceous and now. (A) Aptian–Cretaceous (120 Ma). Plates are to a first approximation moving N–S, associated with the opening of the Tethys seaway. (B) Present. Plates are to a first approximation moving E–W. Figure courtesy of Lisa Gahagan, UT Institute of Geophysics.

6. Conclusions

We have made great advances towards understanding how INSZ occurs and what are the responses in the upper plate. Quantitative understanding of SNSZ is needed, particularly to constrain the conditions under which it can occur and how early stages of lithospheric foundering evolve to become true subduction. Such studies present great opportunities for understanding the significance of ophiolites, the history of plate motions and how the solid Earth system operates.

Acknowledgements

Thanks to T.A. Meckel (U. Texas at Austin Institute of Geophysics) for comments on the MRC, to Tom Shipley (U. Texas at Austin Institute of Geophysics) for helping me understand what material is and is not accreted across accretionary prisms, to Lisa Gahagan (U. Texas at Austin Institute of Geophysics) for plate reconstructions used to generate Fig. 5, and to Mike Gurnis (CalTech) for sharing a preprint that captures the thoughts of his group on how subduction zones nucleate. Thanks to Mike Gurnis, Jim Hawkins and Norm Sleep for thoughtful reviews, and to Trey Hargrove for editorial suggestions. My research on subduction zones would not have been possible without support of the US National Science Foundation through the MARGINS program. This is UTD Geosciences contribution # 1027.

References

[1] D.J. Stevenson, Styles of mantle convection and their influence on planetary evolution, Comptes rendus Gèoscience 335 (2003) 99–111.

[2] R.J. Stern, Subduction Zones, Reviews of Geophysics, 2002, DOI:10.1029/2001RG000108

[3] B.H. Hager, R.J. O'Connell, A simple global model of plate dynamics and mantle convection, Journal of Geophysical Research 86 (B6) (1981) 4843–4867.

[4] G.F. Davies, M.A. Richards, Mantle convection, Journal of Geology 100 (1992) 151–206.

[5] W. Alvarez, Geologic evidence for the plate-driving mechanism: the continental undertow hypothesis and the Australian–Antarctic Discordance, Tectonics 9 (1990) 1213–1220.

[6] D. Forsyth, S. Uyeda, On the relative importance of the driving forces of plate motions, Geophysical Journal of the Royal Astronomical Society 43 (1975) 163–200.

[7] G.F. Davies, On the emergence of plate tectonics, Geology 20 (1992) 963–966.

[8] D.L. Anderson, Plate tectonics as a far-from-equilibrium self-organized system, in: S. Stein, J.T. Freymueller (Eds.), Plate Boundary Zones, vol. 30, American Geophysical Union, Washington DC, 2002, pp. 411–425.

[9] C.P. Conrad, C. Lithgow-Bertelloni, How mantle slabs drive plate tectonics, Science 298 (5591) (2002) 207–209.

[10] E.R. Oxburgh, E.M. Parmentier, Compositional and density stratification in oceanic lithosphere—causes and consequences, Journal of the Geological Society (London) 133 (1977) 343–355.

[11] C. Lithgow-Bertelloni, M.A. Richards, Cenozoic plate driving forces, Geophysical Research Letters 22 (1995) 1317–1320.

[12] M.A. Richards, C. Lithgow-Bertelloni, Plate motion changes, the Hawaiian–Emperor Bend, and the apparent success and failure of geodynamic models, Earth and Planetary Science Letters 137 (1996) 19–27.

[13] S.P. Grand, R.D.V.D. Hilst, S. Widiyantoro, Global seismic tomography; a snapshot of convection in the Earth, GSA Today 7 (1997) 1–7.

[14] H. Kárason, R.D.V.D. Hilst, Constraints on mantle convection from seismic tomography, in: M.R. Richards, R. Gordon, R.D.V.D. Hilst (Eds.), The History and Dynamics of Global Plate Motion, 121, American Geophysical Union, Geophysical Monograph, Washington, D.C, 2000, pp. 277–288.

[15] N.L. Sleep, Evolution of the mode of convection within terrestrial planets, Journal of Geophysical Research 105 (E7) (2000) 17563–17578.

[16] E.M. Moores, Pre-1 Ga (pre-Rodinian) ophiolites: their tectonic and environmental implications, Geological Society of America Bulletin 114 (2002) 80–95.

[17] W. Hamilton, An alternative earth, GSA Today 13 (11) (2003) 4–12.

[18] T.M. Kusky, J.H. Li, R.T. Tucker, The Dongwanzi Ophiolite: complete Archean ophiolite with extensive sheeted dike complex, North China craton, Science 292 (2001) 1142–1145.

[19] A. Kontinen, An early Proterozoic ophiolite: the Jormua mafic–ultramafic complex, northeastern Finland, Precambrian Research 35 (1987) 313–341.

[20] D.J. Scott, H. Helmstaedt, M.J. Bickle, Purtuniq Ophiolite, Cape Smith Belt, northern Quebec, Canada: a reconstructed section of early Proterozoic oceanic crust, Geology 20 (1992) 173–176.

[21] R.J. Stern, P.R. Johnson, A. Kröner and B. Yihas, Neoproterozoic Ophiolites of the Arabian–Nubian shield, in: T. Kusky (Ed.), Precambrian Ophiolites and Related Rocks, Developments in Precambrian Geology, vol. 13, Elsevier, Amsterdam, pp. 95–128.

[22] M. Gurnis, C. Hall, L. Lavier, Evolving force balance during incipient subduction, Geochemistry, Geophysics, Geosystems 5 (2004) DOI:10.1029/2003GC000681.

[23] M. Cloos, Lithospheric buoyancy and collisional orogenesis: subduction of oceanic plateaus, continental margins, island

arcs, spreading ridges, and seamounts, Geological Society of America Bulletin 105 (1993) 715–737.

[24] S.-M. Lee, Deformation from the convergence of oceanic lithosphere into Yap Trench and its implication for early-stage subduction, Journal of Geodynamics 37 (2004) 83–102.

[25] K.A. Hegarty, J.K. Weissel, D.E. Hayes, Convergence at the Caroline–Pacific plate boundary: collision and subduction, in: D.E. Hayes (Ed.), The Tectonic and Geologic Evolution of Southeast Asian Seas and Islands: Part 2, vol. 27, American Geophysical Union, Geophysical Monograph, Washington DC, 1983, pp. 326–348.

[26] P. England, R. Engdahl, W. Thatcher, Systematic variation in the depths of slabs beneath arc volcanoes, Geophysical Journal International 156 (2004) 377–408.

[27] K.S. Krishna, J.M. Bull, R.A. Scrutton, Evidence for multi-phase folding of the central Indian Ocean lithosphere, Geology 29 (2001) 715–718.

[28] D. McKenzie, M.J. Bickle, The volume and composition of melt generated by extension of the lithosphere, Journal of Petrology 29 (1988) 625–679.

[29] S. Mueller, R.J. Phillips, On the initiation of subduction, Journal of Geophysical Research 96 (B1) (1991) 651–665.

[30] J. Toth, M. Gurnis, Dynamics of subduction initiation at preexisting fault zones, Journal of Geophysical Research 103 (B8) (1998) 18053–18067.

[31] A.I. Shemenda, Horizontal lithosphere compression and sub-duction: constraints provided by physical modelling, Journal of Geophysical Research 97 (B7) (1992) 11097–11116.

[32] C.R. Neal, J.J. Mahoney, L.W. Kroenke, R.A. Duncan, M.G. Petterson, The Ontong Java Plateau, in: M.F. Coffin, J.J. Mahoney (Eds.), Large Igneous Provinces: Continental, Oceanic, and Planetary Flood Volcanism, vol. 100, American Geophysical Union, Geophysical Monograph, Washington DC, 1997, pp. 183–216.

[33] M. Cloos, Thrust-type subduction earthquakes and seamount asperities: a physical model for seismic rupture, Geology 20 (1992) 601–604.

[34] P.A. Cooper, B. Taylor, Polarity reversal in the Solomon Islands arc, Nature 314 (1985) 428–430.

[35] E.J. Phinney, P. Mann, M.F. Coffin, T.H. Shipley, Sequence stratigraphy, structure, and tectonic history of the southwestern Ontong Java Plateau adjacent to the North Solomon Trench and Solomon Islands arc, Journal of Geophysical Research 104 (B9) (1999) 20446–20449.

[36] M.G. Petterson, T. Babbs, C.R. Neal, J.J. Mahoney, A.D. Saunders, R.A. Duncan, D. Tolia, R. Magu, C. Qopoto, H. Mahoa, D. Natogga, Geological-tectonic framework of Solomon Islands, SW Pacific: crustal accretion and growth within an intra-oceanic setting, Tectonophysics 301 (1999) 35–60.

[37] J.F. Casey, J.F. Dewey, Initiation of subduction zones along transform and accreting plate boundaries, triple-junction evolution, and forearc spreading centres—implications for ophiolitic geology and obduction, in: I.G. Gass, S.J. Lippard, A.W. Shelton (Eds.), Ophiolites and Oceanic Lithosphere, Geological Society, London Special Publication 13 (1984) 269–290.

[38] C. DeMets, R.G. Gordon, D.F. Argus, S. Stein, Effect of recent revisions to the geomagnetic reversal time scale on estimates of current plate motions, Geophysical Research Letters 21 (20) (1994) 2191–2194.

[39] L.J. Ruff, J.W. Given, C.O. Sanders, C.M. Sperber, Large earthquakes in the Macquarie Ridge Complex: transitional tectonics and subduction initiation, Pure and Applied Geophysics 128 (1989) 72–129.

[40] C. Frohlich, M.F. Coffin, C. Massell, P. Mann, C.L. Schuur, S.D. Davis, T. Jones, G. Karner, Constraints on Macquarie Ridge tectonics provided by Harvard focal mechanisms and teleseismic earthquake locations, Journal of Geophysical Research 102 (B3) (1997) 5029–5041.

[41] J.F. Lebrun, G. Lamarche, J.Y. Collot, J. Deteil, Abrupt strike-slip fault to subduction transition; the Alpine Fault-Puysegur Trench connection, New Zealand, Tectonics 19 (2000) 688–706.

[42] H. Anderson, T. Webb, J. Jackson, Focal mechanism of large earthquakes in the South Island of New Zealand, implications for the accomodation of Pacific–Australia plate motion, Geophysical Journal International 115 (1993) 1032–1054.

[43] G. Lamarche, J.-F. Lebrun, Transition from strike-slip faulting to oblique subduction: active tectonics at the Puysegur Margin, South New Zealand, Tectonophysics 316 (2000) 67–89.

[44] M.A. House, M. Gurnis, P.J.J. Kamp, R. Sutherland, Uplift in the Fiordland Region, New Zealand: implications for incipient subduction, Science 297 (5589) (2002) 2038–2041.

[45] J.-Y. Collot, G. Lamarche, R. Wood, J. Delteil, M. Sosson, J.-F. Lebrun, M.F. Coffin, Morphostructure of an incipient sub-duction zone along a transform plate boundary: Puysegur ridge and trench, Geology 23 (1995) 519–522.

[46] J.T. Wilson, Did the Atlantic close and then re-open? Nature 211 (505) (1966) 676–681.

[47] J.F. Dewey, J.M. Bird, Mountain belts and the new global tectonics, Journal of Geophysical Research 75 (1970) 2625–2647.

[48] S. Cloetingh, R. Wortel, N.J. Vlaar, On the initiation of subduction zones, Pure and Applied Geophysics 129 (1989) 7–25.

[49] K. Regenauer-Lieb, D.A. Yuen, J. Branlund, The initiation of subduction: criticality by addition of water? Science 294 (5542) (2001) 578–580.

[50] S.G. Erickson, Sedimentary loading, lithospheric flexure, and subduction initiation at passive margins, Geology 21 (1993) 125–128.

[51] D.V. Kemp, D.J. Stevenson, A tensile flexural model for the initiation of subduction, Geophysical Journal 125 (1996) 73–94.

[52] O. Eldholm, Magmatic-tectonic evolution of a volcanic rifted margin, Marine Geology 102 (1991) 43–61.

[53] J.P. Brun, M.O. Beslier, Mantle exhumation at passive margins, Earth and Planetary Science Letters 142 (1996) 161–173.

[54] Y. Niu, M.J. O'Hara, J.A. Pearce, Initiation of subduction zones as a consequence of lateral compositional buoyancy contrast within the lithosphere: a petrological perspective, Journal of Petrology 44 (2003) 851–866.

[55] P. Bird, Continental delamination and the Colorado Plateau, Journal of Geophysical Research 84 (B13) (1979) 7561–7571.

[56] L.A. Lastowka, A.F. Sheehan, J.M. Schneider, Seismic evidence for partial lithospheric delamination model of Colorado Plateau uplift, Geophysical Research Letters 28 (7) (2001) 1319–1322.

[57] P. England, G. Houseman, Extension during continental convergence, with application to the Tibetan Plateau, Journal of Geophysical Research 94 (B12) (1989) 17561–17579.

[58] X. Yuan, S.V. Sobolev, R. Kind, Moho topography in the central Andes and its geodynamic implications, Earth and Planetary Science Letters 199 (2002) 389–402.

[59] F. Prat-Chalot, R. Girbacea, Partial delamination of continental mantle lithosphere, uplift-related crust-mantle decoupling, volcanism and basin formation; a new model for the Pliocene–Quaternary evolution of the southern East-Carpathians, Romania, Tectonophysics 327 (2000) 83–107.

[60] Z. Gvirtzman, Partial detachment of a lithospheric root under the southeast Carpathians: towards a better definition of the detachment concept, Geology 30 (2002) 51–54.

[61] M. Ducea, Constraints on the bulk composition and root foundering rates of continental arcs; a California arc perspective, Journal of Geophysical Research B 107 (B11) (2002) DOI: 10.1029/2001JB000643.

[62] M. Jull, P.B. Kelemen, On the conditions for lower crustal convective instability, Journal of Geophysical Research 106 (B4) (2001) 6423–6446.

[63] J.W. Hawkins, Geology of supra-subbduction zones—implications for the origin of ophiolites, in: Y. Dilek, S. Newcomb (Eds.), Ophiolite Concept and the Evolution of Geological Thought, Special Paper, Geological Society of America, Boulder, 2003, pp. 227–268.

[64] J.A. Pearce, Subduction zone ophiolites, in: Y. Dilek, S. Newcomb (Eds.), Ophiolite Concept and the Evolution of Geological Thought, Special Paper, Geological Society of America, Boulder, 2003, pp. 269–294.

[65] J. Pearce, S.J. Lippard, S. Roberts, Characteristics and tectonic significance of supra-subduction zone ophiolites, in: B.P. Kokelaar, M.F. Howells (Eds.), Marginal Basin Geology, Geological Society, London Special Publication 16 (1984) 77–94.

[66] R.J. Stern, N.C. Smoot, A bathymetric overview of the Mariana forearc, The Island Arc 7 (1998) 525–540.

[67] P. Fryer, J.P. Lockwood, N. Becker, S. Phipps, C.S. Todd, Significance of serpentine mud volcanism in convergent margins, in: Y. Dilek, E.M. Moores, D. Elthon, A. Nicolas (Eds.), Ophiolites and Oceanic Crust, Geological Society of America Spec. Pap. 349 (2000) 35–51.

[68] R.J. Stern, M.J. Fouch, S. Klemperer, An overview of the Izu-Bonin-Mariana subduction factory, in: J. Eiler (Ed.), Inside the Subduction Factory, vol. 138, American Geophysical Union, Geophysical Monograph, Washington DC, 2003, pp. 175–222.

[69] S.M. DeBari, B. Taylor, K. Spencer, K. Fujioka, A trapped Philippine Sea plate origin for MORB from the inner slope of the Izu-Bonin trench, Earth and Planetary Science Letters 174 (1999) 183–197.

[70] C.E. Hall, M. Gurnis, M. Sdrolias, L.L. Lavier, R.D. Muellar, Catastrophic initiation of subduction following forced convergence across fracture zones, Earth and Planetary Science Letters 212 (2003) 15–30.

[71] J.C. Lewis, T.B. Byrne, X. Tang, A geologic test of the Kula-Pacific Ridge capture mechanism for the formation of the West Philippine Basin, Geological Society of America Bulletin 114 (2002) 656–664.

[72] B. Taylor, A.M. Goodliffe, The West Philippine Basin and the initiation of subduction, revisited, Geophysical Research Letters 31 (L12602) (2004).

[73] J.A. Tarduno, R.A. Duncan, D.W. Scholl, R.D. Cottrell, B. Steinberger, T. Thordarson, B.C. Kerr, C.R. Neal, F.A. Frey, M. Torii, C. Carvallo, The Emperor Seamounts; southward motion of the Hawaiian Hotpsot plume in Earth's mantle, Science 301 (5636) (2001) 1064–1069.

[74] C.G. Macpherson, R. Hall, Tectonic setting of Eocene boninite magmatism in the Izu-Bonin-Mariana forearc, Earth and Planetary Science Letters 186 (2001) 215–230.

[75] S.H. Bloomer, B. Taylor, C.J. MacLeod, et al., Early arc volcanism and the ophiolite problem: a perspective from drilling in the Western Pacific, in: B. Taylor, J. Natland (Eds.), Active Margins and Marginal Basins of the Western Pacific, American Geophysical Union, Washington D.C., 1995, pp. 67–96.

[76] J.W. Shervais, Birth, death, and resurrection: the life cycle of suprasubduction zone ophiolites, Geochemistry, Geophysics, Geosystems 2 (2001) (Paper 2000GC000080).

[77] R.J. Stern, S.H. Bloomer, Subduction zone infancy: examples from the Eocene Izu-Bonin-Mariana and Jurassic California, Geological Society of America Bulletin 104 (1992) 1621–1636.

[78] E.M. Moores, L.H. Kellogg, Y. Dilek, Tethyan ophiolites, mantle convection, and tectonic "historical contingency"; a resolution of the "ophiolite conundrum", in: Y. Dilek, E. Moores, D. Elthon, A. Nicolas (Eds.), Ophiolites and oceanic crust; new insights from field studies and the Ocean Drilling Program, vol. 349, Geological Society of America, Boulder, 2000, pp. 3–12.

[79] J.H. Natland, J. Tarney, Petrologic Evolution of the Mariana Arc and Back-Arc Basin System—A Synthesis of Drilling Results in the South Philippine Sea, in: D.M.H., et al., (Eds.), Initial Reports of the Deep Sea Drilling Project, vol. 60, U.S. Government Printing Office, Washington DC, 1982, pp. 877–908.

[80] J.W. Hawkins, S.H. Bloomer, C.A. Evans, J.T. Melchior, Evolution of intra-oceanic arc–trench systems, Tectonophysics 102 (1984) 175–205.

[81] E. Bonatti, P.J. Michael, Mantle peridotites from continental rifts to ocean basins to subduction zones, Earth and Planetary Science Letters 91 (1989) 297–311.

[82] I.J. Parkinson, J.A. Pearce, Peridotites from the Izu-Bonin-Mariana forearc (ODP Leg 125); evidence for mantle melting and melt-mantle interaction in a supra-subduction zone setting, Journal of Petrology 39 (1998) 1577–1618.

[83] M.-P. Doin, P. Henry, Subduction initiation and continental crust recycling; the roles of rheology and eclogitization, Tectonophysics 342 (2001) 163–191.

[84] Z. Garfunkel, C.A. Anderson, G. Schubert, Mantle circulation and the lateral migration of subducted slabs, Journal of Geophysical Research 91 (B7) (1986) 7205–7223.

[85] C.R. Ranero, J.P. Morgan, K. McIntosh, C. Refchert, Bending-related faulting and mantle serpentinization at the Middle America trench, Nature 425 (2003) 367–373.

[86] C.R. Ranero, V. Sallares, Geophysical evidence for hydration of the crust and mantle of the Nazca plate during bending at the north Chile trench, Geology 32 (2004) 549–552.

[87] S.M. Peacock, Are the lower planes of double seismic zones caused by serpentine dehydration in subduction oceanic mantle? Geology 29 (2001) 299–302.

[88] J.-Y. Royer, R.G. Gordon, The motion and boundary between the Capricorn and Australian plates, Science 277 (1997) 1268–1274.

Bob Stern received his undergraduate degree at UC Davis and his PhD at Scripps Institution of Oceanography, where he also developed an interest on subduction zones, island arcs, ophiolites and crustal evolution. His research interests for the past 28 years have alternated between studying modern arc systems, especially the Marianas in the Western Pacific, and the formation of juvenile Neoproterozoic crust in NE Africa and Arabia.

ELSEVIER

Reprinted from
Earth and Planetary Science Letters 226 (2004) 1–15

www.elsevier.com/locate/epsl

How life began on Earth: a status report

Jeffrey L. Bada*

Scripps Institution of Oceanography, University of California at San Diego, La Jolla, CA 92093-0212, United States

Received 7 January 2004; received in revised form 16 July 2004; accepted 22 July 2004
Editor: A.N. Halliday

Abstract

There are two fundamental requirements for life as we know it, liquid water and organic polymers, such as nucleic acids and proteins. Water provides the medium for chemical reactions and the polymers carry out the central biological functions of replication and catalysis. During the accretionary phase of the Earth, high surface temperatures would have made the presence of liquid water and an extensive organic carbon reservoir unlikely. As the Earth's surface cooled, water and simple organic compounds, derived from a variety of sources, would have begun to accumulate. This set the stage for the process of chemical evolution to begin in which one of the central facets was the synthesis of biologically important polymers, some of which had a variety of simple catalytic functions. Increasingly complex macromolecules were produced and eventually molecules with the ability to catalyze their own imperfect replication appeared. Thus began the processes of multiplication, heredity and variation, and this marked the point of both the origin of life and evolution. Once simple self-replicating entities originated, they evolved first into the RNA World and eventually to the DNA/Protein World, which had all the attributes of modern biology. If the basic components water and organic polymers were, or are, present on other bodies in our solar system and beyond, it is reasonable to assume that a similar series of steps that gave rise of life on Earth could occur elsewhere.
© 2004 Elsevier B.V. All rights reserved.

Keywords: prebiotic soup; metabolist theory; origin of life; pre-RNA World

1. Introduction

One of the major scientific questions that confront humanity is whether life exists beyond Earth. If the conditions and processes that resulted in the origin of life on Earth are common elsewhere, then it is reasonable to expect that life could be widespread in the Universe. It is generally assumed that there are two fundamental requirements for life as we know it: the presence of liquid water and organic polymers, such as nucleic acids and proteins. Water's unique properties (excellent solvent, exceptionally large liquid temperature range, etc.) make it an ideal medium for chemical reactions to take place. Polymers are needed to carry out the central biological functions of replication and catalysis. Without these vital components, as far as we know, life is impossible.

* Tel.: +1 858 534 4258; fax: +1 858 534 2674.
  E-mail address: jbada@ucsd.edu.

0012-821X/$ - see front matter © 2004 Elsevier B.V. All rights reserved.
doi:10.1016/j.epsl.2004.07.036

The origin and early evolution of life on Earth can be divided into several stages [1–3]: the prebiotic epoch; the transition to primitive biotic chemistry (the pre-RNA World); the evolution of the early biotic chemistry of the pre-RNA World into self-replicating RNA molecules (the RNA World); and the evolution of the RNA World into modern DNA/protein biochemistry (DNA/Protein World) which was a common ancestor of all subsequent life on Earth. The appearance of the first molecular entities capable of multiplication, heredity and variation, which probably occurred in the later part of the pre-RNA World, marked the point of the origin of both life and evolution.

It is only the two end members in this series, the prebiotic epoch and the DNA/Protein World, that we know the most about [3]. We can readily investigate in the laboratory, using a variety of plausible geochemical conditions, the possible routes by which compounds of biological interest could have been produced on Earth. Meteorites can be studied to determine what clues they contain about natural abiotic organic chemistry. The genes and proteins of modern organisms can be dissected in order to ascertain information about their possible origins. The first time that direct evidence of life's existence would have been preserved in the form of physical fossils in ancient rocks occurred in the DNA/Protein World when the compartmentalization of the biochemical machinery by cell-like membrane structures comparable to those used in modern biology likely became widespread. Fossilized structures that resemble single-cell organisms similar to modern day cyanobacteria have been found in rocks formed 3.5 billion years ago (Ga) [4], although whether these structures are indeed ancient fossils or artifacts has recently become controversial [5]. Rocks older than 3.5 Ga have been so extensively altered by metamorphic processes that any molecular or fossilized evidence of earlier life has apparently been largely obliterated (for example, see [6]).

To evaluate how life may have begun on Earth, we must access what the Earth was like during its early history and under what conditions the processes thought to be involved in the origin of life took place. Considerable progress has been made in our knowledge of the early Earth and in how the transition from abiotic to biotic chemistry may have occurred. Nevertheless, there are still enormous gaps in our understanding of how the simple organic compounds associated with life as we know it reacted to generate the first living entities and how these in turn evolved into organisms that left behind actual evidence of their existence in the rock record.

2. Origin of life theories

Two views on how the transition from abiotic organic compounds to autonomous self-replicating molecules capable of evolving by natural selection into ones of increasing efficiency and complexity took place are presently dominant [1]:

(a) *The prebiotic soup theory*: Organic compounds in the primordial oceans, derived from a variety of possible sources, underwent polymerization producing increasingly complex macromolecules, some of which by chance were capable of catalyzing their own self-replication. These simple self-replication entities evolved into increasingly complex ones and eventually into organisms with modern biochemistry.

(b) *The metabolist theory*: A primitive type of "metabolic life" characterized by a series of self-sustaining reactions based on monomeric organic compounds made directly from simple constituents (CO_2, CO) arose in the vicinity of mineral-rich hydrothermal systems. According to this theory, at first, "life" did not have any requirement for informational molecules. As the system of self-sustaining reactions evolved in complexity, genetic molecules were somehow incorporated in order for metabolic-based life to develop into biochemistry as we know it.

Besides these two dominant theories, there have also been numerous suggestions that life began elsewhere and was transported to Earth (for example, see [7]), but this only shifts the problem of the origin of life to a different location.

According to the modern version of the prebiotic soup theory [1], organic compounds derived from "homegrown" chemical synthetic reactions on Earth and the infall of organic rich material from space accumulated in the primordial oceans. These compounds then underwent further reactions in the primal broth, producing ones with increasing molecular complexity. Some of these reactions took place at interfaces of mineral deposits with primitive ocean water, while others occurred when the primitive ocean constituents were concentrated by various mechanisms, such as evaporation in shallow water regions or the formation of eutectic brines produced during the freezing of parts of the oceans.

From the assortment of simple organic compounds in the primitive oceans, geochemical processes next resulted in the synthesis of polymeric molecules. As the variety of polymers (oligomers) that were assembled from the simple monomers by polymerization processes became more varied, some by chance acquired functions, such as the ability to catalyze other reactions. With the rise of catalytic molecules, increasingly complex macromolecules were produced and eventually by chance molecules with the ability to catalyze their own imperfect replication appeared. Although these first replicators at first probably represented only a tiny fraction of the large array of macromolecules, with the ability to catalyze their own replication, they would have soon become increasingly more abundant. This would have marked the transition from purely abiotic chemistry to primitive biochemistry. These first self-replicating molecular entities began the evolutionary cascade that next led to the RNA World and then to the DNA/Protein World that had all the characteristics of modern biochemistry. The chance aspects of this scenario have been used to argue that the prebiotic soup theory is flawed and unscientific [8], but chance events have shaped the course of life's evolution several times throughout its history, with the Cretaceous/Tertiary impact-induced extinction event 65 million years ago (Ma) being one striking example.

In contrast to the prebiotic soup theory, the metabolist theory claims that life at its beginning was nothing more than a continuous chain of sulfide mineral catalyzed self-sustaining chemical reactions with no requirement for genetic information [8–11]. This theory has recently become popular, as some researchers have questioned the validity of the prebiotic soup theory. 'Metabolic life' is rightfully referred to as "Life as we don't know it" [11]: life as we know it is based on both chemistry *and information*. In an attempt to incorporate informational molecules into the metabolist scheme, it has recently been suggested that an elaborate cascade of metabolic reactions entrained within sulfide minerals around hydrothermal vents developed all the way to RNA molecules and even primitive cells [12]. However, given the transient, short-lived nature of hydrothermal systems, the plausibility of this process under geochemical conditions seems questionable.

In principle, self-sustaining autotrophic reactions conceivably could have arisen in any type of environment as long as the reactant/product molecules survived long enough to continue to be part of the overall reaction chain. Of the various reaction schemes that have been proposed, however, none have been demonstrated to be autocatalytic with one possible exception [13]. The exception is the formose reaction, where the formation of a diverse variety of sugars from formaldehyde in the presence of alkaline catalysts apparently involves an autocatalytic cycle that can result in the continuous autocatalytic synthesis of sugars as long as there is an unlimited supply of formaldehyde [13].

Advocates of the metabolist theory generally favor hydrothermal environments (for example, see [9,14]). However, even if the metabolic-type reaction schemes that have been proposed were feasible, they would not have been unique to hydrothermal temperatures. Reactions that take place rapidly at elevated hydrothermal temperatures would also occur at the more moderate temperatures characteristic of the overall surface of the Earth, albeit at slower rates. The central issue is which temperature regime is more geochemically relevant. A germane example is petroleum, which is mainly produced by a series of reactions that take place as buried sedimentary organic matter is subjected to geochemical processing over time scales of several million years. Although petroleum can be produced at hydrothermal temperatures of 300 to 350 °C in periods as short as 100 years [15], the majority of the Earth's petroleum was formed at peak temperatures of ~120 °C, which demonstrates the dominance of the lower temperature regimes at least in the geochemical processes associated with petroleum formation [16].

The potential importance of autocatalytic reactions should not be underestimated, however. Self-sustaining reaction chains could have played an important role in enriching the prebiotic soup in molecules that were perhaps not readily synthesized by other abiotic reactions or which were unstable and thus need to be continuously and rapidly synthesized in order to be available for subsequent reactions. In this sense, the prebiotic soup and metabolist theories are synergistic and complementary.

3. The primitive Earth

During the final planetismal accretionary phase of the Earth and especially in the period immediately following the moon forming impact event at 4.51 Ga, the Earth's surface was likely covered with a liquid rock, or magma, ocean [17]. Any water at the surface would have been present as steam in the atmosphere. These high temperatures would have incinerated any organic compounds, derived from whatever sources. Thus, in its earliest history, the Earth was likely devoid of an organic carbon reservoir, especially in comparison to the modern Earth where 20–30% of the surficial carbon is present as organic matter.

Although the rock record of the Earth prior to ~3.9 Ga is not preserved, oxygen isotopic analyses of the 4.3 to 4.4 Ga detrital zircons indicate that temperatures had apparently decreased to the point that liquid water was present on the Earth's surface about 100–200 million years after accretion [18,19]. Earth's water is thought to have been derived mainly from degassing of hydrated minerals initially present in the mantle and to a lesser extent from the infall of asteroids and comets [20]. If the Earth's water did indeed mainly come from degassing of the mantle as the early Earth underwent differentiation, then most of the water on the surface today (10^{21} l) was probably released early in Earth's history. Although the composition of early ocean water is not known, if all the salt deposits presently stored on the continents along with that present in saline groundwaters was originally in the oceans, the salt content of the primitive oceans could have been nearly twice that today [21].

Based on the ancient zircon evidence, Earth during the period 4.4 to 4.0 Ga may have been cool enough to allow for the presence of extensive liquid water oceans for long periods [22]. Based on the lunar cratering record, during this same period, the Earth was apparently relatively free of ocean vaporizing or sterilizing impacts [23]. When liquid water started to accumulate, temperatures would have become cool enough to allow for the survival of organic compounds derived from various sources. It was thus likely that during this presumably relatively quiescent interval between 4.4 and 4.0 Ga, some of the key steps in the origin of life may have occurred (see Fig. 1).

The Sun is estimated to have been ~30% less luminous than today during the early history of the

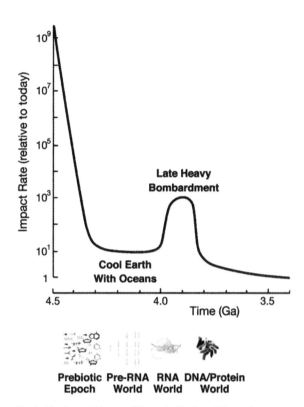

Fig. 1. The impact history of the early Earth and the various stages involved in the origin and evolution of life. Based on [2,22].

Earth, which gives rise to the so-called faint young Sun paradox [24]: without a significantly enhanced greenhouse effect relative to the modern Earth, global surface temperatures would have been −40 °C because of a planetary albedo near that of ice [25]. Consequently, the Earth could have become a permanently frozen planet early in its history. Although it was first suggested nearly 25 years ago that atmospheric CO_2 levels 10^3 to 10^4 times modern levels provide one solution to the frozen Earth dilemma, recent considerations of the early carbon cycle suggest that before extensive tectonic recycling of crustal sediments became common, most of the carbon on the Earth's surface would have remained buried in the crust and mantle as calcium carbonate [26]. Thus, CO_2 may not have been present in the atmosphere at levels adequate to prevent global glaciation, unless other greenhouse gases were present. Apparently, CO_2 levels were insufficient to prevent global glaciation at around 2.45 to 2.22 Ga and at ~600 Ma when the Sun was ~17% and ~6%, respectively, less luminous than today [24,25]. Without enhanced greenhouse warming associated with higher CO_2 levels, the widespread occurrence of an ocean ice cover could have been common in the early history of the Earth [27]. The presence of methane and ammonia in the atmosphere provides a potential solution to the faint early Sun/frozen over early ocean problem (for example, see [28]). However, these gases would have been prone to rapid UV photolysis and thus might not have accumulated to sufficiently high concentrations to prevent ocean freezing.

Even if the Earth's early oceans were totally ice covered, the oceans would not have completely frozen [27]. Heat from the Earth's interior leaking through the oceanic crust would have provided a heat source to the deep ocean beneath the ice layer. A similar heat source today provides sufficient basal heat to produce large subsurface lakes such as Lake Vostok at the base of the Antarctic ice sheet. Using a simple one-dimensional heat flow model and assuming a heat flow of ~3 times the present day value, it has been estimated that the ice thickness on an ice covered early ocean would have been ~300 ± 100 m.

Bolide impacts could have played a role in melting the iced-over primitive ocean. Bolide induced melting could have been especially important during the 3.8 to 4.0 Ga 'late heavy bombardment' period [29] and helped promote the transition to a more permanently ice-free ocean. Bolide impacts could have also, however, had some unwanted dire consequences with respect to the transition to an ice free early Earth. Impacts may have generated an opaque equatorial debris ring that, because of its shadow, caused reduced solar insolation at low latitudes [30]. This in turn could have triggered the onset of glaciations and the return of global ice.

It was under these uncertain, chaotic, seemingly adverse and tumultuous conditions that the ingredients for life somehow accumulated and the first primitive life forms emerged on the Earth.

3.1. The Prebiotic Epoch

In 1953, Stanley L. Miller demonstrated the ease by which important biomolecules, such as amino and hydroxy acids, could be synthesized under what were viewed at the time as plausible primitive Earth conditions (see [31] for a summary of this classic experiment). A key aspect of the experiment was the formation of hydrogen cyanide (HCN), aldehydes and

ketones produced during the sparking of the reduced gases H_2, CH_4 and NH_3. The formation of these reagents suggested that the compounds were produced by the Strecker–Cyanohydrin reaction, first discovered in 1850 by the German chemist Adolph Strecker. The actual synthesis takes place in aqueous solution, implying that on the early Earth, amino acids could have been produced in bodies of water, provided the necessary reagents were present. Only α-hydroxy acids are formed in the absence of ammonia. Thus, the concentration of ammonia in the primitive oceans would have been critical in determining whether amino acids would be have been synthesized by this process.

HCN is a critical reagent. Not only is it a central component of the Strecker reaction, but polymerization of HCN itself, even in -20 and -78 °C HCN-rich brines [32], generates important molecules, such as glycine, adenine and guanine. HCN is unstable and hydrolyzes to ammonia and formate, a reaction that is very rapid at elevated temperatures (half-life at 100 °C is ~1 day at neutral pH [33]). Based on estimates of the rates of HCN production and hydrolysis, it has been estimated that the steady-state concentration of HCN in the primitive oceans would have been about 2×10^{-5} M at 0 °C and only 4×10^{-12} M at 100 °C at pH 7 [33]. This result strongly suggests that in order for HCN to play a significant role in prebiotic chemistry on the early Earth, temperatures at the time must have been cool.

The hydrolysis of HCN provides a potential source of ammonia in the primitive oceans even if ammonia was absent in the early atmosphere. Ammonia would have been continuously produced in the oceans by HCN hydrolysis, provided there was a continuous source of HCN, which remains uncertain, however. Ammonia may have also been injected into the early oceans by hydrothermal vent discharges [34], although most of the ammonia detected in modern hydrothermal systems is likely derived from the high-temperature decomposition of biologically produced organic matter rather than being abiotic in origin. In addition, ammonia could have been produced in the early oceans by the ferrous iron catalyzed reduction of nitrite [35]. If the primitive oceans were cool and more acidic than today, the ammonia would have been mainly dissolved in the ocean present as NH_4^+. This implies that even if there was adequate NH_4^+ dissolved in the oceans to support prebiotic reactions, the atmospheric ammonia levels may have been too low to provide for sufficient greenhouse warming to keep the early Earth's surface temperature above freezing.

One often-overlooked aspect of the Miller experiment is that the main product was oily goo. With a methane-rich atmosphere, oily material would have been produced in huge quantities on the early Earth, forming an oil slick that would have unimaginable on the Earth today. Oily material could have formed a protective layer on the primitive ocean surface that allowed not only for molecules to be protected from destruction by the sun's ultraviolet light [36], but also may have helped promote the condensation of simple monomeric compounds into polymers by acting as an anhydrous solvent [37]. In addition, the oily layer could have decreased the vapor pressure of water and thus the OH radical concentration in the atmosphere. As a consequence, the atmospheric lifetimes of reducing gases, such as methane and ammonia, could have been substantially increased [37].

Since the classic Miller experiment, numerous researchers have demonstrated that a large assortment of organic molecules can be synthesized using a variety of gaseous mixtures and energy sources (for example, see [38,39]). Most of the molecules that play an essential role in modern biochemistry, such as amino acids, nucleobases, sugars, etc., have been synthesized under plausible geochemical conditions. The conditions employed have ranged from the highly reducing conditions first used by Miller to less reducing mixtures containing CO and CO_2 [40]. However, with neutral atmospheric mixtures containing CO_2 and N_2, the yields of amino acids and other essential organic compounds is vanishingly low.

Many geoscientists today doubt that the primitive atmosphere had the highly reducing composition used by Miller in 1953. Although reducing conditions may not have existed on a global scale, localized high concentrations of reduced gases may have existed around volcanic eruptions, especially in hot-spot island-arc systems that may have been common on the early Earth. Whether reducing volcanic gases would have been dominant in these systems on the early Earth would depend on the oxidation state of the early mantle, which could have been more reducing than today [41]. The localized release of reduced gases by volcanic eruptions on the early Earth would likely have been immediately exposed to intense lightning (see Fig. 2), which is commonly associated with

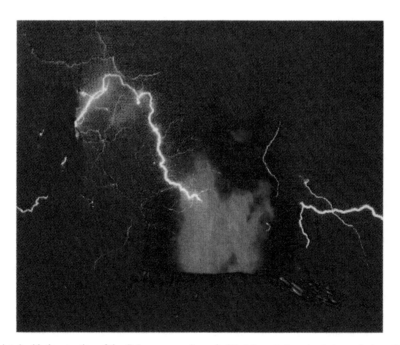

Fig. 2. Lightning associated with the eruption of the Galunggung volcano in West Java, Indonesia photographed on October 16, 1982 (taken from [42]).

volcanic eruptions today [42,43]. With present day volcanic gas mixtures, NO is the main product [44], but with more reducing mixtures containing H_2, CH_4 and N_2, acetylene, HCN and other prebiotic reagents would have been produced [45]. Thus, in localized volcanic plumes, prebiotic reagents may have been produced, which after washing out of the atmosphere could have become involved in the synthesis of organic molecules. Island-arc systems may have been particularly important in localized Strecker-type syntheses because the reagents could have rained out into tidal areas where they could be concentrated by evaporation or periodic freezing.

An alternative to direct Earth-based syntheses is that the organic compounds needed for the origin of life may have come from extraterrestrial sources, such as interplanetary dust particles (IDPs), comets, asteroids and meteorites [46]. It is well established that carbonaceous meteorites contain a wide assortment of organic compounds [47], including ones, such as amino acids and nucleobases, that play a critical role in biochemistry. The major organic component in carbonaceous chondrites is complex macromolecular material similar to that produced in the spark discharge experiment. The input of this type of material from space could have contributed to the organic goo that may have covered parts of the early Earth and oceans.

Whether extraterrestrial organic material was efficiently delivered intact to the Earth, however, remains an uncertain issue. Because of the high temperatures associated with large impacts, most organic compounds originally present in the bolide should be destroyed, although robust molecules, such as fullerenes, may survive intact [48]. Bolide impacts may have had a beneficial effect on prebiotic chemistry, however: large amounts of methane could have been produced and which in turn could have yielded substantial amounts of HCN by photodissociation reactions [49]. This could have resulted in the bolide impact induced episodic synthesis of some key biomolecules by both the Strecker reaction and direct HCN polymerization.

Discussions of exogenous delivery have focused on IDPs because they represent the largest source of extraterrestrial material after large bolide impacts [46]. However, IDPs can be heated to temperatures of 1000 to 1500 °C during atmospheric deceleration, which could cause extensive decomposition of any organic compounds present. In experiments designed to simulate the atmospheric entry of Murchison-like IDPs [50], it was found that a large fraction of the amino acid glycine vaporized (sublimed) and survived when the grains where heated to 550 °C under partial vacuum. However, other amino acids present in Murchison did not sublime and were completely destroyed. These results suggest that sublimation of glycine present in IDPs may provide a way for this amino acid to survive atmospheric entry heating whereas all other amino acids apparently are destroyed. This is consistent with amino acid analyses of micrometeorites collected from Antarctic ice [51]. These results indicate that <5% of the micrometeorites contain endogenous amino acids. If amino acids where originally present in IDPs at Murchison-like levels, apparently only a small fraction of the amino acids escapes decomposition during atmospheric entry.

It is now generally assumed that the inventory of organic compounds on the early Earth would have been derived from a combination of both direct Earth-based syntheses and input from space. The simple abiotic monomeric organic compounds derived from these sources would have accumulated in the early oceans as well as other bodies of water and provided the raw material for the subsequent reactions.

4. The transition to primitive biotic chemistry

Polymers composed of at least 20–100 monomeric units (mers) are thought to be required in order to have any primitive catalytic and replication functions [2,3]. Thus, polymerization processes taking place on the primitive Earth must have been capable of producing polymers of at least this minimum size.

Polymerization is a thermodynamically unfavorable process. In order to overcome this problem, the selective adsorption of monomers onto mineral surfaces has been suggested as one means of promoting polymerization and this process that has been demonstrated in the laboratory using a variety of simple compounds and activated monomers [52,53]. The potential importance of mineral-assisted catalysis is demonstrated by the montmorillonite-promoted polymerization of activated adenosine and uridine derivatives producing 25–50 mer oligonucleotides [53], the general length range considered necessary for primitive biochemical functions. Absorption onto surfaces involves the formation of weak noncovalent van der Waals interactions and thus the mineral-based concentration process and subsequent polymerization would be most efficient at cool temperatures [54,55].

As the length of polymers formed on mineral surfaces increases, they tend to be more firmly bound to the mineral [56]. In order for these polymers to be involved in subsequent interactions with other polymers or monomers, they would need to be released. This could be accomplished by warming the mineral although this would also tend to hydrolyze the absorbed polymers. Polymers could also be released by concentrated salt solutions [52,56], a process that could take place in tidal regions during evaporation or freezing of seawater.

The direct concentration of dilute solutions of monomers could be accomplished by evaporation and by eutectic freezing of dilute aqueous solutions. The evaporation of tidal regions and the subsequent concentration of their organic constituents have been proposed in the synthesis of a variety simple organic molecules (for example, see [57]). Eutectic freezing of dilute reagent solutions has also been found to promote the synthesis of key biomolecules [32]. It has been shown that the freezing of dilute solutions of activated amino acids at $-20\,^{\circ}C$ yields peptides at higher yields than in experiments with highly concentrated solutions at 0 and 25 $^{\circ}C$ [58]. In addition, recent studies have shown that eutectic freezing is especially effective in the nonenzymatic synthesis of oligonucleotides [59].

Salty brines could have played a role in the polymerization of amino acids and perhaps other important biopolymers as well. Short peptides have been synthesized using concentrated NaCl solutions containing Cu (II) and 40–50 mM amino acids [60]. Clay minerals, such as montmorillonite, apparently promote the reaction. Again, the evaporation of tidally flushed lagoons or the freezing of the primitive oceans could have produced the concentrated salty brines needed to promote this salt-induced polymerization process.

Hydrothermal systems may have been sites for the formation of short peptides. Because peptide bond formation becomes more favorable at higher temperatures, this reaction is one of the exceptions to the low temperature usually dominates "rule" discussed earlier. Experiments using glycine and other amino acids carried out under simulated high-temperature hydro-thermal conditions have been shown to produce peptides containing up to 6 amino acid units [61,62]. However, the initial amino acid concentrations in these experiments are unrealistically high (0.1 M) and at more dilute concentrations, the peptide yields would be expected to decrease dramatically. Amino acids rapidly decompose and peptide bonds are rapidly hydrolyzed at elevated temperatures, so long-term survival and steady-state concentration of peptides at elevated temperatures are problematic. This is particularly important with respect to the residence time at high temperature during circulation through hydrothermal systems. The model experiments published to date use short exposure times (i.e., 1+ h) at high temperatures compared to 1–30+ years associated with actual hydrothermal systems [63]. However, autocatalytic cycles, which might take place at temperatures lower than the peak hydrothermal temperatures of ~350 to 400 $^{\circ}C$, could have been important in maintaining a modest steady-state concentration of short peptides [64].

As polymerized molecules increased in length and became more complex, some of these began to fold into configurations that could bind and interact with other molecules. Primitive catalysts that promoted a variety of reactions could have thus arisen. Some of these catalytic reactions may have assisted in making the polymerization process more efficient. As the variety of polymeric combinations increased, a large library of random sequences would have been generated. By chance, some of these polymers acquired the ability to catalyze their own imperfect self-replication. Although these self-replicating molecules at first may have been scarce in the overall pool of polymers, because of their ability to catalyze their own replication, they would have become increasingly more abundant and soon dominated. There are now several known examples of self-replicating molecular systems that have been studied in the laboratory and these provide examples of the types of molecular systems that could have given rise to early self-replicating entities [65].

The appearance of the first molecular entities capable of replication, catalysis and multiplication would have marked the origin of both life and evolution. At a minimum, the first living molecular entities must have had the following properties: they could make copies of themselves, although the replication was not exact so mutants that had some sort of selective advantage might be generated; they could either make 'activated' molecules that could be used in the replication process or utilize mineral surfaces to promote replication; and they must 'live' long enough to ensure that they survive long enough to be replicated. The challenge is to determine the best candidate system that best fits this minimalist definition.

Because of the huge number of possible random combinations of nucleotides from sugars, phosphate and nucleobases, it is unlikely that a RNA molecule capable of catalyzing its own self-replication arose spontaneously [2]. In addition, the ribose component of RNA is very unstable making its presence in the prebiotic milieu unlikely. Rather than RNA, some type of simpler self-replicator must have come first and several possible contenders have been suggested [2]. It is generally assumed that the first molecular self-replicating living entities must have had the capacity to store information and thus were nucleic acid based, although the component nucleobases and the backbone that held the polymer together were not necessarily the same as those in modern RNA and DNA. Possible candidates include nucleic acid analogues, such as peptide nucleic acid (PNA), where the backbone consists of linked amino acid derivatives, such as *N*-

(2-aminoethyl)glycine or AEG (the nucleobases are attached by an acetic acid linkage to the amino group of glycine) and threose nucleic acid (TNA), where the backbone is made up of L-threose connected by 3′, 2′ phosphodiester bonds.

PNA is attractive because its backbone is achiral (lacking handedness) which eliminates the need for the selection of chirality before the time of the origin of life. Its components, AEG and nucleobases linked to acetic acid, have been produced under simulated prebiotic conditions [66]. However, PNA is susceptible to an *N*-acyl migration reaction producing a rearranged PNA. This problem could be minimized, however, by blocking the N-terminal position by acetylation, for example.

Based on an extensive study of sugar-based nucleic acids, TNA appears to be superior with respect to its base-pairing attributes, especially with RNA, compared to other possible sugar-based nucleic acids [67,68]. The tetrose sugar in TNA could have been synthesized during the reaction cascade that takes place during the formose reaction. The 4-carbon sugars threose and erythrose could have be readily synthesized by the dimerization of glycolaldehyde, which in turn could have been produced from the dimerization of formaldehyde (see Reaction (1)):

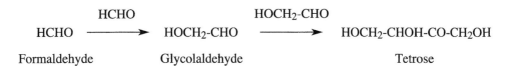

Reaction 1: The formation of tetrose sugars from the dimerization of formaldehyde and glycolaldehyde. This is only part of the overall formose series of reactions and both hexoses and pentoses would have been produced by subsequent reactions.

However, in order to avoid the production of a large array of 5- and 6-carbon competing sugars, the reaction would need to somehow be quenched. TNA suffers from the chirality quandary associated with all sugar-based nucleic acid backbones. Although the presence of a 4-carbon sugar in TNA reduces this problem to 2 sugars and 4 stereoisomers, it remains a formable challenge to demonstrate how oligonucleotides composed of only L-threose could be preferentially synthesize under prebiotic conditions.

It is possible that PNA preceded TNA and in fact assisted in the transition to TNA-based replicating entities. As stated above, the selection of the chiral sugar component of TNA would have required some sort of selection process to be in operation. The incorporation of chiral sugar nucleotides at the end of a PNA chain, that could have occurred simply by chance, can induce chirality into a nucleic acid produced by PNA-induced oligomerization [69]. PNA could have thus assisted in conveying the critically important biological property of chirality into polymers near the time of the origin of life. This possibility potentially solves an ongoing dilemma about whether the origin of chirality occurred before the origin of life, or whether its origin occurred during the evolution of early living entities.

Regardless of the type of nucleic acid-like analogue, or other type of replicator system, that was used by the first self-replicating entities, polymer stability and survival would have been of paramount importance. Nucleic acids in general have very short survival times at elevated temperatures. The half-life for cleavage of the phosphodiester bonds in RNA has been estimated to be <1000 years at 0 °C [70], <1 day at pH 8, 35 °C [68] and <1 s at 150 °C [71]. DNA is more stable but is still completely fragmented in minutes at 250 °C, in an hour at 150 °C and in 10^3 to 10^4 years at 0 °C [70,72]. The amino acyl bonds in tRNAs, which are involved in biological protein synthesis, are hydrolyzed in ~10 s at 100 °C [73]. In contrast to elevated temperatures, stabilities would be greatly enhanced at lower temperatures.

The stability of TNA is apparently similar to that of DNA [68]. The stability of PNA has been partly investigated and provided the N-acyl migration reaction can be minimized, the amide linkage in PNA should have a stability at neutral pH similar to that of peptide bonds in proteins [74]. This suggests that in environments with temperatures of around 25 °C, its survival time would be in the range of 10^4–10^5 years.

Salty brines may have played a role in early nucleic acid survival. The stabilities of several tRNAs were significantly increased in 1–2 M NaCl solution in comparison to that in pure water [75]. The stability of DNA also increases with increasing salt concentration (Fig. 3). If this trend is applicable to other nucleic acid analogues, such as PNA and TNA, then salt solutions would have provided a protective environment that could have enhanced the survival of early self-replicating molecular entities.

Stability limitations suggest that the origin of simple nucleic acid-based living entities would likely only have been feasible in cool, perhaps salty environments on the early Earth. The first living entities that arose, regardless of the process, may not have survived the high temperatures generated by subsequent bolide impacts, however (see Fig. 1). Simple self-replicating entities may have originated several times before

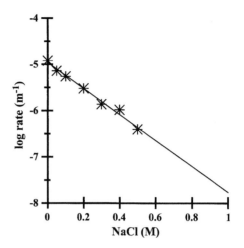

Fig. 3. The rate (m^{-1}=minutes^{-1}) of depurination of adenine in DNA at pH 4.5 and 60 °C as a function of sodium chloride concentration. The rate for guanine is 2–3 times faster (taken from [74]).

conditions became tranquil enough for periods sufficiently long to permit their survival and evolution into more advanced life.

4.1. The transition to modern biochemistry

The evolution of the first living molecular living entities into ones based entirely on RNA (the RNA World) would have been the next step in the evolution towards modern biochemistry. RNA has been found to be an all-in-one molecule that cannot only store information but also catalyze reactions [2]. Laboratory-based "test tube evolution" experiments have demonstrated that catalytic RNA molecules (ribozymes) have the capacity to carry out a wide range of important biochemical reactions [2,76]. The RNA World could have had a large repertoire of catalytic RNA molecules perhaps functioning in concert with one another. Although the complex series of reactions needed to permit multiplication, genetic transfer and variation required in the RNA World has so far not been demonstrated in the laboratory, optimism remains because of the relative immaturity of this area of research [2,76].

The invention of protein synthesis and the encapsulation of reaction machinery needed for replication may have taken place during the RNA World. Four of the basic reactions involved in protein biosynthesis are catalyzed by ribozymes and it has been noted that the complimentary nature of these reactions is not likely accidental but rather suggestive that they had a common origin most likely in the RNA World [77]. If this was the case, then the origin of a primitive nucleobase code used for protein biosynthesis had its origin in the RNA World although the bases used in the early code could have been different from the ones used today [78].

RNA molecules adsorbed onto clays, such as montmorillonite, which can catalyze the formation of RNA oligomers, can be encapsulated into fatty acid vesicles whose formation in turn is accelerated by the clay [79]. By incorporating additional fatty acid micelles, these vesicles can grow and divide while still retaining a portion of their contents needed to support RNA replication. In this manner, some of the basic machinery needed for RNA self-replication could have been compartmentalized into prototype cells.

By the time RNA-based life appeared on Earth, the supply of simple abiotic organic compounds derived from the sources discussed above had likely greatly diminished. Many of the components of the primordial soup would have been converted into polymers including those associated with living entities and thus the raw materials needed to sustain primitive life had become largely exhausted. This implies that the origin of simple metabolic-like pathways must have arisen at this point in order ensure a supply of the components needed to sustain the existence of the primitive living entities. This is where the self-staining autotrophic reactions discussed in the "metabolist theory" could have played an important role. The metabolic pathways needed to produce essential components required by primitive living entities were perhaps originally nonenzymatic or semienzymatic autocatalytic processes that later became fine tuned as ribozymatic and protein-based enzymatic processes became dominant [80,81]. The amount of organic carbon produced via the autotrophic fixation of CO_2, CH_4 and simple organic compounds such as formic and acetic acids by metabolic processes in the RNA World could have far exceeded the amounts of organic compounds remaining, or still being supplied, from either homegrown processes or extraterrestrial sources under the best conditions. At this point, the reservoir of organic material present on the Earth shifted from one totally characterized by compounds of abiotic origin to one made up of components synthesized by early biotic processes.

The main limitation in the RNA World would have been the extreme instability of RNA. This implies that RNA molecules must have been very efficient in carrying out self-replication reactions in order to maintain an adequate inventory of molecules needed for survival. The instability of RNA could have been the primary reason for the transition to the DNA/Protein World where, because of the increased stability of the genetic molecules, survival would have been less dependent on polymer stability. In the RNA World, ribozymes may have arisen that could catalyze the polymerization of DNA and in this manner, information stored in RNA could be transferred to the more stable DNA [2]. Because of their superior stability, much longer DNA oligomers could have accumulated and this provided for an enhanced storage capacity of information that could be passed on to the next generation of living entities. In addition, using test-tube evolution strategies, deoxyribozymes, the DNA analogues of ribozymes, have been discovered [82]. This may imply that some DNA molecules inherited catalytic properties from ancestors in the RNA World. Before long, RNA which once played the singular role of replication and catalysis was replaced by the more efficient and robust DNA/Protein World wherein RNA was demoted to a role of messenger/transcriber of DNA stored information needed for protein biosynthesis.

Although DNA is more stable than RNA, it is still rapidly degraded at elevated temperatures. In addition, protein enzymes denature rapidly at elevated temperatures. This must have at least initially limited the environments where DNA/Protein-based life could survive for any significant period of time and, as was the case for other earlier nucleic-acid-based living entities, survival would have been the most favorable under cool conditions. Nevertheless, several researchers have advocated high temperatures, especially those associated with hydrothermal vent systems, as the environment where DNA/Protein-based entities first arose. Proponents for a high temperature transition cite the fact that the universal tree of extant life appears to be rooted in hyperthermophilic organisms. Thus, if the last universal common ancestor (LUCA) to all modern biology was a hyperthermophile, then it is concluded that the first DNA/Protein-based life must have arisen in a similar type of environment. However, this argument is flawed for several reasons. First, the recognition that the deepest branches in rooted universal phylogenies are occupied by hyperthermophiles is controversial and does not provide by itself conclusive proof of a high-temperature origin of DNA/ Protein-based life [83]. In addition, lateral gene transfer of thermoadaptative traits has apparently greatly compromised the genetic record present in modern organisms, which makes any conclusions about the environment where the DNA/Protein World originated questionable [84]. In addition, an analysis of protein sequences has found only one enzyme, reverse gyrase, that is specific to hyperthermophiles; other proteins are apparently not ancestral to these organisms and are likely simply heat-adapted versions of those present in cooler temperature organisms [85]. Even if the LUCA was a hyperthermophile, there are alternative explanations for their basal distribution, such as the possibility that hyperthermophily is an evolutionary relic from early Archean high-temperature regimes associated with severe bolide impact events during the late heavy bombardment period (see Fig. 1). In this latter case, heat-loving DNA/Protein-based life was simply the major survivor of an impact-induced catastrophe that destroyed the bulk of the early DNA/Protein World.

5. Future challenges

Although the overall scenario presented here provides a framework of the processes involved in the appearance of life on Earth, there are still many details that need to be fully elucidated. One of the most challenging areas is determining what types of polymerization processes could have given rise to nucleic acid-based molecules, especially ones with catalytic properties. For example, although appealing as possible candidates for the first self-replicating molecular living entities, both PNA and TNA have negative aspects, mainly the lack of any demonstrated oligerization process for efficiently producing these nucleic acid analogues under plausible prebiotic conditions. In addition, the reactions needed to permit multiplication, genetic transfer and variation required in PNA, TNA and RNA Worlds have not been achieved in the laboratory and this represents a formable challenge. The time of the origin of chirality, although discussed here as arising during early evolution of life, remains largely unknown.

Perhaps the biggest uncertainty is whether a metabolic-like set of self-sustaining chemical reactions can be considered 'alive' and whether this type of

chemistry preceded information-based chemistry associated with life as we know it. This prospect is highly debatable and although there are compelling reasons that genetic informational molecules were a vital component of the first living entities on Earth, there is considerable polarization on this issue [86]. As discussed here, metabolic-type reactions likely contributed important molecules during the prebiotic epoch and thus could have had a central role in the processes that gave rise to the origin of life. However, whether "metabolic life" could have truly existed by itself and preceded life as we know it is controversial. This issue has important implications with respect to searching for life beyond Earth. How would we recognize simple metabolic-based life elsewhere and how would we go about testing to determine if it was compatible or hazardous to life as we know it on Earth?

6. Conclusions

Much remains unknown about the actual processes and sequence of events that gave rise to the first living entities on Earth. Someday, an artificial molecular-based self-replicating entity will probably be created in the laboratory and this will allow us to study some of the attributes of primitive "living" systems. However, there is no guarantee that this will in any way be representative of the types of early living entities that may have first appeared on the early Earth.

The exploration of extraterrestrial worlds may provide some of the missing information about how life began on Earth and its earliest stages of evolution. There are compelling reasons to believe that the core organic components used by living organisms on Earth, polymers made of amino and nucleic acids, would be part of biochemistry elsewhere, although the exact structural makeup of these key molecules may be different [87]. If the abiotic chemistry and its subsequent evolution into primitive living molecular entities discussed here is widespread, we may find examples of some of the stages associated with this process perhaps on Mars, Europa or on some of the extra-solar Earth-like planets that may abound in the Universe.

Searching for signs of extraterrestrial life in our solar system should be fairly straightforward because we can send spacecraft to directly explore promising bodies and if positive results are obtained, eventually return samples to Earth for direct analyses. Several missions by ESA and NASA are being planned to just this, and hopefully, in the not-too-distant future, we may know whether life on Earth is unique to our solar system. Detecting life on Earth-like planets outside our solar system will likely remain a daunting challenge for some time, however [88].

References

[1] J.L. Bada, A. Lazcano, Origin of life—some like it hot, but not the first biomolecules, Science 296 (2002) 1982–1983.

[2] G.F. Joyce, The antiquity of RNA-based evolution, Nature 418 (2002) 214–221.

[3] C. de Duve, A research proposal on the origin of life, Orig. Life Evol. Biosph. 33 (2003) 559–574.

[4] J.W. Schopf, The Cradle of Life: The Discovery of the Earth's Earliest Fossils, Princeton University Press, 1999, 336 pp.

[5] J.M. Garcia-Ruiz, S.T. Hyde, A.M. Carnerup, A.G. Christy, M.J. van Kranendonk, N.J. Welham, Self-assembled silica–carbonate structures and detection of ancient microfossils, Science 302 (2003) 1194–1197.

[6] M.A. van Zullen, A. Lepland, G. Arrhenius, Reassessing the evidence for the earliest traces of life, Nature 418 (2002) 627–630.

[7] W.N. Napier, A mechanism for interstellar panspermia, Mon. Not. R. Astron. Soc. 348 (2004) 46–51.

[8] G. Wächtershäuser, The origin of life and its methodological challenge, J. Theor. Biol. 187 (1997) 483–494.

[9] G. Wächtershäuser, The case for a hyperthermosphilic, chemolithoautotrophic origin of life in an iron–sulfur world, in: J. Wiegel, M.W.W. Adams (Eds.), Thermophiles: The Keys to Molecular Evolution and Origin of Life? Taylor & Francis, London, 1998, pp. 47–57.

[10] R. Shapiro, A replicator was not involved in the origin of life, IUBMB Life 49 (2000) 173–176.

[11] G. Wächtershäuser, Life as we don't know it, Science 289 (2000) 1307–1308.

[12] W. Martin, M.J. Russell, On the origins of cells: a hypothesis for the evolutionary transitions from abiotic geochemistry to chemoautotrophic prokaryotes, and from prokaryotes to nucleated cells, Philos. Trans. R. Soc. Lond., B (2003) 59–85.

[13] L.E. Orgel, Self-organizing biochemical cycles, Proc. Natl. Acad. Sci. U. S. A. 97 (2000) 12503–12507.

[14] G.D. Cody, N.Z. Boctor, R.M. Hazen, J.A. Brandes, H.J. Morowitz, H.S. Yoder, Geochemical roots of autotrophic carbon fixation: hydrothermal experiments in the system citric acid, H_2O-($\pm$FeS)-($\pm$NiS), Geochim. Cosmochim. Acta 65 (2001) 3557–3576.

[15] K.A. Kvenvolden, J.B. Rapp, F.D. Hostettler, J.D. King, G.E. Claypool, Organic geothermometry of petroleum from Escanaba Trough, offshore northern California, Adv. Org. Geochem. 13 (1987) 351–355.

[16] R. di Primio, B. Horsfield, M.A. Guzman-Vega, Determining the temperature of petroleum formation from the kinetic properties of petroleum asphaltenes, Nature 406 (2000) 173–176.

[17] N.H. Sleep, K. Zahnle, P.S. Neuhoff, Initiation of clement surface conditions on the earliest Earth, Proc. Natl. Acad. Sci. U. S. A. 98 (2001) 3666–3672.

[18] S.A. Wilde, J.W. Valley, W.H. Peck, C.M. Graham, Evidence from detrital zircons for the existence of continental crust and oceans on the Earth 4.4 Gyr ago, Nature 409 (2001) 175–178.

[19] S.J. Mojzsis, T.M. Harrison, R.T. Pidgeon, Oxygen-isotope evidence from ancient zircons for liquid water at the Earth's surface 4300 Myr ago, Nature 409 (2001) 178–181.

[20] N.F. Dauphas, F. Robert, B. Marty, The late asteroidal and cometary bombardment of Earth as recorded in water deuterium to protium ratio, Icarus 148 (2000) 508–512.

[21] L.P. Knauth, Salinity history of the Earth's early ocean, Nature 395 (1998) 554–555.

[22] J.W. Valley, W.H. Peck, E.M. King, S.A. Wilde, A cool early Earth, Geology 30 (2002) 351–354.

[23] G. Ryder, Mass flux in the ancient Earth–Moon system and benign implications for the origin of life on Earth, J. Geophys. Res. 107 (E4) (2002) 13 (art. no. 5022).

[24] E. Tajika, Faint young Sun and the carbon cycle: implications for the Proterozoic global glaciations, Earth Planet. Sci. Lett. 214 (2003) 443–453.

[25] P.F. Hoffman, D.P. Schrag, The snowball Earth hypothesis: testing the limits of global change, Terra Nova 14 (2002) 129–155.

[26] N.H. Sleep, K. Zahnle, Carbon dioxide cycling and implications for climate on ancient Earth, J. Geophys. Res. 106 (2001) 1373–1399.

[27] J.L. Bada, C. Bigham, S.L. Miller, Impact melting of frozen oceans on the early Earth: implications for the origin of life, Proc. Natl. Acad. Sci. U. S. A. 91 (1994) 1248–1250.

[28] C. Sagan, C. Chyba, The early faint sun paradox: organic shielding of ultraviolet-labile greenhouse gases, Science 276 (1997) 1217–1221.

[29] R. Schoenberg, B.S. Kamber, K.D. Collerson, S. Moorbath, Tungsten isotope evidence from ~3.8 Gyr metamorphosed sediments for early meteorite bombardment of the Earth, Nature 418 (2002) 403–405.

[30] P.J. Fawcett, M.B.E. Boslough, Climatic effects of an impact-induced equatorial debris ring, J. Geophys. Res. 107 (D15) (2002) 18 (art. no. 4231).

[31] A. Lazcano, J.L. Bada, The 1953 Stanley L. Miller experiment: fifty years of prebiotic organic chemistry, Orig. Life Evol. Biosph. 33 (2003) 235–242.

[32] M. Levy, S.L. Miller, K. Brinton, J.L. Bada, Prebiotic synthesis of adenine and amino acids under Europa-like conditions, Icarus 145 (2000) 609–613.

[33] S. Miyakawa, H.J. Cleaves, S.L. Miller, The cold origin of life: A. Implications based on the hydrolytic stabilities of hydrogen cyanide and formamide, Orig. Life Evol. Biosph. 32 (2002) 195–208.

[34] J.A. Brandes, N.Z. Boctor, G.D. Cody, B.A. Cooper, R.M. Hazen, H.S. Yoder, Abiotic nitrogen reduction on the early Earth, Nature 395 (1998) 365–367.

[35] D.P. Summers, Sources and sinks for ammonia and nitrate on the early Earth and the reaction of nitrite with ammonia, Orig. Life Evol. Biosph. 29 (1999) 33–46.

[36] H.J. Cleaves, S.L. Miller, Oceanic protection of prebiotic organic compounds from UV radiation, Proc. Natl. Acad. Sci. U. S. A. 95 (1998) 7260–7263.

[37] F.P.R. Nilson, Possible impact of a primordial oil slick on the atmospheric and chemical evolution, Orig. Life Evol. Biosph. 32 (2002) 247–253.

[38] S.L. Miller, The endogenous synthesis of organic compounds, in: A. Brack (Ed.), The Molecular Origins of Life: Assembling Pieces of the Puzzle, Cambridge University Press, Cambridge, UK, 1998, pp. 59–85.

[39] J.D. Sutherland, J.N. Whitfield, Prebiotic chemistry: a bioorganic perspective, Tetrahedron 53 (1997) 11493–11527.

[40] S. Miyakawa, H. Yamanashi, K. Kobayashi, H.J. Cleaves, S.L. Miller, Prebiotic sysnthesis from CO atmospheres: Implications for the origins of life, Proc. Natl. Acad. Sci. U. S. A. 99 (2002) 14628–14631.

[41] D. Canil, H.St.C.O. O'Neill, D.G. Pearson, R.L. Rudnick, W.F. McDonough, D.A. Carswell, Ferric iron in peridotites and mantle oxidation states, Earth Planet. Sci. Lett. 123 (1994) 205–220.

[42] J.A. Katili, A. Sudradjat, Galunggung: the 1982–1983 Eruption, Volcanological Survey of Indonesia, Directorate General of Geology and Mineral Resources, Department of Mines and Energy, Republic of Indonesia, 1984, p. 3.

[43] S.R. McNutt, C.M. Davis, Lightning associated with the 1992 eruptions of Crater Peak, Mount Spurr Volcano, Alaska, J. Volcan. Geotherm. Res. 102 (2000) 45–65.

[44] R. Navarro-González, M.J. Molina, L.T. Molina, Chemistry of Archean volcanic lightning, in: M. Akaboshi, N. Fujii, N. Navarro-González (Eds.), Role of Radiation in the Origin and Evolution of Life, Kyoto University Press, Kyoto, Japan, 2000, pp. 121–141.

[45] A. Segura, R. Navarro-González, Experimental simulation of early Martian volcanic lightning, Adv. Space Res. 27 (2001) 201–206.

[46] C.F. Chyba, C. Sagan, Endogenous production, exogenous delivery, and impact-shock synthesis of organic molecules: an inventory for the origins of life, Nature (1992) 125–132.

[47] O. Botta, J.L. Bada, Extraterrestrial organic compounds in meteorites, Surv. Geophys. 23 (2002) 411–467.

[48] L. Becker, R.J. Poreda, T.E. Bunch, Fullerenes: an extraterrestrial carbon carrier phase for noble gases, Proc. Natl. Acad. Sci. U. S. A. 97 (2000) 2979–2983.

[49] Y. Sekine, S. Sugita, T. Kadono, T. Matsui, Methane production by large iron meteorite impacts on the early Earth, J. Geophys. Res. 108 (E7) (2003) 11 (art. no. 5070).

[50] D.P. Glavin, J.L. Bada, Survival of amino acids in micrometeorites during atmospheric entry, Astrobiology 1 (2001) 259–269.

[51] D.P. Glavin, G. Matrajt, J.L. Bada, A search for extraterrestrial amino acids in Antarctic micrometeorites: implications for the exogenous delivery of organic compounds, Adv. Space Sci. 33 (2004) 106–113.

[52] A.R. Hill, C. Böhler, L.E. Orgel, Polymerization on the rocks: negatively charged α-amino acids, Orig. Life Evol. Biosph. 28 (2001) 235–243.

[53] J.P. Ferris, Montmorillonite catalysis of 30–50 mer oligonucleotides: laboratory demonstration of potential steps in the origin of the RNA World, Orig. Life Evol. Biosph. 32 (2002) 311–332.

[54] S.J. Sowerby, C.-M. Mörth, N.G. Holm, Effect of temperature on the adsorption of adenine, Astrobiology 1 (2001) 481–487.

[55] R. Liu, L.E. Orgel, Polymerization of β-amino acids in aqueous solution, Orig. Life Evol. Biosph. 28 (1998) 47–60.

[56] L.E. Orgel, Polymerization on the rocks: theoretical introduction, Orig. Life Evol. Biosph. 28 (1998) 227–234.

[57] K.E. Nelson, M.P. Robertson, M. Levy, S.L. Miller, Concentration by evaporation and the prebiotic synthesis of cytosine, Orig. Life Evol. Biosph. 31 (2001) 221–229.

[58] R. Liu, L.E. Orgel, Efficient oligomerization of negatively-charges β-amino acids at -20 °C, J. Am. Chem. Soc. 119 (1997) 4791–4792.

[59] A. Kanavarioti, P.A. Monnard, D.W. Deamer, Eutectic phases in ice facilitate nonenzymatic nucleic acid synthesis, Astrobiology 1 (2001) 271–281.

[60] B.M. Rode, Peptides and the origin of life, Peptides 20 (1999) 773–786.

[61] E. Imai, H. Honda, K. Hatori, A. Brack, K. Matsuno, Elongation of oligopeptides in a simulated submarine hydrothermal system, Science 283 (1999) 831–833.

[62] Y. Ogata, E. Imai, H. Honda, K. Hatori, K. Matsuno, Hydrothermal circulation of seawater through hot vents and contribution of interface chemistry to prebiotic synthesis, Orig. Life Evol. Biosph. 30 (2000) 527–537.

[63] D. Kadko, D.A. Butterfield, The relationship of hydrothermal fluid composition and crustal residence time to the maturity of vent fluids on the Juan de Fuca Ridge, Geochim. Cosmochim. Acta 62 (1998) 659–668.

[64] C. Huber, W. Eisenreich, S. Hecht, G. Wächtershäuser, A possible primordial peptide cycle, Science 301 (2003) 938–940.

[65] A. Robertson, A.J. Sinclair, D. Philp, Minimal self-replication systems, Chem. Soc. Rev. 29 (2000) 141–152.

[66] K.E. Nelson, M. Levy, S.L. Miller, Peptide nucleic acids rather than RNA may have been the first genetic molecule, Proc. Natl. Acad. Sci. U. S. A. 97 (2000) 3868–3871.

[67] A. Eschenmoser, Chemical etiology of nucleic acid structure, Science 284 (1999) 2118–2124.

[68] A. Eschenmoser, The TNA-family of nucleic acid systems: properties and prospects, Orig. Life Evol. Biosph. 34 (2004) 277–306.

[69] I.A. Kozlov, L.E. Orgel, P.E. Nielsen, Remote enantioselection transmitted by an achiral peptide nucleic acid backbone, Angew. Chem. Int. Ed. 39 (2000) 4292–4295.

[70] J.L. Bada, Biogeochemistry of organic nitrogen compounds, in: B.A. Stankiewics, P.F. van Bergen (Eds.), Nitrogen-Containing Macromolecules in the Bio- and Geospheres, ACS Symposium Series, vol. 707, American Chemical Society, Washington DC, 1999, pp. 64–73.

[71] V. Marcano, P. Benitez, L. Fajardo, E. Palacios-Prü, Stability of ribonucleic acid in protective environments of alkanes $\geq n$-C_{18}—results from experiments in the laboratory, in: J. Chela-Flores, T. Owen, F. Raulin (Eds.), First Steps in the Origin of Life in the Universe, Kluwer Academic Publishers, Dordrecht, The Netherlands, 2001, pp. 99–102.

[72] D.P. Glavin, M. Schubert, J.L. Bada, Direct isolation of purines and pyrimidines from nucleic acids using sublimation, Anal. Chem. 74 (2002) 6408–6412.

[73] V.G. Stepanov, J. Nyborg, Thermal stability of aminoacyl–tRNAs in aqueous solutions, Extremopiles 6 (2002) 485–490.

[74] X.S. Wang, Stability of Genetic Informational Molecules under Geological Conditions, PhD thesis, Scripps Institution of Oceanography, University of California at San Diego, 1998, pp. 200.

[75] M. Tehei, B. Franzetti, M.C. Maurel, J. Vergne, C. Hountondji, G. Zaccai, The search for traces of life: the protective effect of salt on biological mancromolecules, Extremophiles 6 (2002) 427–430.

[76] D.P. Bartel, P.J. Unrau, Constructing an RNA World, Trends Chem. Biol. 9 (1999) M9–M13.

[77] R.K. Kumar, M. Yarus, RNA-catalyzed amino acid activation, Biochemistry 40 (2001) 6998–7004.

[78] M.P. Robertson, S.L. Miller, Prebiotic synthesis of 5-substituted uracils: a bridge between the RNA World and the DNA-protein world, Science 268 (1995) 702–705.

[79] M.M. Hanczyc, S.M. Fujikawa, J.W. Szostak, Experimental models of primitive cellular compartments: encapsulation, growth, and division, Science 302 (2003) 618–622.

[80] H.J. Cleaves, S.L. Miller, The nicotinamide biosynthetic pathway is a by-product of the RNA World, J. Mol. Evol. 52 (2001) 73–77.

[81] H.J. Morowitz, J.D. Kostelnik, J. Yang, G.D. Cody, The origin of intermediary metabolism, Proc. Natl. Acad. Sci. U. S. A. 97 (2000) 7704–7708.

[82] A. Sreedhara, Y. Li, R. Breaker, Ligating DNA with DNA, J. Am. Chem. Soc. 126 (2004) 3454–3460.

[83] C. Brochier, H. Phillippe, Phylogeny: a non-hyperthermophilic ancestor for bacteria, Nature 417 (2002) 244.

[84] W.F. Doolittle, The nature of the univerisal ancestor and the evolution of the proteome, Curr. Opin. Biol. 10 (2000) 355–358.

[85] P. Forterre, A hot story from comparative genomics: reverse gyrase is the only hyperthermophile-specific protein, Trends Genet. 18 (2002) 236–238.

[86] A. Pross, Causation and the origin of life, metabolism or replication first? Org. Life Evol. Biosphere 34 (2004) 307–321.

[87] N.P. Pace, The universal nature of biochemistry, Proc. Natl. Acad. Sci. U. S. A. 98 (2001) 805–808.

[88] D.J. Des Marais, M.O. Harwit, K.W. Jucks, J.F. Kasting, D.N.C. Lin, J.I. Lunine, J. Schneider, S. Seager, W.A. Traub, N.J. Woolf, Remote sensing of planetary properties and biosignatures on extrasolar terrestrial planets, Astrobiology 2 (2002) 153–181.

Jeffrey L. Bada is professor of Marine Chemistry and Director of the NASA Specialized Center of Research and Training (NSCORT) in Exobiology at the Scripps Institution of Oceanography, University of California at San Diego. Prof. Bada's research deals with the geochemistry of amino acids, organic cosmogeo-chemistry, the sources and stability of organic compounds on the primitive Earth and other solar system bodies, the origin of homochirality on Earth and the detection of possible remnants of ancient life on Mars both by in situ analyses on the planet, and from the study of Martian meteorites. Dr. Bada has revived sublimation as an extraction technique for organic molecules from natural samples and has played a pioneering role in the development of the Mars Organic Analyzer (MOA) instrument package that is designed to search for amino acids and other organic compounds directly on the surface of Mars.

Reprinted from
Earth and Planetary Science Letters 225 (2004) 253–269

www.elsevier.com/locate/epsl

Early Earth differentiation

Michael J. Walter[a],*, Reidar G. Trønnes[b,c]

[a]*Department of Earth Sciences, University of Bristol, Wills Memorial Building, Queen's Road, Bristol BS8 1RJ, United Kingdom*
[b]*Nordic Volcanological Institute, University of Iceland, Natural Sciences Building, 101 Reykjavík, Iceland*
[c]*Geological Museum, UNM, University of Oslo, P.O. Box 1172 Blindern, N-0318 Oslo, Norway*

Received 5 January 2004; received in revised form 24 June 2004; accepted 9 July 2004

Editor: A.N. Halliday

Abstract

The birth and infancy of Earth was a time of profound differentiation involving massive internal reorganization into core, mantle and proto-crust, all within a few hundred million years of solar system formation (t_0). Physical and isotopic evidence indicate that the formation of iron-rich cores generally occurred very early in planetesimals, the building blocks of proto-Earth, within about 3 million years of t_0. The final stages of terrestrial planetary accretion involved violent and tremendously energetic giant impacts among core-segregated Mercury- to Mars-sized objects and planetary embryos. As a consequence of impact heating, the early Earth was at times partially or wholly molten, increasing the likelihood for high-pressure and high-temperature equilibration among core- and mantle-forming materials. The Earth's silicate mantle harmoniously possesses abundance levels of the siderophile elements Ni and Co that can be reconciled by equilibration between iron alloy and silicate at conditions comparable to those expected for a deep magma ocean. Solidification of a deep magma ocean possibly involved crystal–melt segregation at high pressures, but subsequent convective stirring of the mantle could have largely erased nascent layering. However, primitive upper mantle rocks apparently have some nonchondritic major and trace element refractory lithophile element ratios that can be plausibly linked to early mantle differentiation of ultra-high-pressure mantle phases. The geochemical effects of crystal fractionation in a deep magma ocean are partly constrained by high-pressure experimentation. Comparison between compositional models for the primitive convecting mantle and bulk silicate Earth generally allows, and possibly favors, 10–15% total fractionation of a deep mantle assemblage comprised predominantly of Mg-perovskite and with minor but geochemically important amounts of Ca-perovskite and ferropericlase. Long-term isolation of such a crystal pile is generally consistent with isotopic constraints for time-integrated Sm/Nd and Lu/Hf ratios in the modern upper mantle and might account for the characteristics of some mantle isotope reservoirs. Although much remains to be learned about the earliest formative period in the Earth's development, a convergence of theoretical, physical, isotopic and geochemical arguments is beginning to yield a self-consistent portrait of the infant Earth.
© 2004 Published by Elsevier B.V.

Keywords: planetary accretion; core formation; magma ocean; element partitioning; mantle geochemistry

* Corresponding author. Tel.: +44 117 954 5378; fax: +44 117 925 3385.
E-mail address: M.J.Walter@bristol.ac.uk (M.J. Walter).

0012-821X/$ - see front matter © 2004 Published by Elsevier B.V.
doi:10.1016/j.epsl.2004.07.008

1. In the beginning

An unfortunate consequence of the Earth's unruly beginnings is that we have essentially no direct rock record of its birth and early childhood. Yet it was during that earliest stage when much of the Earth's personality was shaped. The first 860 Ma of the Earth's lifetime are generally referred to as the 'Hadean', a stage during which the Earth underwent major episodes of differentiation of a magnitude that will unlikely ever be repeated. The metallic core separated from the silicate mantle, and the silicate mantle may have developed much of its internal character.

A detailed sketch of the Earth's birth can now be drawn on the basis of stochastic and hydrodynamic modeling of solar nebula processes. The picture that emerges depicts a violent infancy. The formation of a solar nebula starts with gravitational contraction of a localized dense region in an interstellar molecular cloud, and a supersonic shock event, probably from a nearby supernova or perhaps an AGB star, triggered cloud collapse [1,2]. Within the inner, or 'terrestrial', region of the nebular disc, sticky primordial dust particles coagulate to form a large population of meter- to kilometer-sized objects [3]. Gravity and gas-drag began to affect the larger objects, driving a period of runaway growth lasting on the order of 10^6 years, the product of which includes a vast array of planetesimals and planetary embryos [4,5]. A final phase of coalescence, completed on a time scale of 10^8 years, occurs mainly by infrequent albeit massive impacts among larger objects, resulting in a few terrestrial planets that sweep up the flotsam and jetsam. Importantly, such modeling predicts that the lion's share of a terrestrial planet's mass and energy is acquired during this late-impact stage [6–8]. Vindication of the standard planetary accretion model comes from hydrodynamic simulations which indicate that giant impacts can account for the origin and low density (iron-poor) of the Moon, the spin angular momentum of the Earth–Moon system, as well as the Earth's rotation [8–12].

Giant impacts involve a vast amount of potential and kinetic energy. Physical models show that the energy buried in the proto-Earth from a giant collision with a Mars-sized impactor, such as in a 'moon-forming' event, would have been sufficient to raise interior temperatures by thousands of degrees, causing high degrees of melting and a global magma ocean [13,14]. During its accretion period, the proto-Earth likely endured several massive impacts and may have experienced transient magma oceans. The birth of Earth was violent and hot. How did such an angry young planet grow and differentiate into the seemingly well-adjusted, mature planet we know today?

2. Core formation

The formation of a metallic core is the most dramatic differentiation event in a terrestrial object involving segregation of dense, iron-rich metal alloy from magnesium-rich silicate. Understanding the core formation process provides insight into accretion and the earliest formative period of a planet, and in recent years, answers have begun to emerge to key questions regarding the timing, physical conditions and mechanisms of core formation.

2.1. When do planetary cores form?

Parent/daughter fractionation of short- and long-lived radionuclides during core formation generates isotopic signals that constrain the timing of accretion and core formation in terrestrial objects (see Table 1

Table 1
Radiogenic isotope systems

Parent–Daughter	Half-life (years)	$D_{parent}^{met/sil}/D_{daughter}^{met/sil}$[a]	$D_{parent}^{CP/RM}/D_{daughter}^{CP/RM}$[b]
Short-lived			
^{26}Al–^{27}Mg	0.7×10^6	~1	~1
^{146}Sm–^{142}Nd	103×10^6	~1	<1
^{182}Hf–^{182}W	9×10^6	≪1	>1
Long-lived			
^{87}Rb–^{87}Sr	49×10^9	~1	>1
^{147}Sm–^{143}Nd	106×10^9	~1	<1
^{176}Lu–^{176}Hf	36×10^9	~1	<1
^{238}U–^{206}Pb	4.5×10^9	≪1	>1
^{235}U–^{207}Pb	0.8×10^9	≪1	>1

[a] $D_{parent}^{met/sil}/D_{daughter}^{met/sil}$ = the ratio of partition coefficients, $D^{metal/silicate}$, during metal–silicate equilibration for parent and daughter elements in radiogenic isotope systems.

[b] $D_{parent}^{CP/RM}/D_{daughter}^{CP/RM}$ = the ratio of partition coefficients, $D^{CP/RM}$, during fractionation of a deep mantle crystal pile, CP, from residual mantle, RM, for parent and daughter elements.

for review of some relevant isotope systems) [15–17]. The short-lived Hf–W isotope system places severe restrictions on the timing of core formation in planet-esimals [18,19]. Iron-rich cores apparently formed in some objects within about 3 million years of t_0, a time frame that provides important constraints on plausible core formation mechanisms in growing terrestrial bodies.

In contrast, analyses of Earth samples reveal a much later, or more protracted, core formation. The W isotopic composition of mantle silicates suggests core formation at ~30 million years after t_0, but this model 'age' requires a wholesale core–mantle reequilibration event at that time [18,19]. But it is unlikely that such a global event ever happened, because Pb and Xe isotopes indicate more than a factor of two longer core formation time scales [20]. Defining a core formation rate for the Earth depends on the degree to which silicate and iron equilibrate between impactors and the proto-Earth, especially during a giant moon-forming impact that might have disturbed the cosmochemically volatile U–Pb system relative to the refractory Hf–W system [20]. Defining an absolute 'age' of core formation may have little real meaning for a continuous accretion process, but in a broad sense, the radioisotope data prescribe that the bulk of the Earth's core formed on a 10^7- to 10^8-year time scale, generally consistent with accretion time scales predicted by stochastic modeling.

2.2. How do planetary cores form?

Common wisdom has held that molten iron alloy does not 'wet' silicate grain boundaries, rendering the silicate effectively impermeable. For this reason, core formation in planetary bodies has been thought to require extensive silicate melting, enabling dense molten metal droplets to sink through silicate melt [21–23]. To achieve high degrees of silicate melting in a planetesimal requires prodigious heat, and the most important early source of energy may have been the decay of short-lived radioactive ^{26}Al ($t_{1/2}$~0.7 Ma), which is known to have been active in the solar disc [24]. Models of the thermal evolution of a growing planetesimal indicate that for an object large enough to retain heat efficiently (r~30 or 40 km), the thermal evolution is dependent on both the timing of accretion

and the initial ^{26}Al/^{27}Al abundance ratio [25–27]. These models show that bodies with ^{26}Al/^{27}Al greater than about 1×10^{-5} and which accreted within a few ^{26}Al half-lives of t_0 would have heated internally to temperatures sufficient to melt iron alloy and even silicate, as illustrated in Fig. 1. High initial ^{26}Al/^{27}Al ratios, as for example the canonical initial ^{26}Al/^{27}Al ratio of 5×10^{-5} that is based on measurements of calcium–aluminum inclusions (CAI) in meteorites [28], would clearly have produced sufficient heat to melt planetesimals. However, recent measurements on carbonaceous chondrites reveal much lower initial ratios in some cases, of the order 5×10^{-6} [29], indicating that radiogenic heating may not have been sufficient to cause partial melting and implying a larger role for impact heating among planetesimals [29].

Recent experimental results indicate that excess molten iron alloy over a percolation threshold of only about 5 vol.% can create permeability in a solid silicate matrix [27]. The metal content in planetary objects may generally exceed 10 vol.%, so objects heated sufficiently to melt iron alloy could have experienced an incipient albeit inefficient stage of core formation upon melting of iron alloy. Modeling indicates that melting and segregation of iron alloy could have occurred within a few million years of the origin of the solar system, consistent with the short time frame allotted by tungsten isotopes (Fig. 1). Efficient removal of molten iron alloy to a planetary core may have required heating to well above the silicate solidus, such as would have occurred for high initial ^{26}Al/^{27}Al ratios and rapid accretion or as a consequence of shock-heating among small, porous planetesimals [29]. Alternatively, efficient removal of iron alloy may have been aided by heat advection in the silicate, as recent experiments show that plastic deformation of the silicate matrix promotes connec-tivity in an otherwise impermeable silicate, possibly even at low melt fractions [32].

2.3. How did the Earth's core form?

The prevailing evidence indicates early core for-mation in planetesimals. The Earth acquired most of its mass from giant impacts with Mercury- to Mars-sized objects during late-stage accretion, and these objects, as well as the embryonic Earth, are likely to

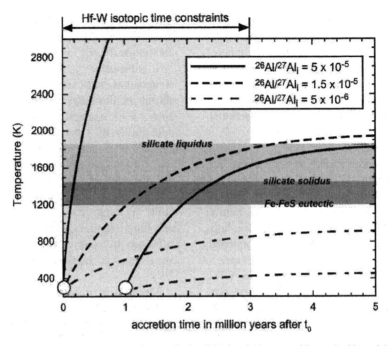

Fig. 1. Thermal models for post-accretional heating of a planetary body of CI chondritic composition and with a minimum radius of 30 km by decay of short-lived [26]Al. The silicate/metal proportion is 85:15 (vol.%) and the initial temperature is assumed to be 255 K. Heating trajectories are shown for a canonical initial [26]Al/[27]Al ratio of 5×10^{-5} (solid lines) [28], as well as for lower ratios of 1.5×10^{-5} (dashed lines) and 5×10^{-6} (long–short dashed lines). Estimated 1 atm melting intervals in the Fe–FeS system (dark shaded region) and Allende silicate (medium shaded region) are shown [30,31]. Heating paths emanating from 0 and 1 Myr show the effects of accretion time. The W isotopic composition of iron meteorites indicates core formation within about 3 m.y. of solar system formation in planetesimals (light shaded region). Both the initial [26]Al/[27]Al ratio and the timing of accretion have large effects on the rate of heating [26,28], but bodies that accreted within 1 Myr of t_0 may generally have reached the melting points of iron alloy and silicate within 3 Myr.

have had preformed cores [33]. What happens to the metallic core of an impactor during the violent collision with the proto-Earth?

Hydrodynamic simulations of giant impacts indicate a dependence on the angular momentum of the collision but generally show that while some of the impactor core may merge directly with the Earth's proto-core, much of the impactor core would be ejected into low orbit and would fall rapidly back to the Earth [9]. Models are too low in resolution to trace the fate of core materials in detail, but if the iron falls back to the Earth in large fragments, then dense metallic blobs should sink rapidly to the base of the magma ocean produced by the impact. If the entire silicate mantle were molten, the metal blobs would sink directly to the proto-core with little equilibration, but if the base of the magma ocean were solid silicate, then metal might pool until gravitational instability permits movement to the core in large diapirs [22].

Alternatively, metal from the impactor and perhaps even the Earth's proto-core may have become highly fragmented and emulsified in the magma ocean, in which case small metallic droplets could equilibrate rapidly with silicate melt. The fate of the cores of the impactor and proto-Earth during a giant impact and the process of emulsification are poorly understood dynamical problems needing attention.

Direct evidence of core formation is preserved in the siderophile ('iron-loving') element geochemistry of mantle rocks. Siderophile elements partition nearly wholesale into metal during core formation, but trace amounts remain in the silicate. Element partitioning is sensitive to the physical conditions of metal–silicate equilibrium (e.g., P, T, X), and if core formation were a singular, equilibrium event, then the conditions of equilibrium can be deduced from the pattern of siderophile element depletion in mantle rocks.

Compositional models that are based on the assumption that refractory elements are present in chondritic relative proportions in bulk Earth have yielded a generally consistent picture regarding the degree to which siderophile elements are depleted in the mantle, as shown on Fig. 2 [34–36]. The so-called highly siderophile elements (HSE) such as Au and the Pt-group elements exist as a group in near-chondritic relative proportions in the Earth's mantle, meaning that whatever process established their abundances did not fractionate them one from another. Chondritic relative HSE cannot be reconciled with measured low-pressure and -temperature metal/silicate partitioning behavior, which show that within the group, these elements have highly variable siderophile character, with partition coefficients varying by as much as 10^8 [35]. Furthermore, the HSE are present at levels that exceed expectations for metal segregation at shallow mantle conditions by many orders of magnitude. It has been suggested that metal–silicate equilibration at higher pressures and temperatures might potentially explain HSE abundances, but the limited experimental data do not support that contention [37]. The standard model is that core formation effectively stripped all the HSE from the mantle, and subsequently, HSE abundances were raised to observed values by the final 1% of Earth accretion comprised of meteoric materials with chondritic proportions of HSE [38]. Alternatively, back-mixing of a small amount outer-core materials might also have established HSE abundances in the mantle [39]. In either case, HSE abundances cannot be used as a signal of core–mantle equilibration.

In contrast to the HSE, the abundances of moderately and slightly siderophile elements (MSE and SSE, Fig. 2) would not have been significantly altered by the late addition of meteoric material and so are potentially much more informative about the conditions of core formation. Of these elements, Ni and Co have provided the most powerful constraints on core formation because their abundances in the mantle are most accurately and precisely known, and their partitioning behavior has been relatively well studied experimentally over a wide range of conditions. A fascinating feature of mantle rocks is that within uncertainty, Ni and Co are present in essentially chondritic relative proportions, both having apparently been depleted similarly during core formation (Fig. 2). A relatively complete experimental database that describes Ni and Co partitioning between molten silicate and molten iron alloy at pressures up to ~40 GPa and over a range of high

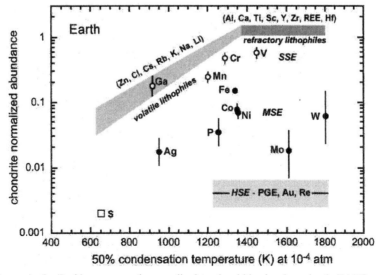

Fig. 2. Elemental abundances in the Earth's upper mantle normalized to chondritic abundance levels [34,35] and plotted relative to 50% condensation temperatures in a vacuum (10^{-4} atm), which is used as a proxy for a nebular condensation sequence. The generalized trends of refractory lithophile and volatile elements are shown as shaded regions. Slightly siderophile elements (SSE) are plotted as open circles and moderately siderophile elements (MSE) as solid circles. The highly siderophile elements (HSE) have nearly constant abundance as a group, and abundance uncertainties are shown for simplicity by a shaded region.

temperatures (e.g., ~2000 to 3000 K) and oxygen content now exists. Experiments show that at constant oxygen content, an increase in pressure and/or temperature causes both Ni and Co to become less siderophile, but at different rates, so that they can have similar partitioning behavior under certain conditions as required by mantle rocks. [40–42].

Fig. 3 shows an example of how experimental partitioning data can be used to deduce a set of conditions at which metal and silicate might last have equilibrated in the Earth. The diagram, which is constructed on the basis of our own parameterizations to high-pressure and -temperature experimental data (see figure caption), shows the locus of pressures and temperatures that satisfies the condition that the ratio of the Ni and Co exchange partition coefficients (K_D) is ~1.1 during metal–silicate equilibrium. In this model, the relative and absolute abundances of Ni and Co in the Earth's upper mantle are satisfied simultaneously at ~40 GPa and ~2800 K. Modeling like this is used as primary evidence for metal–silicate equilibration in a deep magma ocean [40,41,52],

plausibly linked to large-scale melting during one or more giant impacts during late-stage accretion.

Partitioning models have been developed for other MSE and SSE (e.g., Ga, P, W, and Mo), and a signature of high-pressure and -temperature metal–silicate equilibration is interpreted for these elements as well, although their mantle abundances and partitioning behavior are far less well constrained than for Ni and Co [35,41,52]. It is well to remember that limitations inherent to the partition coefficient parameterization scheme, due for example to correlated effects among variables or nonlinear effects, make extrapolation outside the range of the data precarious at best and misleading at worst. Further constraints on plausible metal–silicate equilibrium conditions require collection of internally consistent sets of partitioning data over a wide pressure, temperature and compositional range for each element in order to more accurately deduce the effects of each variable (e.g., P, T, X) in isolation. A solidly anchored thermodynamic understanding of how intensive properties effect partitioning is needed in order to reliably

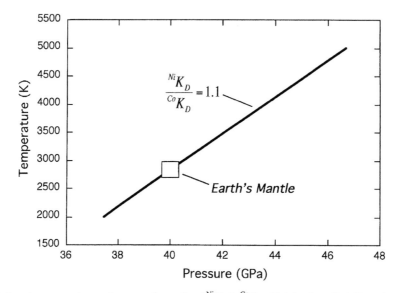

Fig. 3. The locus of solutions in pressure–temperature space for a ratio of $^{Ni}K_D$ to $^{Co}K_D$ of 1.1:1, where K_D is the molten Fe alloy/molten silicate exchange partition coefficient, $K_D = \left(^{met}X_M / {}^{sil}X_{MO}\right) \div \left(^{met}X_{Fe} / {}^{sil}X_{FeO}\right)$, where M=Ni or Co and assuming that Fe, Ni and Co are in a divalent oxidation state. The large square shows the 'unique' pressure–temperature condition that satisfies the absolute Ni and Co K_Ds required to reproduce the observed abundances of these elements in the Earth's mantle. The diagram is constructed based on parameterizations to high-pressure and -temperature experimental partitioning data [40,42–51]. Data are regressed to the equation, $\ln K_D = \alpha 1/T + \beta P/T + \chi \ln(1-X_S) + \delta \ln(1-X_C) + k$, where T is temperature in Kelvin, P is pressure in GPa, X_S and X_C are the mol fractions of sulfur and carbon in the iron alloy and k is a constant. Values for fitted coefficients with standard uncertainties are as follows: Ni: α=9887 (853), β=−181 (16), χ=0.17 (0.19), δ=1.03 (0.5) and k=−0.12 (0.42), n=68, R^2=0.90; Co: α=6847 (1113), β=−70 (15), χ=0.78 (0.17), δ=0.58 (0.38) and k=−0.53 (0.48), n=44, R^2=0.75.

extrapolate to conditions outside those encompassed by the experimental database.

Presuming that a signal of high-pressure and -temperature equilibration is robust, interpreting exactly what the signal means is a matter of conjecture. In the simplest model, metal segregates from a magma ocean, ponds at the ocean floor, equilibrates essentially isobarically and eventually sinks to the core in diapirs or perhaps by percolation through high-pressure silicate phases [33,41]. This model is probably unrealistic because numerical calculations show that large blobs of metal require several orders of magnitude more time to equilibrate than plausible magma ocean crystallization time scales [53]. Another possibility is that emulsified metal precipitates out of the magma ocean in much the same manner as rain out of clouds [22]. Small metal droplets would equilibrate rapidly with silicate melt as they descend, and deduced metal–silicate equilibrium conditions would reflect the average of a polybaric, polythermal process [42,53].

Geochemically based core formation modeling has typically presumed a singular metal–silicate equilibration event (single-stage). However, accretion theory leads us to expect numerous large impacts among core-segregated objects, each of which could have left an indelible imprint on the siderophile element content of the mantle (multistage). For example, it is conceivable that a large impactor could have been a fragment of a previously differentiated object with overall nonchondritic siderophile element ratios (e.g., Mercury). Another major unknown factor is the degree of metal–silicate reequilibration between impactor and target materials. Whether or not current siderophile element abundances in the mantle reflect a single, final giant impact in which effectively all the silicate and metal were re-homogenized and equilibrated or reflect a multistage history of metal–silicate partial equilibration is a difficult question to assess. Reconstructing the details of a multi-event history of metal segregation is a daunting and perhaps intractable prospect.

3. Mantle differentiation in the Hadean

Large-scale melting of Earth is unavoidable in the standard model of planetary accretion. Deposition of impact energy when a planetary embryo swallows up planetesimals during late-stage accretion would result in magma ocean formation. Repeated impacts could sustain a magma ocean especially in the presence of a blanketing atmosphere [54]. A giant collision of a 'Mars-sized' impactor with proto-Earth is the prevailing theory for the origin of the Moon [9,11,12], and such an impact could deliver sufficient energy to melt the entire planet [13].

Hydrodynamic impact models indicate that shock-induced impact melting would create a melt pond within the impact crater, which upon isostatic readjustment would spread like a blanket over the surface, forming a deep magma ocean [14]. The extent and depth of the magma ocean depend on many factors including the impactor/target mass ratio, impact velocity and initial temperatures of the objects [14]. How would a magma ocean cool and crystallize? Would crystal fractionation occur and be preserved? Does the modern mantle provide physical or geochemical evidence that informs us about magma ocean differentiation processes?

3.1. How would a magma ocean crystallize?

Cooling of a peridotite magma ocean would eventually lead to crystallization from the bottom up because liquidus crystallization curves have dT/dP slopes that are steep relative to an adiabat in a convecting magma ocean, as shown schematically in Fig. 4. Experimental melting studies show that crystallization in the deep mantle ($>\sim$700 km) would be dominated by Mg-perovskite, with minor amounts of Ca-perovskite and ferropericlase, whereas crystallization at shallower levels would be dominated by olivine and its high-pressure polymorphs, majorite garnet, and lesser amounts of pyroxenes [55–57].

Numerical simulations show that the magma ocean solidification process is sensitive to many factors, including surface temperature, cooling rate, crystal nucleation and growth rates and melt and crystal-mush viscosities [54,58,59]. Surface temperature and cooling rate are highly dependant on whether a rock vapor or steam atmosphere is present during cooling [54]. In the absence of an atmosphere, heat radiation to space is very efficient and the deeper portion of a magma ocean cools rapidly (10^3 years). The upper portion of the magma ocean (e.g., transition zone and

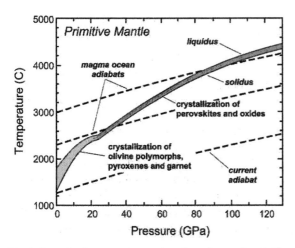

Fig. 4. Generalized pressure–temperature phase diagram for a fertile peridotite mantle [55–57]. Schematic magma ocean adiabats illustrate that crystallization would proceed from the bottom up. At pressures greater than about 25 GPa, crystallization would involve Mg-perovskite, Ca-perovskite and (Mg.Fe)-oxide, and at lower pressures, it would invovle olivine polymorphs, garnet and pyroxenes.

shallow upper mantle) could remain hot and molten much longer due to the depression of the solidus at pressures less than about 25 Gpa (Fig. 4), especially if a 'chill-crust' forms at the surface—heat radiation to space from the surface would be efficient enough to cool and crystallize a surface layer of the magma ocean (D. Stevenson personal communication). If an atmosphere persists, it provides very effective thermal insulation and cooling time scales are much longer (10^6–10^7 years) even for deeper portions of the magma ocean.

Hydrodynamic models indicate that vigorous convection during early stages of crystallization might effectively prevent accumulation of crystals if the melt–solid viscosity contrast is sufficiently low [58,59]. The viscosity contrast depends critically on the physical properties of melt at high pressures and the nucleation and growth rates of the crystals, properties that are poorly known for deep mantle phases. The potential for crystal fractionation in a deep magma ocean remains a matter of speculation [54].

An initially stratified mantle after magma ocean solidification would have important chemical and dynamic consequences for mantle evolution. Nearly 4.5 billion years of solid-state convection subsequent

to solidification may have largely erased nascent stratification. However, if lower mantle phases are intrinsically dense and stiff, they could form layers, or 'deep crystal piles', that are resistant to convective mixing.

3.2. Is the modern mantle stratified?

The extent, scale and morphology of mantle chemical heterogeneity has long been a hotly debated topic in mantle geophysics, geodynamics and geochemsitry, with implications for the nature of mantle convection [60–65]. Geochemical and isotope heterogenities in mantle derived magmas dictate distinctive geochemical reservoirs, evoking models for chemical layering [60]. Global seismic discontinuities at ~410 and 660 km image mineralogic boundaries separating the 'upper' mantle from the 'transition zone' and 'lower' mantle. These boundaries can be well ascribed to phase transitions based on known phase relations and elastic properties of high-pressure and -temperature mineral assemblages in fertile mantle peridotite [64,66,67]. Seismic tomography indicates that subducting slabs penetrate deep into the lower mantle [68], at least in some regions, and return flow of material from the deep mantle is implicated. The mixing efficiency of mantle convection is linked to the driving forces of subducting slabs and the vorticity or toroidal motion induced by rotation and strike-slip motion of the surface plates [69,70]. Whole-mantle convection, even when moderated by the 660-km phase transition and a high-viscosity lower mantle, is expected to generally erase large-scale heterogeneities over the age of the Earth. Moreover, mixing was probably considerably more vigorous when the Earth was young and hot. These considerations make it likely that the modern mantle is not chemically stratified in a gross sense.

Yet stable chemical domains in the deeper reaches of the lower mantle remain a distinct possibility. Seismic observations and theoretical models have indicated the possible presence of dense, compositionally distinct regions in the lowermost mantle [71,72]. The depth of the upper boundary of this 'layer' is thought to be laterally and temporally highly variable [72]. Indeed, numerical simulations suggest that seismic observations might best be explained if dense material is restricted to discontinuous piles rather than a continuous, undulating boundary [73].

Massive crystallization differentiation in the Hadean mantle, possibly involving flotation of olivine in the upper mantle, concentration of majorite at transition zone depths or fractionation of lower mantle phases, should have been largely or even completely homogenized by the Earth's convective engine. However, it remains conceivable that some fraction of a crystal pile deposited at the base of the mantle could have become rheologically isolated from the main convecting regime or perhaps slowly eroded as the convecting mantle interacts with the stiffer crystalline mass. The pressure and temperature dependance of densitiy and viscocity is not well known for perovskite and oxide phases at deep mantle conditions. In lieu of direct validation or repudiation of the plausibility for long-term isolation of a deep mantle layer on mineral–physical grounds, it is of interest to investigate what kind of geochemical signal would be expected in the convecting mantle should such a layer actually exist.

3.3. What is the composition of the primitive convecting mantle?

Crystal fractionation induces chemical changes in an evolving magma, and isolation of a crystal pile in the deep mantle would impart a chemical imprint on the remaining residual mantle. The search for a geochemical signal involves a compositional comparison among the original magma ocean (i.e., bulk silicate Earth), the isolated crystal pile and the modern convecting portion of the mantle. To properly constrain a magma ocean crystallization model, reliable compositional estimates for each of these reservoirs are required.

Meteorite compositions, either as direct analogues or through deductions from compositional trends, form the basis of most bulk Earth and bulk silicate Earth compositional models [34,74–77]. The relative constancy of cosmochemically refractory element ratios (e.g., Ca/Al, Al/Ti, Sc/REE) in primitive chondritic meteorites of all flavors helps justify the assumption that refractory elements were delivered to the Earth en masse in primitive solar nebula proportions [34,36]. Correlations among refractory lithophile elements in chondritic meteorites and selected sets of primitive upper mantle rocks have been interpreted to indicate that the bulk silicate Earth has chondritic relative abundances of these elements (such as Ca, Al, Ti, Sc, Zr and REE) [34,74–77].

The crux of the early mantle differentiation issue is whether or not the modern convecting mantle (plus crust) has the same composition as the bulk silicate Earth. Model primitive upper mantle compositions, which may be representative of the convecting mantle, have been constructed based directly on primitive peridotite compositions [78–80], mixtures of depleted peridotite and basalt ('pyrolite') [81] and melt extraction trends in residual peridotites [82]. Fig. 5 shows a set of chondrite-normalized major and trace refractory lithophile element ratios that are most accurately known based on literature assessments of primitive upper mantle composition. Common among these models are superchondritic Mg/Si and Ca/Al ratios by about 20% and 15%, respectively, whereas the Al/Mg ratio is generally subchondritic. Ratios involving Mg and Si may be unreliable as geochemical indicators, however, because they are not constant among various chondrite groups, varying by as much as 20% relative and indicating that a cosmochemical process fractionated these elements prior to Earth accretion [76]. In contrast, the Ca/Al ratio is essentially constant within uncertainty among nearly all chondrite groups, consistent with the highly refractory nature of these elements. Ca/Al is presumably very near the chondritic value in the bulk silicate Earth, in which case the apparently superchondritic ratio in the primitive upper mantle could be a consequence of mantle differentiation [80,82].

Among trace elements Sc, Yb and Sm abundances show a relatively small scatter in primitive mantle rocks, and elemental ratios among these and major elements are typically within about 10–20% of chondritic ratios. Overall, the pattern exhibited in Fig. 5 is one of a primitive convecting mantle with some elemental ratios that are close to chondritic, but with others that may have been fractionated to various degrees away from initially chondritic values.

3.4. How would crystal fractionation in the deep mantle affect mantle chemistry?

Crystal–melt separation can be a very efficient mechanism for fractionating major and trace element

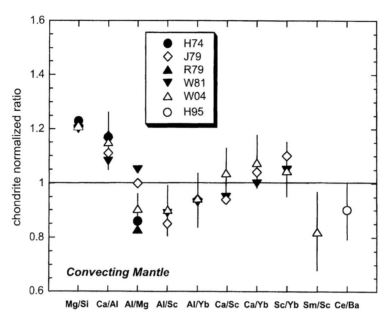

Fig. 5. Elemental ratios in primitive upper mantle rocks normalized to CI chondritic values [34]. Data sources are H74 [78], J79 [79], R79 [83], W81 [84], W04 [80] and H95 [85]. Error bars are best estimates of the mean at the 95% confidence level from the data compilation of [80], with the exception of the Ce/Ba ratio which is based on data from [85].

ratios. Some elements concentrate in the melt phase and are incompatible in minerals (e.g., K, Na, Ba, LREE), whereas others are compatible minerals (e.g., Mg, Ni, Cr). Elemental fractionations are commonly used to retrace the crystallization history of differentiated lavas that erupted at the Earth's surface and in the same way can be used to detect signals of magma ocean fractionation. Mantle convection may have efficiently destroyed initial layering throughout much of the mantle, in which case crystal fractionation in the deep mantle would be the most likely mechanism for imposing long-term geochemical effects. Peridotite crystallization studies show that a deep mantle crystal assemblage would be comprised of Mg-perovskite, Ca-perovskite and ferropericlase [55,56,86]. Although crystallization experiments have only been made at pressures corresponding to the uppermost lower mantle (~25 to 35 GPa), Mg-perovskite is expected to be the first and proportionally dominant phase to crystallize throughout most of the lower mantle, followed down temperature by ferropericlase and Ca-perovskite as the magma ocean cools and solidifies.

Mg-perovskite has a subchondritic Mg/Si ratio, so it is not surprising that removal of this phase in the lower mantle has been envisioned as a way to elevate the Mg/Si ratio in the upper mantle [87]. With this impetus, early partitioning studies concentrated on how Mg-perovskite fractionates major and trace element ratios. Measuring reliable partition coefficients at extreme pressures and temperatures proved challenging, but a general consensus emerged that only very minor amounts of Mg-perovskite fractionation in the lower mantle could be accommodated given near-chondritic refractory element ratios in primitive upper mantle [88,89]. The severe geochemical restrictions imposed by these studies have generally been accepted as evidence against magma ocean crystal fractionation [63,88,89] and have even been used to argue that a magma ocean likely never existed [90].

There are factors, however, that preclude an outright dismissal of at least some fractionation of deep mantle phases. The first of these is that the convecting mantle can be interpreted to have some nonchondritic refractory elemental ratios, most notably the Ca/Al ratio, and these fractionations beg for an explanation (Fig. 5). Second, partition coefficients for all phases were not well known in early magma ocean fractionation models. A recent spate of experiments has vastly improved our knowledge of partitioning among deep mantle phases with some remarkable new findings [80,91–93]. For example, it has recently been revealed that Ca-perovskite can accommodate copious amounts of trace elements, so that even a small amount in a fractionating assemblage can exert considerable leverage on certain elemental ratios [92,93]. Furthermore, in a surprising number of cases, Ca-perovskite fractionates elemental ratios in a sense opposite to that of Mg-perovskite [80].

The exact proportions in which deep mantle phases might have been removed from a crystallizing magma ocean is not known with any certainty, but simple major element mass balance dictates an assemblage dominated by Mg-perovskite with small amounts of Ca-perovskite and ferropericlase [80]. By way of example, Fig. 6 shows how the elemental ratios from Fig. 5 change in two scenarios for deep mantle crystallization, which are based on our own modeling [80]. One involves pure Mg-perovskite fractionation, and the other fractionation

dominated by Mg-perovskite but with minor amounts of Ca-perovskite and ferropericlase. In these examples, the amount of crystal fractionation is chosen to generate a Ca/Al ratio in the residual magma ~15% greater than the chondritic value in order to match the mean estimate of this ratio in the primitive convecting mantle. In one model, 10% of the magma ocean crystallizes as pure Mg-perovskite, whereas in the other model, 15% of the magma ocean crystallizes as an assemblage comprised of 93% Mg-perovskite, 3% Ca-perovskite and 4% ferropericlase.

The Mg/Si ratio is nominally unchanged in these models, reinforcing the conclusion that the high mantle Mg/Si ratio is either inherited from accretionary materials or that Si was sequestered into the core [75,76,83]. Another feature is that, Mg/Si notwithstanding, the models do not produce elemental fractionations that are grossly discordant with observed element ratios in convecting upper mantle, and it is intriguing that the multiphase deep mantle crystal assemblage produces elemental fractionations

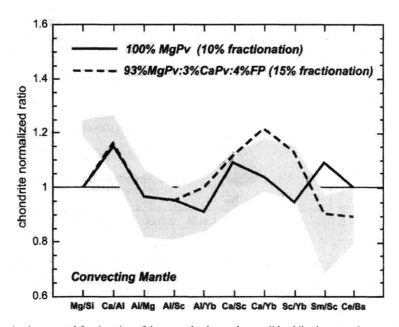

Fig. 6. Two models showing how crystal fractionation of deep mantle phases changes lithophile element ratios away from initially chondritic values in the mantle. The solid line is for the fractionation of 10% of pure Mg-perovskite. The dashed line is for the fractionation of 15% of an assemblage consisting of a mixture of 93% Mg-perovskite, 3% Ca-perovskite and 4% ferropericlase. Partition coefficients for Mg-perovskite and ferropericlase are from [80]. Partition coefficients for Ca-perovskite are based on the modeling of [80] for a coexisting melt with 4 wt.% CaO, based on a parameterization of data from [92,93]. Primitive mantle elemental ratios are as in Fig. 5.

that resemble closely those observed in convecting mantle.

Geochemical models such as these are not uniquely required by phase equilibrium constraints, and there are plenty of possible fractionation scenarios involving realistic combinations of deep mantle phases that do not produce chemical fractionations that match well with convecting mantle [80]. Model outcomes are especially sensitive to the amount of Ca-perovskite in the assemblage, as well to the chosen partitioning models for both perovskite phases [80,91–93]. Much needs to be learned about the pressure, temperature and compositional effects on element partitioning at deeper mantle conditions, as all experiments to date are at conditions valid only for the very top of the lower mantle.

Given these caveats, however, it can at least be stated that geochemical modeling indicates that a reservoir of ultra-high-pressure phases constituting something like 10% to 15% of the mantle is plausible.

Such a reservoir, if evenly distributed, would extend about 500 km above the core mantle boundary but might more likely exist as mountainous, discontinuous crystal piles [73]. If such a reservoir has existed for nearly 4.5 Ga, then an obvious place to look for a signal is in radiogenic isotope systems with parent/daughter ratios that are affected by crystal–melt fractionation.

3.5. How would crystal fractionation in the deep mantle effect mantle isotopes?

The isotopic composition of radiogenic daughter nuclides in modern mantle derived magmas that erupted as volcanic rocks at mid-oceanic ridges and ocean islands yield information about time-integrated changes in the parent/daughter ratios in their mantle source rocks. Both long- and short-lived radionuclides provide probes into early mantle differentiation processes (see Table 1). For example, Fig. 7 shows

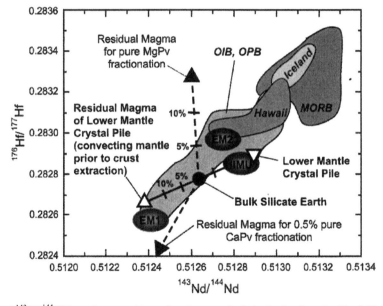

Fig. 7. The $^{176}Hf/^{177}Hf$ and $^{143}Nd/^{144}Nd$ isotopic compositions of modern, mantle-derived volcanic rocks. The fields for mid-ocean ridge basalt (MORB), ocean island basalt (OIB), ocean plateau basalts (OPB), Hawaii and Iceland, as well as the compositions of bulk silicate Earth and mantle isotopic components (enriched mantle I, EMI; enriched mantle II, EMII; high U/Pb, HIMU) are taken from [94,95]. Models are shown for the time-integrated effects on the residual magma (i.e., residual mantle) of the removal and isolation from the bulk silicate Earth of deep mantle phases in a Hadean magma ocean. The dashed trends show the time-integrated isotopic compositions in the residual magma for removal of up to 15% of pure Mg-perovskite or 0.5% removal of pure Ca-perovskite. The solid hatched line shows isotopic compositions in the lower mantle crystal pile and residual magma (convecting mantle before crust extraction) for removal of up to 15% of an assemblage composed of 93% Mg-perovskite, 3% Ca-perovskite and 4% ferropericlase, as described in the caption in Fig. 6. In the fractionation models, isotopic growth occurs in a chondritic reservoir for the first 30 Ma. At 4.536 Ga ago, Sm/Nd and Lu/Hf are fractionated due to crystal removal in the deep mantle.

$^{143}Nd/^{144}Nd$ versus $^{176}Hf/^{177}Hf$ for modern oceanic volcanic rocks. Isotopic compositions form a relatively tight array that passes through an estimated model chondritic bulk silicate Earth reservoir [94]. Correlated positive and negative deviations from bulk silicate Earth are interpreted to reflect mantle components that are relatively radiogenic (e.g., the 'depleted' MORB source) and unradiogenic (e.g., the 'enriched' OIB sources).

Fig. 7 shows models demonstrating the effect that removal and isolation of deep mantle phases in the early Hadean could have on the isotopic composition of the residual magma, which would now be the modern convecting mantle [80]. Mg-perovskite crystallization produces a marked increase in Lu/Hf, whereas Sm/Nd remains essentially constant. More than about 5% removal of pure Mg-perovskite in the early Hadean would yield a modern convecting mantle with a bulk isotopic composition outside the observed mantle array, a fact used previously as a compelling argument against a buried reservoir of pure Mg-perovskite [95]. Ca-perovskite has considerable leverage on element ratios during crystal fractionation because Lu and REE are extremely compatible in this phase and it has a very high Lu/Hf ratio. Removal of even as little as ~0.2% Ca-perovskite in isolation would move convecting mantle outside its observed field, but importantly, in the opposite sense to Mg-perovskite fractionation (Fig. 7).

Removal of deep mantle phases in the proportions 93% Mg-perovskite, 3% Ca-perovskite and 4% ferropericlase, as in the model shown in Fig. 6, produces a reduction in both Lu/Hf and Sm/Nd in the residual magma from initially chondritic values. Separation of ~15% of this crystalline assemblage in the early Hadean would produce subsequent isotopic compositions in the residual magma (e.g., convecting mantle) that remain internal to the mantle array (Fig. 7) [80]. Furthermore, the Nd and Hf isotopic compositions of the residual mantle are remarkably akin to the modern Enriched Mantle I (EMI) reservoir. The complementary crystal pile at the base of the lower mantle would grow an isotopic composition akin to the modern HiMu reservoir (high U/Pb), at least on this isotope diagram. Interestingly, available partition coefficients, although not well known, make it conceivable that a deep crystal pile could have low

Rb/Sr and elevated U/Pb relative to chondrite [80,91–93], the residual magma acquiring complementary ratios. In such case, the correspondence in Fig. 7 between the isotopic characteristics of the deep crystal pile and residual magma with HiMu and EMI, respectively, might translate to Sr and Pb isotopes as well.

After separation of the crystal pile, the convecting portion of the mantle would have become differentiated further by permanent separation of continental crust material, perhaps initially by late-stage melt expulsion, a considerable portion of which might have been extracted within a few hundred million years of t_0 based on fossil ^{142}Nd observed in Archean rocks [96,97]. This, together with continued continental crust formation, would differentiate the convecting mantle into an incompatible element depleted reservoir ('MORB mantle') and enriched continental crust [98].

3.6. Speculations on deep mantle differentiation

As the juvenile Earth grew during the first few tens of millions of years, at some time it would have reached a size large enough for perovskite phases to become stable at the base of the silicate mantle. Accretion models lead one to expect that it is probabilistically favorable that the size threshold for perovskite stability would have been breached during a giant impact between the proto-Earth and a large impactor. Besides the large mass gain in the proto-Earth, the energy deposited in the impact could have been sufficient to substantially melt the silicate mantle (as we have seen, evidence that such a magma ocean stage might have occurred comes from the abundances of Ni and Co in the mantle). Subsequent cooling and solidification of the silicate magma ocean might have involved the fractional removal of deep mantle perovskite and oxide phases. Further impact accretion of the Earth to its final size may have involved more than one magma ocean stage, each possibly involving deep mantle crystal fractionation. Vigorous convection in the silicate mantle of a hot, young Earth may have reworked and homogenized much of the mantle, but some amount of dense, deep mantle crystalline phases might have remained rheologically stable at the base of the deepening mantle. Indeed, proto-crystalline piles of deep mantle phases may have been eroding at various rates since primordial crystallization. Compositional models involving moderate amounts (e.g., 10% to 15%) of an isolated ultra-high-pressure crystal fraction can produce elemental and isotopic characteristics of the residual magma that are generally consistent with those observed in primitive upper mantle rocks.

4. Perspectives

In spite of the missing rock record documenting differentiation in the infant Earth, important information has been gathered from diverse fields including observational and theoretical astrophysics, cosmochemistry and high-pressure experimental geophysics and geochemistry. What is most intriguing given the present knowledge of early Earth differentiation is the apparent convergence of several different approaches regarding the timing of events. Accretion models predict that giant impacts should occur during the late stages of accretion on a time scale of 10^7–10^8 years after t_0 [6]. Radiogenic isotopes yield an 'age' of core formation on similar time scales, of the order of 50 million years after t_0, that could be related to the origin of the Moon and magma ocean formation [18–20]. Siderophile element abundances in mantle rocks apparently preserve a signature of high-pressure–temperature equilibration [52]. Complete crystallization of the magma ocean could occur on a 10^7 year time scale [54]. Long- and short-lived Hf and Nd isotopes indicate very early differentiation in the mantle also on a time scale of 10^7–10^8 years after t_0 [96]. Advances in our understanding will likely continue to come from a multifaceted approach.

Advances in numerical modeling and laboratory experiments at high pressures and temperatures will facilitate increased understanding of the physicochemical processes of accretion, terrestrial core formation and magma ocean formation and differentiation. Higher resolution hydrodynamic modeling is needed to address the fate of metal cores during giant impacts and particularly whether metal emulsification is a likely prospect. Determination of the physical properties of metals, minerals and melts at deep mantle conditions is needed to better constrain numerical modeling of the processes of magma ocean formation, cooling, convection and crystallization in the magma ocean, as well as metal segregation from silicate. A more comprehensive and accurate database on experimentally determined partitioning coefficients for a range of elements between minerals and melts in silicate-dominated systems and between Fe-dominated metal and silicate is also needed.

High-pressure experiments within the stability field of lower mantle materials are challenging. The trade-off between sample volume and pressure applies to multi-anvil experiments using sintered diamond anvils, as well as to diamond anvil cell experiments. Experiments covering the entire lower mantle pressure regime are required in order to realistically constrain early Earth differentiation processes. This is especially important in terms of crystallization in a deep magma ocean. Currently, it is possible to obtain relatively stable and long-duration melting experiments with relatively low thermal gradients across the sample capsule up to about 25 GPa using conventional multi-anvil configurations. A future challenge is to develop corresponding thermal stability at the higher pressures generated by diamond anvils.

Acknowledgements

We extend thanks to R. Carlson, J. Chambers, A. Corgne, M. Drake, T. Elliot, D. Frost, A. Halliday, K. Hirose, E. Ito, B. McDonough, E. Nakamura, D. Presnall, K. Righter, D. Rubie, E. Takahashi, J. Wade, B. Wood and T. Yoshino, whom, for better or worse, and usually without consent, all contributed through conversation and otherwise to the particular portrait of the Earth's childhood depicted herein. Reviews by M. Drake, D. Stevenson and an anonymous reviewer are much appreciated and helped improve the manuscript.

References

[1] P.N. Foster, A.P. Boss, Triggering star formation with stellar ejecta, Astrophys. J. 468 (1996) 784–796.

[2] C.M.O. Alexander, A.P. Boss, R.W. Carlson, Review: cosmochemistry—the early evolution of the inner solar system: a meteoric perspective, Science 293 (2001) 64–68.

[3] S.J. Weidenschilling, The origin of comets in the solar nebula: a unified model, Icarus 127 (1997) 290–306.

[4] G.W. Wetherill, G.R. Stewart, Formation of planetary embryos: effects of fragmentation, low relative velocity, and independent variation of eccentricity and inclination, Icarus 106 (1993) 190–209.

[5] E. Kokubo, S. Ida, Formation of protoplanets and planetesimals in the solar nebula, Icarus 143 (1996) 15–27.

[6] G.W. Wetherill, Provenance of the terrestrial planets, Geochim. Cosmochim. Acta 58 (1994) 4513–4520.

[7] J.E. Chambers, G.W. Wetherill, Making the terrestrial planets: N-body integrations of planetary embryos in three dimensions, Icarus 136 (1998) 304–327.

[8] R.M. Canup, C.B. Agnor, Accretion of the terrestrial planets and the Earth–Moon system, in: R.M. Canup, K. Righter (Eds.), Origin of the Earth and Moon, The University of Arizona Press, Tucson, 2000, pp. 113–132.

[9] A.G.W. Cameron, Higher-resolution simulations of the giant impact, in: R.M. Canup, K. Righter (Eds.), Origin of the Earth and Moon, The University of Arizona Press, Tucson, 2000, pp. 133–144.

[10] J.J. Lissauer, L. Dones, K. Ohtsuki, Origin and evolution of terrestrial planet rotation, in: R.M. Canup, K. Righter (Eds.), Origin of the Earth and Moon, The University of Arizona Press, Tucson, 2000, pp. 101–112.

[11] R. Canup, E. Asphaug, Origin of the Moon in a giant impact near the end of Earth's formation, Nature 412 (2001) 708–712.

[12] R. Canup, Simulations of a late lunar-forming impact, Icarus 168 (2004) 433–456.

[13] W. Benz, G.W. Cameron, Terrestrial effects of the giant impact, in: H.E. Newsom, J.H. Jones (Eds.), Origin of the Earth, Oxford University Press, New York, 1990, pp. 61–67.

[14] W.B. Tonks, H.J. Melosh, Magma ocean formation due to giant impacts, J. Geophys. Res. 98 (1993) 5319–5333.

[15] D.C. Lee, A.N. Halliday, Hafnium–tungsten chronometry and the timing of terrestrial core formation, Nature 378 (1995) 771–774.

[16] C.L. Harper, S.B. Jacobsen, Evidence for [182]Hf in the early solar system and constraints on the timescale of terrestrial core formation, Geochim. Cosmochim. Acta 60 (1996) 1131–1153.

[17] A.N. Halliday, D.-C. Lee, S.B. Jacobsen, Tungsten isotopes, the timing of metal–silicate fractionation, and the origin of the Earth and Moon, in: R.M. Canup, K. Righter (Eds.), Origin of the Earth and Moon, The University of Arizona Press, Tucson, 2000.

[18] Q. Yin, S.B. Jacobsen, K. Yamashita, J. Blichert-Toft, P. Telouk, F. Albarede, A short timescale for terrestrial planet formation from Hf–W chronometry of meteorites, Nature 418 (2002) 949–952.

[19] J. Kleine, C. Munker, K. Mezger, H. Palme, Rapid accretion and early core formation on asteroids and the terrestrial planets from Hf–W chronometry, Nature 418 (2002) 952–955.

[20] A.N. Halliday, Mixing, volatile loss and compositional change during impact-driven accretion of the Earth, Nature 427 (2004) 505–509.

[21] W.G. Minarik, F.J. Ryerson, E.B. Watson, Textural entrapment of core-forming melts, Science 272 (1996) 530–533.

[22] D.J. Stevenson, Fluid dynamics of core formation, in: H.E. Newsom, J.H. Drake (Eds.), Origin of the Earth, Oxford University Press, New York, 1990, pp. 231–249.

[23] T. Rushmer, W.G. Minarik, G.J. Taylor, Physical processes of core formation, in: R.M. Canup, K. Righter (Eds.), Origin of the Earth and Moon, The University of Arizona Press, Tucson, 2000, pp. 227–244.

[24] I.D. Hutcheon, R. Hutchinson, Evidence from the Semarkona ordinary chondrite for [26]Al heating of small planets, Nature 337 (1989) 238–241.

[25] A. Ghosh, H.Y. McSween, A thermal model for the differentiation of Asteroid 4 Vesta, Icarus 134 (1998) 187–206.

[26] R. Merk, D. Breuer, T. Spohn, Numerical modeling of [26]Al-induced radioactive melting of asteroids considering accretion, Icarus 159 (2002) 183–191.

[27] T. Yoshino, M.J. Walter, T. Katsura, Core formation in planetesimals triggered by permeable flow, Nature 422 (2003) 154–157.

[28] G.J. MacPherson, A.M. Davis, E.K. Zinner, The distribution of [26]Al in the early solar system—a reappraisal, Meteorology 30 (1995) 365–386.

[29] T. Kunihiro, A.E. Rubin, K.D. McKeegan, J.T. Wasson, Initial [26]Al/[27]Al in carbonaceous-chondrite chondrules: too little [26]Al to melt asteroids, Geochim. Cosmochim. Acta 68 (2004) 2947–2957.

[30] Y. Fei, C.M. Bertka, L. Finger, High-pressure iron-sulfide compound, Fe_3S_2, and melting relations in the Fe–FeS system, Science 275 (1997) 1621–1623.

[31] C.B. Agee, J. Li, M.C. Shannon, S. Circone, Pressure–temperature phase diagram for the Allende meteorite, J. Geophys. Res. 100 (1995) 17725–17740.

[32] D. Bruhn, N. Groebner, D.L. Kohlstedt, An interconnected network of core-forming melts produced by shear deformation, Nature 403 (2000) 883–886.

[33] S.R. Taylor, S.R. Norman, Accretion of differentiated planetesimals to the Earth, in: H.E. Newsom, J.H. Jones (Eds.), Origin of the Earth, Oxford University Press, New York, 1990, pp. 29–44.

[34] W.F. McDonough, S.-S. Sun, The composition of the Earth, Chem. Geol. 120 (1995) 223–253.

[35] M.J. Walter, H.E. Newsom, W. Ertel, A. Holzheid, Siderophile elements in the earth and moon: metal/silicate partitioning and implications for core formation, in: R.M. Canup, K. Righter (Eds.), Origin of the Earth and Moon, The University of Arizona Press, Tucson, 2000, pp. 265–290.

[36] M.J. Drake, K. Righter, Detrermining the composition of the Earth, Nature 416 (2002) 39–44.

[37] A. Holzheid, P. Sylvester, H.S.C. O'Neill, D.C. Rubie, H. Palme, Evidence for a late chondritic veneer in the Earth's mantle from high-pressure partititoning of palladium and platinum, Nature 406 (2000) 396–399.

[38] H.S. O'Neill, The origin of the Moon and the early history of the Earth—a chemical model: 2. The Earth, Geochim. Cosmochim. Acta 55 (1991) 1159–1172.

[39] D. Walker, Core participation in mantle geochemistry: Geochemical Society Ingerson Lecture, GSA Denver, October 1999, Geochim. Cosmochim. Acta 64 (2000) 2911–2987.

[40] J. Li, C. Agee, Geochemistry of mantle–core differentiation at high pressure, Nature 381 (1996) 686–689.

[41] K. Righter, M. Drake, G. Yaxley, Prediction of siderophile element metal–silicate partition coefficients to 20 GPa and 2800 C: the effects of pressure, temperature, oxygen fugacity and silicate and metallic melt compositions, Phys. Earth Planet. Inter. 100 (1997) 115–134.

[42] J. Li, C.B. Agee, The effect of pressure, temperature, oxygen fugacity and composition on partitioning of nickel and cobalt between liquid Fe–Ni–S alloy and liquid silicate: implications for the Earth's core formation, Geochim. Cosmochim. Acta 65 (2001) 1821–1832.

[43] D. Walker, L. Norby, J.H. Jones, Superheating effects on metal–silicate partitioning of siderophile elements, Science 262 (1993) 1858–1861.

[44] V.J. Hillgren, M.J. Drake, D.C. Rubie, High-pressure and high-temperature experiments on core–mantle segregation in an accreting Earth, Science 264 (1994) 1442–1445.

[45] Y. Thibault, M.J. Walter, The influence of pressure and temperature on the metal–silicate partition coefficients of nickel and cobalt in a model C1 chondrite and implications for metal segregation in a deep magma ocean, Geochim. Cosmochim. Acta 59 (1995) 991–1002.

[46] E. Ohtani, H. Yurimoto, S. Seto, Element partitioning between metallic liquid, silicate liquid, and lower-mantle minerals: implications for core formation of the Earth, Phys. Earth Planet. Inter. 100 (1997) 97–114.

[47] E. Ito, T. Katsura, T. Suzuki, Metal silicate partitioning of Mn, Co and Ni at high-pressures and high-temperatures and implications for core formation in a deep magma ocean, in: M.H. Manghnani, T. Yagi (Eds.), Properties of Earth and Planetary Materials at High Pressure and Temperature, American Geophysical Union, Washington, DC, 1998, pp. 215–225.

[48] M.A. Bouhifd, A.P. Jephcoat, The effect of pressure on partitioning of Ni and Co between silicate and iron-rich metal liquids: a diamond-anvil cell study, Earth Planet. Sci. Lett. 209 (2003) 245–255.

[49] D. Jana, D. Walker, The influence of silicate melt composition on distribution of siderophile elements among metal and silicate liquids, Earth Planet. Sci. Lett. 150 (1997) 463–472.

[50] D. Jana, D. Walker, The impact of carbon on element distribution during core formation, Geochim. Cosmochim. Acta 61 (1997) 2759–2763.

[51] D. Jana, D. Walker, The influence of sulfur on partitioning of siderophile elements, Geochim. Cosmochim. Acta 61 (1997) 5255–5277.

[52] K. Righter, Metal–silicate partitioning of siderophile elements and core formation in the early Earth, Ann. Rev. Earth Planet. Sci. 31 (2003) 135–174.

[53] D.C. Rubie, H.J. Melosh, J.E. Reid, C. Liebske, K. Righter, Mechanisms of metal–silicate equilibration in the terrestrial magma ocean, Earth Planet. Sci. Lett. 205 (2003) 239–255.

[54] Y. Abe, Thermal and chemical evolution of the terrestrial magma ocean, Phys. Earth Planet. Int. 100 (1997) 27–39.

[55] E. Ito, T. Katsura, A. Kubo, M. Walter, Melting experiments of mantle materials under lower mantle conditions with implica-tion to fractionation in magma ocean, Phys. Earth Planet. Inter. 143–144 (2004) 397–406.

[56] R.G. Tronnes, D.J. Frost, Peridotite melting and mineral–melt partitioning of major and minor elements at 22–24.5 GPa, Earth Planet. Sci. Lett. 97 (2002) 117–131.

[57] J. Zhang, C. Herzberg, Melting experiments on anhydrous peridotite KLB-1 from 5.0 to 22.5 GPa, J. Geophys. Res. 99 (B9) (1994) 17729–17742.

[58] W.B. Tonks, H.J. Melosh, The physics of crystal settling and suspension in a turbulent magma ocean, in: H.E. Newsom, J.H. Jones (Eds.), Origin of the Earth, Oxford University Press, New York, 1990, pp. 151–174.

[59] V.S. Solamotov, D.J. Stevenson, Suspension in convective layers and style of differentiation of a terrestrial magma ocean, J. Geophys. Res. 98 (1993) 5375–5390.

[60] A. Zindler, S. Hart, Chemical geodynamics, Annu. Rev. Earth Planet. Sci. 14 (1986) 493–571.

[61] D.L. Anderson, Chemical composition of the mantle, J. Geophys. Res. 88 (1983) B41–B52.

[62] C.R. Bina, P.G. Silver, Constraint on lower mantle composi-tion and temperature from density and bulk sound velocity profiles, Geophys. Res. Lett. 17 (1990) 1153–1156.

[63] R.W. Carlson, Mechanisms of Earth differentiation—conse-quences for the chemical structure of the mantle, Rev. Geophys. 32 (1994) 337–361.

[64] I. Jackson, S.M. Rigden, Composition and temperature of the Earth's mantle: seismological models interpreted through exper-imental studies of Earth materials, in: I. Jackson (Ed.), The Earth's Mantle: Composition, Structure and Evolution, Cambridge University Press, Cambridge, UK, 1998, pp. 405–460.

[65] A.W. Hofmann, Sampling mantle heterogeneity through oceanic basalts: isotopes and trace elements, in: R.W. Carlson (ed.), The Mantle and Core, vol. 2, pp. 61–102; Treatise on Geochemistry, in: H.D. Holland, K.K. Turekian (eds.), Elsevier-Pergamon, Oxford, 2003.

[66] G.R. Helffrich, B.J. Wood, The Earth's mantle, Nature 412 (2001) 501–507.

[67] C.R. Bina, Seismological constraints upon mantle composition, in: R.W. Carlson (ed.), The Mantle and Core, vol. 2, pp. 39–60; Treatise on Geochemistry, in: H.D. Holland, K.K. Turekian (eds.) Elsevier–Pergamon, Oxford, 2003.

[68] B.L.N. Kennett, R.D. van der Hilst, Seismic structure of the mantle: from subduction zone to craton, in: I. Jackson (Ed.), The Earth's Mantle: Composition, Structure and Evolution, Cam-bridge University Press, Cambridge, UK, 1998, pp. 381–404.

[69] P.E. van Keken, S.J. Zhong, Mixing in a 3D spherical model of present-day mantle convection, Earth Planet. Sci. Lett. 171 (1999) 533–547.

[70] P.E. van Keken, E.H. Hauri, C.J. Ballentine, Mantle mixing: the generation, preservation, and destruction of chemical hetero-geneity, Annu. Rev. Earth Planet. Sci. 30 (2002) 493–525.

[71] R.D. van der Hilst, H. Karason, Compositional heterogeneity in the bottom 1000 kilometers of Earth's mantle: toward a hybrid convection model, Science 283 (1999) 1885–1888.

[72] L.H. Kellogg, B.H. Hager, R.D. van der Hilst, Composi-tional stratification in the deep mantle, Science 283 (1999) 1881–1884.

[73] P.J. Tackley, Strong heterogeneity caused by deep mantle layering, Geochem. Geophys. Geosyst. 3 (2002) (10.1029/2001GC000167).

[74] S.R. Hart, A. Zindler, In search of a bulk-Earth composition, Chem. Geol. 57 (1986) 247–267.

[75] C.J. Allegre, J.P. Poirier, E. Humbler, A.W. Hofmann, The chemical composition of the Earth, Earth Planet. Sci. Lett. 134 (1995) 515–526.

[76] H.S.C. O'Neill, H. Palme, Composition of the silicate Earth: implications for accretion and core formation, in: I. Jackson (Ed.), The Earth's Mantle: Composition, Structure and Evolution, Cambridge University Press, Cambridge, UK, 1998, pp. 3–126.

[77] H. Palme and H.S.C. O'Neill, Cosmochemical estimates of mantle composition, in: R.W. Carlson (ed.), The Mantle and Core, vol. 2, pp. 1–38; Treatise on Geochemistry, in: H.D. Holland, K.K. Turekian (eds.), Elsevier-Pergamon, Oxford, 2003.

[78] R. Hutchison, The formation of the Earth, Nature 250 (1974) 556–558.

[79] E. Jagoutz, H. Palme, H. Baddenhausen, K. Blum, K. Cendales, G. Dreibus, B. SPettel, V. Lorenz, H. Wanke, The abundances of major, minor and trace elements in the Earth's mantle as derived from primitive ultramafic nodules, Poceedings of the 10th Lunar and Planetary Science Conference, 1979, pp. 2031–2050.

[80] M.J. Walter, E. Nakamura, R. Tronnes, D. Frost, Experimental constraints on crystallization differentiation in a deep magma ocean, Geochim. Cosmochim. Acta (2004).

[81] A.E. Ringwood, Composition and Petrology of the Earth's Mantle, McGraw-Hill, New York, 1975, 618 pp.

[82] M.J. Walter, Melt extraction and compositional variability in mantle lithosphere, in: R.W. Carlson (ed.), The Mantle and Core, vol. 2, pp. 363–394; Treatise on Geochemistry, in: H.D. Holland, K.K. Turekian (eds.), Elsevier-Pergamon, Oxford, 2003.

[83] A.E. Ringwood, Origin of the Earth and Moon, Springer-Verlag, New York, 1979, 295 pp.

[84] H. Wanke, Constitution of terrestrial planets, Philos. Trans. R. Soc. Lond. A 303 (1981) 287–302.

[85] A.N. Halliday, D.-C. Lee, S. Tommasini, G.R. Davies, C.R. Paslick, J.G. Fitton, D.E. James, Incompatible trace elements in OIB and MORB and source enrichment in the sub-oceanic mantle, Earth Planet. Sci. Lett. 133 (1995) 379–395.

[86] E. Ito, E. Takahashi, Melting of peridotite at uppermost lower-mantle conditions, Nature 328 (1987) 514–516.

[87] C.B. Agee, D. Walker, Mass balance and phase density constraints on early differentiation of chondritic mantle, Earth Planet. Sci. Lett. 90 (1988) 144–156.

[88] T. Kato, A.E. Ringwood, T. Irifune, Experimental determination of element partitioning between silicate perovskites, garnets and liquids: constraints on early differentiation of the mantle, Earth Planet. Sci. Lett. 89 (1988) 123–145.

[89] E.A. McFarlane, M.J. Drake, D.C. Rubie, Element partitioning between Mg-perovskite, magnesiowustite, and silicate melt at conditons of the Earth's mantle, Geochim. Cosmochim. Acta 58 (1994) 5161–5172.

[90] J.H. Jones, H. Palme, Geochemical consequences on the origin of the Earth and Moon, in: R.M. Canup, K. Righter (Eds.), Origin of the Earth and Moon, The University of Arizona Press, Tucson, 2000, pp. 197–216.

[91] A. Corgne, B. Wood, Silicate perovskite–melt partitioning of trace elements and geochemical signature of a deep perovskitic reservoir, Geochim. Cosmochim. Acta (2004).

[92] A. Corgne, B.J. Wood, $CaSiO_3$ and $CaTiO_3$ perovskite–melt partitioning of trace elements: implications for gross mantle differentiation, Geophys. Res. Lett. (2002) (2001GL014398).

[93] K. Hirose, N. Shimizu, W.V. Westrenen, Y. Fei, Trace element partitioning in Earth's lower mantle, Physics of Earth and Planetary Interiors 146 (2003) 249–260.

[94] J. Blichert-Toft, F. Albarede, The Lu–Hf isotope geochemistry of chondrites and the evolution of the mantle–crust system, Earth Planet. Sci. Lett. 148 (1997) 243–258.

[95] V.J.M. Salters, W.M. White, Hf isotopic constraints on mantle evolution, Chem. Geol. (1998) 447–460.

[96] G. Caro, B. Bourdon, J.-L. Birck, S. Moorbath, $^{146}Sm–^{142}Nd$ evidence from Isua metamorphosed sediments for early differentiation of the Earth's mantle, Nature 423 (2003) 428–432.

[97] M. Boyet, J. Blichert-Toft, M. Rosing, M. Storey, P. Telouk, F. Albarede, ^{142}Nd evidence for early Earth differentiation, Earth Planet. Sci. Lett. 214 (2003) 427–442.

[98] A.W. Hofmann, Chemical differentiation of the Earth: the relationship between mantle, continental crust, and oceanic crust, Earth Planet. Sci. Lett. 90 (1988) 297–314.

Michael J. Walter is a Reader in Earth Sciences at the University of Bristol. He received a BS degree in Geology from the University of Nebraska, Omaha, and a PhD degree in Geology from the University of Texas, Dallas. His research in experimental geochemistry and geophysics focuses on mineral–melt phase equilibria and element partitioning, the origin of magmas and their residues, geochemistry of iron alloys and mineral physics in order to obtain a better understanding of the origin and evolution of the Earth and terrestrial planets.

Reidar G. Trønnes is an associate professor at the Geological Museum, University of Oslo. He has conducted petrologic and geochemical research at institutions in Norway, Canada, Iceland, Germany and Japan. His research includes experimental studies of the high-pressure melting relations of komatiite and peridotite to further constrain the early differentiation of the Earth. His general research interests are focused on planetary evolution and dynamics.

Reprinted from
Earth and Planetary Science Letters 224 (2004) 1–17

www.elsevier.com/locate/epsl

Chondrules

B. Zanda*

Laboratoire d'Etudes de la Matière extraterrestre MNHN-UMS2679 CNRS 61, rue Buffon 75005- Paris, France
Department of Geological Sciences, Rutgers University, P.O. Box 1179, Piscataway, NJ 08855-1179, USA

Received 17 June 2003; received in revised form 5 May 2004; accepted 5 May 2004

Abstract

Chondrules are submillimeter spheres that constitute up to 80% of the volume of the most primitive meteorites. That they result from the solidification of a melt in low-gravity in the early solar system has been known for nearly two centuries, but the conditions of their formation and their significance still elude our understanding. It has been variously proposed that they both predate and postdate planet formation and some have suggested that they are the product of the planet-forming process itself. There is mounting evidence that rather than resulting from a trivial event (or series of events) which melted only a small fraction of solids in the disk, chondrule formation significantly transformed the original material present in the early solar system and contributed to the chemical and isotopic compositions of the first planets. The only meteorites that preserved the chemical composition and isotopic signatures of the earliest solar system solids are the CI chondrites that contain no preserved chondrules and probably had very few, if any. All other chondrites have experienced various levels of metal/silicate and refractory/volatile fractionation that may have resulted from chondrule formation, although a number of researchers argue that these fractionations existed before chondrules and probably resulted from nebular-wide condensation. The current most popular mechanisms for forming chondrules in a nebular setting are radiation emitted by the protosun in the X-wind setting or shock waves propagated in the protoplanetary disk. In the latter case, chondrule formation may have contributed to the first stages of accretion, which would have helped preserve the chemical complementarity between chondrules and matrix. It is important that the chemical and isotopic properties, and even the petrology, of chondrules be reassessed in order to allow the development of chondrule formation models that better fit these constraints.
© 2004 Elsevier B.V. All rights reserved.

Keywords: chondrules; chondrites; formation of the solar system; solar nebula; early chemical fractionations

1. Introduction

1.1. Chondrules and chondrites

The most common kind of meteorites that have been observed to fall on the Earth are *chondrites*. These are like no rocks formed on Earth and there now exists a sizeable body of data to indicate that they formed from dust and debris from the solar nebula that escaped being incorporated into planets. As such, they provide unique insights into processes operating in the circumstellar disk from which the planets formed at the very start of the solar system. Chondrites are so named because they usually contain large amounts of small (at most milli-meter-sized) spherules called *chondrules*.

Although these are the most abundant objects in chondrites, the origin of chondrules is extremely

* Laboratoire d'Etudes de la Matière extraterrestre MNHN-UMS2679 CNRS 61, rue Buffon 75005- Paris, France. Tel.: +33-1-4079-3542; fax: +33-1-4079-5772.
E-mail address: zanda@mnhn.fr (B. Zanda).

0012-821X/$ - see front matter © 2004 Elsevier B.V. All rights reserved.
doi:10.1016/j.epsl.2004.05.005

Jargon Box

Calcium–Aluminum-rich inclusions (CAIs)

CAIs are particles up to a centimeter in size found in chondrites. They are made of minerals containing refractory elements such as Ca, Al and Ti. CAIs are the oldest objects of our solar system which they allowed us to date. They are considered to be either the first objects condensed within the solar system or the only residue of the wholesale evaporation which took place during its formation. A fraction of the CAIs have been melted and their relationship with chondrules is unclear. They differ in compositions but a few intermediate objects exist. They also appear older than chondrules. It is still unknown whether the event(s) that melted some of them is related to chondrule formation.

Chondrites

The meteorites that hit the Earth fall into two broad categories. "Differentiated meteorites" come from asteroids which were melted (like the larger planets) and they sample different layers from these bodies: the eucrites (most of which are basalts) are crust samples while the iron meteorites are core samples. Primitive meteorites, on the other hand, come from bodies which were never melted after their formation in the early solar system and they have preserved characteristics acquired in the solar accretion disk. They are called "chondrites" because their most striking feature is that they are made of up to 80% chondrules.

Three main categories of chondrites were recognized early on: the ordinary (OCs, most common), enstatite (ECs, so reduced that some lithophile elements such as Mn or Ca are found in sulfides) and carbonaceous chondrites (CCs), in which organic matter was detected and a large fraction of which are oxidized (containing no metal but magnetite). We now know that these three categories encompass different domains in the oxygen isotopes diagram shown in Fig. 9 (above, on and below the TFL). These three categories are also divided into subcategories: EH and EL for the ECs depending on their high (H) or low (L) content of Fe; H, L and LL for OCs depending both on their total Fe and their metal content (Fig. 8). A fourth group, apparently related to LLs but even more oxidized, also exists: Rumurutiites (Rs). The subcategories of OCs and even ECs are fairly closely related both in terms of mineralogy and of oxygen isotopes. The CCs are much more diverse and the most of the groups are named after a type meteorite: CI (Ivuna); CM (Murray); CV (Vigarano); CO (Ornans); CR (Renazzo); CK (Karoonda); CH (extremely High in metal).

Chondrules

Chondrules are submillimeter particles found in chondrites of which they can comprise most of the mass. Chondrules are often spherical and are believed to be derived from liquids crystallized in low gravity in the early solar system. Chondrules have a large variety of textures and compositions but they fall into two main chemical categories shown in Fig. 2 and defined in its caption.

---

controversial and has generated more debate than any other feature of meteorites. Their formation may play a major role in determining the composition of the inner planets (e.g., [1]). Now scientists are at the point of being able to propose new theories that are being stimulated by the exciting new information being obtained from astronomical observations of disks around young solar mass stars coupled with astrophysical modelling.

1.2. Formation of the Sun and solar system

The formation of stars and planets is now better understood but the occurrence of widespread melt droplets, chondrules, is not a feature that has been predicted in such work. It is likely that the gravitational collapse of a dust-gas cloud was triggered by the blast from a supernova, which injected newly synthesized isotopes into the future solar system. The cloud material spiralled inwards, forming a Sun and circumstellar disk from which planetary material accreted. The inner regions of this protoplanetary disk, or solar nebula,

must have been hot enough to evaporate all solids, but we do not know how extensive or long-lived was the heating in this region. Material passing from the disk into the growing Sun would have encountered its magnetosphere and it has been proposed that it would have been divided into an inflow and an outflow known as the X-wind [2]. Much mass would have been lost in bipolar outflows but some particles may have decoupled from the X-wind and fallen back into the disk. Whether or not the X-wind model is correct, there were numerous energetic processes within the disk, as well as at its inner edge, some of which may have led to the formation of chondrules by melting solids or by condensing vaporized solids.

1.3. Controversies about chondrules

Linking chondrules with such new theories for disk behavior based on observational astronomy and astrophysics may well be the best way to determine their origin. Three years ago, at the LPS Conference in Houston, a prominent meteoriticist gave a highly controversial plenary lecture [3], later featured in Science [4], in which he expressed his doubts as to whether continuing to accumulate data on chondrules would ever lead to a better understanding of their origin. This is because of the lack of a grand theory into which these data could be fitted. Indeed, the diversity of models that have been proposed to account for chondrule formation (e.g., Fig. 1) is a testimony as to our present lack of understanding of their origin and our need for a framework for understanding the physical and chemical properties of the first solar system bodies. However, as pointed out by Hewins [5], the lack of agreement between meteoriticists about the detailed interpretation of chondrule petrology and geochemistry is part of the reason for our inability to develop such a unifying theory. As suggested by Wood [6], it is important to reevaluate the data and the paradigms on chondrule formation, taking into account more recent developments. However, contrary to Wood [3] and Kerr [4], a strong case can be made that the acquisition of new data taking full advantage of recently developed techniques remains essential, especially on key issues such as formation ages, irradiation signatures and exchanges with the surrounding gas.

Progress will also come from the refinement of astrophysical models based on new astronomical observations that are more closely tailored to the detailed physical and chemical properties of chondrules.

The present paper briefly reviews the main characteristics of chondrules and the relevant properties of chondrites, the leading arguments concerning the origin of chondrules and the nature of their precursor material, and the approaches that presently appear the most promising to unravel this mystery.

2. Key properties of chondrules

2.1. Petrography and chemistry of chondrules

Chondrules are small particles of silicate material that experienced melting before their incorporation into chondritic parent bodies. Particles with different compositions such as *Calcium–Aluminum-rich Inclusions (CAIs)* and basaltic fragments are not considered chondrules [5]. Ideal chondrules are spherical as they solidified from liquid droplets, but most chondrules are not spheres; that is, a significant fraction is

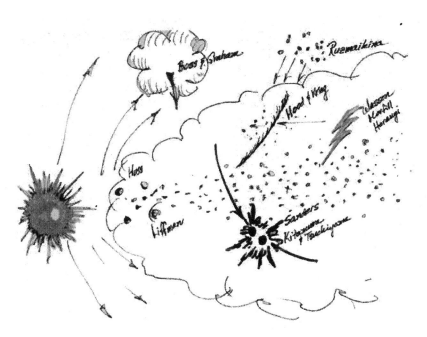

Fig. 1. This cartoon was sketched by the astrophysicist P. Cassen at the "Chondrule Conference" in 1994 as an up-to-date summary of chondrule formation models. The profusion of energy sources makes it a wonder that any material in the disk escaped melting (e.g., the chondritic matrix and the fluffy CAIs). Note that the details of the various models have evolved significantly since.

present as fragments (Fig. 2), some are molded around one another (Fig. 3a) while others were insufficiently melted for surface tension to make them round (Fig. 3b).

Figs. 2 and 3 illustrate some of the key properties of chondrules that any model of chondrule formation must account for the variability of their shapes, textures and compositions, starting with the existence of two main chemical classes, one of which is volatile-rich. Another key point is the evidence for extended or successive episodes of heating in the form of melted rims around some chondrules and relicts from earlier generations of objects. These relicts are usually found as grains (Fig. 4), but chondrules from CO and CV chondrites may also contain isotopic geochemical signatures derived from CAIs [7] or amoeboid olivine aggregates [8].

2.2. Isotopic data

More evidence pertaining to the chondrule formation mechanism stems from isotopic measurements in chondrules. One intriguing problem is that no or only

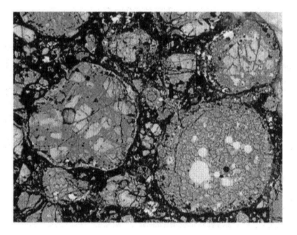

Fig. 2. This reflected light image of a primitive ordinary chondrite shows that it looks like a "sediment" of subspherical particles, chondrules, embedded in a fine-grained matrix loaded with chondrule debris. Chondrules can have highly variable sizes and textures (e.g., [5]), but the main distinction is chemical: volatile-depleted type I chondrules are reduced and often metal-bearing (lower-right), whereas metal is absent from volatile-rich type II chondrules in which Fe is oxidized within the silicates or in sulfides (left). Type II chondrules tend to be coarser-grained than type Is and are often also larger. Chondrules of the two types also have variable Mg/Si contents: some of them are dominantly olivine-rich while others are pyroxene-rich. (Field: 2.3 × 3 mm).

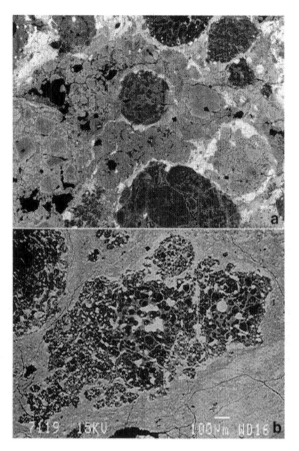

Fig. 3. In some primitive ordinary chondrites, chondrules can be molded around one another (a) indicating that they must have come together while the deformed chondrule was still above its glass transition temperature. In primitive carbonaceous chondrites, chondrules are often contorted (b) and appear like an aggregation of several subunits together with some dust. Some of the subunits and the chondrule as a whole were insufficiently melted to adopt an overall droplet shape. [Field width: 2 mm (a); 3 mm (b)]. On these BSE images, reduced silicates (type I chondrules) appear dark while oxidized material (type II chondrules and matrix) appears light grey, and metal and sulfides are white. These pictures give an idea of the differences between ordinary and carbonaceous chondrites in which chondrules may have (slightly?) different origins. Carbonaceous chondrites have an abundant matrix and mostly type I chondrules, whereas type II chondrules are largely dominant in ordinary chondrites.

very minor isotopic mass fractionations have been found in chondrules so far, even in volatile-depleted ones (e.g., [9,10]). Yet Rayleigh distillation would be expected if chondrules experienced free evaporation. As will be discussed below, this could either indicate that chondrule heating was extremely brief or that the

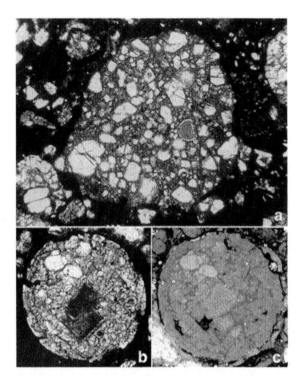

Fig. 4. These two chondrules from a primitive ordinary chondrite contain a relict from a small broken chondrule (a) and several grains from an FeO-rich chondrule which became "dusty" with minuscule beads of Fe produced by the reduction of their FeO when they were included in an FeO-poor chondrule. The dusty grains are visible both in transmitted light (b) and in reflected light (c) where the metal beads on the polished surface can be distinguished. [Field width: 2.2 mm (a); 0.8 mm (b and c)].

environment in which they melted significantly differed from the canonical solar nebula, e.g., in the partial pressures of the lithophile elements.

Another key question is that of the age of chondrules. Based on the decay of extinct radionuclides such as ^{26}Al, ^{53}Mn and ^{129}I, Swindle et al. [11] suggested that the formation of chondrules took place up to several million years after that of CAIs, the earliest solar system objects. The chronological significance of the data for each of these nuclide systems has been the subject of debate. Pb–Pb isochron ages of individual chondrules from a CR chondrite have recently been measured by Amelin et al. [12] and compared to similar ages obtained in CAIs of a CV chondrite. The chondrules were formed at 4564.7 ± 0.6 Ma, 2.5 ± 1.2 My after the refractory inclusions (an age difference in agreement with that based on ^{26}Al). This

indicates that chondrules could not have been formed in the earliest stage of infall and rapid accretion of the solar nebula as advocated by Wood [6]. Moreover, some asteroids and planetary embryos may have been already accreted by the time chondrules were melted and incorporated in chondritic asteroids [13] as recent calculations show planetary embryos forming in much less than 1 million years [14].

Heterogeneities in the spallogenic nuclides ^{11}B and ^{7}Li in chondrules were recently reported by Robert and Chaussidon [15] and were interpreted to result from the irradiation of chondrule precursor material by high energy (MeV) particles (mostly protons, but also alphas and possibly ^{3}He nuclei) emitted by the Sun during its T-Tauri phase. Assuming these results are confirmed, they would constitute the first unequivocal proof of the early irradiation of solar system material because, unlike other short-lived nuclides such as ^{26}Al and ^{53}Mn, these nuclides cannot be formed in supernovae or other stars. The presence of these nuclides offers little additional constraint on the age of chondrules (since the T-Tauri phase of the Sun may have lasted up to a few million years). It would, however, place some constraints on the location where chondrules formed if the irradiating particles originated in the Sun (rather than in Galactic Cosmic Rays as suggested by Desch et al. [16]). MeV particles could only have penetrated as far as 3 AU, the present location of the asteroid belt, if the nebula was "thin" (i.e., low density). Chondrules, however, are unlikely to have formed at the ultralow pressures (below 10^{-8} atm) prevailing in a thin nebula. If the nebula was "thick", and chondrules were irradiated by the Sun, they must have been formed closer to the Sun and then transported out to the asteroid belt.

2.3. Experimental simulations

Experimental petrology also yields information on the chondrule forming process. Textures of chondrules were reproduced successfully about 15 years ago (e.g., [17]) allowing us to derive estimates of peak temperatures and cooling rates. Most chondrules appear to have been heated to about 1500–1600 °C [18] although their liquidus temperatures vary enormously (~ 1200–1900 °C). Their cooling rates ranged from about 10 to 1000 °C/h, much slower than the radiative cooling of isolated spherules into free space, and much faster than global nebular cooling [5]. This is considered evidence that chondrules were produced in large quantities embedded in hot gas [19]. More recent experiments have focused on reproducing the chemical and isotopic properties of chondrules. Yu et al. [20] showed that the oxygen isotopes of a chondrule could result from an exchange between its precursor material and the ambient gas, while in [18,21], they demonstrated that the volatile content of type II chondrules were incompatible with their being formed in the canonical solar nebula. On the other hand, Cohen et al. [22] showed that type I chondrules could have formed from the severe degassing of type II material and Cohen [23] established that such degassing need not result in Rayleigh distillation, as the presence of a volatile-rich atmosphere in the chondrule forming region would have been sufficient to prevent such fractionation. Suppression of isotopic fractionation by exchange during evaporation has also been modelled by Alexander [24].

3. Relevant properties of chondrites

3.1. Petrographic evidence

It would be hard to discuss the origin and significance of chondrules without taking into account some of the key properties of chondrites. As can be seen in (Figs. 2, 3 and 5), chondrites are an assemblage of chondrules and interstitial matrix in highly variable proportions. While chondrules (and chondrule fragments) may constitute up to 80% of the volume of some ordinary chondrites (Fig. 3a), their abundances were probably lower in carbonaceous chondrites (Fig. 3b) and close to zero in CI chondrites. Their sizes (Fig. 5), compositions and textures also vary between chondrite classes. Ferroan and droplet shaped chondrules are much more frequent in ordinary chondrites than in carbonaceous chondrites (Fig. 3). The composition and nature of the fine-grained matrix also varies within an individual chondrite and from one chondrite to another [25]. While all chondrites experienced some degree of aqueous alteration and/or metamorphism, chondrites that have the most matrix appear to be most altered to phyllosilicates, whereas the least altered matrix consists of a complex mixture of material from different sources, including presolar grains, chondrule fragments and condensate material [25].

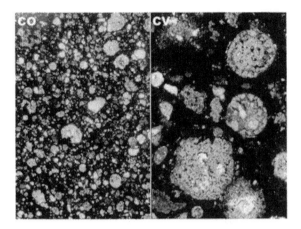

Fig. 5. Transmitted light view of PTS from a CO chondrite and a CV chondrite at the same scale (field 7 × 4.6 mm). Particles that make up chondrites (chondrule, chondrule fragments, metal grains and CAIs) appear to have a very restricted size distribution within a given chondrite class but to differ very significantly from class to class, ranging from an average of 20 μm in CHs and 150 μm in COs up to about 1 mm in CVs. The abundance of matrix also differs significantly ranging from 5% in CHs up to >99% in CIs. In Cos, the matrix amounts on the average to ~ 34% and in CVs to ~ 40% [66].

Apart from the fact that they contain chondrules, the most striking characteristics of chondrites are that their compositions are close to that of the Sun and that metallic minerals tend to be associated with chondrules in carbonaceous chondrites (Fig. 2 and 3b), and are interstitial between chondrules in ordinary chondrites.

3.2. CI chondrites: the solar system reference

The chondrites that most closely mimic the composition of the solar system are the CI chondrites (Fig. 6), which have the peculiarity of containing no or no surviving chondrule and consist almost entirely of highly altered matrix. CI chondrites contain no metal and are so oxidized that magnetite is present. They contain abundant presolar grains and their Cr [27] and Mo [28] were both discovered to consist of two complementary isotopically anomalous components, which appear also to be present in the matrix of other carbonaceous chondrites [27].

It has been argued that the extensive parent-body alteration undergone by CI chondrites was responsible for the destruction of the chondrules and CAIs they

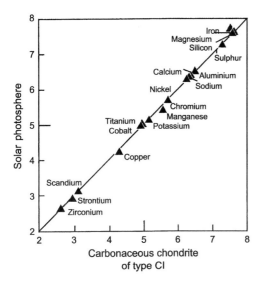

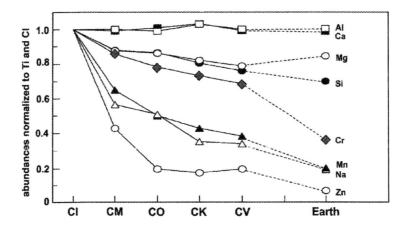

Fig. 6. With the exception of the most volatile elements (H, rare gases, C and O), CI chondrite compositions exactly match those of the solar photosphere which is representative of the sun, i.e., 99.99% of the mass of the solar system. This agreement holds for major and trace elements (after [26]—note scale is exponents).

material, like primitive chondrite matrix, is unlikely to have ever been melted or vaporized and recondensed in the nebula. This explains why CIs are the only meteorites that retained a solar composition and why they are so extensively altered compared to other chondrites: a greater amount of ice would have survived in a chondrule-free region to alter the silicate minerals. This idea receives some support from the detection by [32] in chondrule pyroxenes of water enriched in D up to three times above the values of the matrix and of the chondrule mesostasis. That is, the water in the mesostasis was probably introduced during low temperature alteration, but the water in the pyroxenes must have a different origin, implying that chondrules were formed in a water-rich environment or from a water-bearing material.

3.3. Chemical fractionations among chondrite classes

All other chondrite classes are chemically fractionated with respect to CIs (carbonaceous chondrites are shown in Fig. 7) and the variability of their Fe content and of its oxidation state has long been used as the classification reference (Fig. 8). As will be discussed below, chondrule formation involves both evaporation/condensation and redox processes and has the ability to generate the metal that is absent from CIs. A comparison of the chemical and isotopic properties of the chondrule-bearing chondrites with those of CIs thus suggests that chondrule formation may have played an

originally contained (e.g., [29]). The presence of relict crystals fallen from chondrules or CAIs [29,30] however does not prove that more than 1% of CI material was ever melted. On the contrary, the presolar grain concentrations, the heterogeneities in the Cr and Mo isotopes and the isotopic compositions of organic material [31] all indicate that the vast bulk of CI

Fig. 7. All chondrite classes (and the Earth's mantle) exhibit chemical fractionations compared to CIs and the Sun. In carbonaceous chondrites, the more volatile elements are more fractionated, suggesting the fractionation resulted from a high temperature process involving evaporation and/or condensation (from [33]).

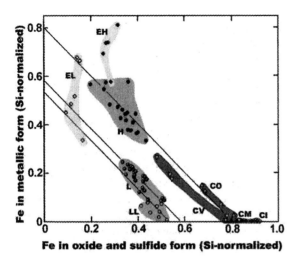

Fig. 8. Chondrite classes were originally defined based on their total Fe content and its oxidation state. In this modified version of the Urey and Craig diagram (after [26]), total Fe is constant along the diagonals, increasing from lower left to upper right. CIs have no metal although only EHs have more total Fe. As for their other chemical properties and the abundance of their chondrules, CMs are intermediate between CIs and the other carbonaceous chondrite groups, having less total Fe and a small amount of metal. EH chondrites are enriched in Fe and in volatiles in general with respect to CIs.

essential role in establishing those properties and generating the chondrite classes. This hypothesis was originally proposed for the volatile/refractory element fractionation as the "two-component" model of Larimer and Anders [34] and is still supported by the complementarity that apparently exists between chondrules and their surrounding matrix [35–37]. That is, when chondrules are more refractory, the matrix tends to be either more abundant or more volatile-enriched, allowing the bulk chondrite to have a bulk composition that is close to CI and significantly less fractionated than each of the individual constituents.

3.4. Oxygen isotopes

The various chemical chondrite classes, the Earth–Moon system and the planet Mars all have distinct oxygen isotopic signatures that cannot be explained by mass dependent fractionation alone (Fig. 9). A mixing from at least two different reservoirs lying on a slope 1 line (i.e., containing different amounts of ^{16}O) is required as well. Since their first discovery by Clayton

et al. [38] almost 30 years ago, the nature and significance of these distinct oxygen reservoirs have been a matter of a heated debate. Recent observations seem to shed new light on these issues and allow us to generate the various reservoirs from the mixing of a limited number of end-members. Following Clayton and Mayeda [39], Young and Russell [40] demonstrated that alteration of CAI minerals shifted their isotopic compositions towards heavier oxygen along mass-dependent fractionation lines whereas unaltered minerals fall along a slope 1 line passing through the ordinary chondrite groups (Fig. 9). This allows us to distinguish two types of vector on the oxygen isotopic diagram, related to different physical processes: mass-dependent fractionation towards the right, which is related to alteration (and may be proportional to the amount of matrix in the material [45]) and mass-independent effects along the ~ 1 slope line "$Y+R$" which remains to be explained. Results by Luck et al. [41] however indicate that this mass-independent fractionation may derive from the process that fractionated volatiles from refractories in chondrites: (1) ^{16}O excesses are coupled with ^{63}Cu excesses both in carbonaceous and (along a distinct trend) in ordinary chondrites [41]; (2) there is a negative correlation between the ^{63}Cu excesses of the various chondrite groups and their ratios of moderately volatile elements such as Mn [41], Na and K [46] relative to the refractory element Al. As discussed below, the process that fractionated volatiles from refractories could be closely related to CAI/chondrule formation since CAIs are enriched in ^{16}O whereas the most volatile-rich chondrules are the most depleted.

3.5. Clasts in chondrites as a clue to their formation environment

Other noteworthy properties of chondrites include the fact that they contain lithic clasts that might have originated from differentiated asteroidal bodies as advocated by Hutchison [47]. Together with the presence of numerous chondrule fragments, this has been taken as evidence for the occurrence of collisions and fragmentation before the final accretion of chondrites, which was therefore suggested to have taken place in a planetary environment [47]. This interpretation appears consistent with the metamorphic Pb–Pb age of the oldest dated chondrites (4.563 Ma [48])

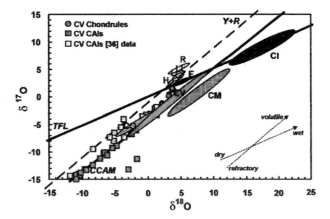

Fig. 9. For convenience, the small variations of oxygen isotopic compositions are described in terms of their deviation in parts per thousand from a terrestrial standard, Standard Mean Ocean Water (SMOW):

$$\delta^{17}O = \{[(^{17}O/^{16}O)_{sample}/(^{17}O/^{16}O)_{SMOW}] - 1\} \times 10^3$$

$$\delta^{18}O = \{[(^{18}O/^{16}O)_{sample}/(^{18}O/^{16}O)_{SMOW}] - 1\} \times 10^3.$$

Each chemical class of chondrites has its own oxygen signature (with some amount of overlap between classes). For simplicity, only the major carbonaceous chondrite classes are displayed here and EH and EL chondrites are grouped as "E". Two types of isotopic fractionations are involved: mass-dependent and mass-independent. Mass-dependent fractionations are usually the result of temperature, coordination or kinetic effects and move the samples along a slope 1/2 line. Except for some very special cases, all terrestrial samples are thus restricted to the 1/2 slope line labelled "terrestrial fractionation line" (TFL), and each chondrite class is characterized by its $\Delta^{17}O$ (the excess of $\delta^{17}O$ relative to the terrestrial fractionation line). In meteorites, mass-independent effects are also present, the origin of which is still not properly understood. Most CAIs and chondrules from CV chondrites lie along a slope ∼ 1 line known as "carbonaceous chondrite anhydrous mineral" (CCAM) which was suggested to result from the addition of pure ^{16}O to these samples [38]. Ordinary chondrites chondrules all fall in the OC domain, i.e., above the TFL. A new understanding of this diagram is emerging. Following the suggestion of [39], Young and Russell [40] recently showed that unaltered minerals from CAIs lie on a slope 1 line ("«Y + R»") going through the ordinary chondrite groups, whereas altered minerals are shifted to the right (along a slope 1/2 line). The $\Delta^{17}O$ of the chondrite groups (which reflects their projection along the TFL onto the «Y + R» line) are related with their ^{63}Cu excesses which are, in turn, correlated with their Mn/Al ratios [41]. Data sources: Clayton and Mayeda [42], Clayton [43], Kalleyman et al. [44] and Young and Russell [40].

which compares to that of eucrites (4.56 Ma [49]). This theory is, however, not widely accepted.

4. The origin of chondrules: their precursors and links with other chondritic components

The question of the origin of chondrules is probably the most tantalizing one in meteoritics. As discussed in [3], we are still in search of a grand theory and an embarrassing number of open questions remain despite the wealth of accumulated data. These questions fall into two closely connecting broad categories: the nature of the material from which the chondrule liquids were generated and the physical mechanism responsible for the melting event. To discuss these issues, the basic observations presented above will be connected with additional recent significant observations.

4.1. The canonical view

It is first necessary to summarize the ideas on chondrules that were most commonly accepted until recently, although some appear to have little other grounding than having been repeated over time, as pointed out by Wood [6]. In the canonical view, equilibrium condensation (resulting from the cooling of an initially entirely vaporized inner nebula of solar composition) generated small solid grains. These grains then aggregated together into small dustballs which were heated to high temperatures by an as yet unidentified heating mechanism. Most researchers agreed that chondrules had formed as closed systems without losses to, or gains from, the nebula (see [6] for references), hence requiring a flash heating mechanism in order to avoid interactions with the nebular gas.

According to this view, chondrule formation had little effect on the chondritic matter aside from textural, and therefore no connection to the chemical and isotopic differences between CIs and other chondrite classes summarized above. As discussed in the following sections, however, a growing number of researchers are challenging this view and arguing for open system behavior based on different lines of evidence.

4.2. Forming chondrules from microdroplets and dust

Wood [6] considered that creating a population of dustballs in the first place was far from straightforward even if this had been commonly ignored. Although this problem does no longer appear critical [50,51], the alternative favored by Wood [6] remains useful to explain the peculiar textures of chondrules in carbonaceous chondrites as the one in Fig. 3b. To account for these, Wood [6] suggested forming chondrules from an accumulation of hot microdroplets and dust, although he recognized that one of the problems with this mechanism is the scarcity of preserved microchondrules in the matrices of primitive chondrites. These might however have been lost and accumulated elsewhere, for example in CH chondrites, as these consist almost exclusively of microchondrules and debris with an average size of 20 μm.

4.3. Forming chondrules by condensing liquid

Condensation of liquids from an enriched gas constitutes another alternative that has been receiving growing attention. Ebel and Grossman [52] showed that liquids with compositions close to those of chondrules could be stable in local environments with high total pressures or high partial pressures of the relevant elements. This hypothesis provides a simple explanation to the absence of Rayleigh distillation induced isotopic fractionations in chondrules. In addition, Krot [53] argues that the extreme variety of compositions of chondrules in CHs (which range from highly refractory-enriched to highly refractory-depleted) can only be ascribed to fractional condensation. It is interesting to note that this hypothesis is not exclusive of the microdroplet mechanism favored by Wood [6] as the condensation might even have been to microdroplets which subsequently would have agglomerated, as he suggested [6].

4.4. Forming chondrules from earlier generations of chondrules

Yet another possibility advocated by Alexander [54] is that chondrules were made from the recycling of earlier generations of chondrules, based on their chemical compositions and on the evidence for successive episodes of melting. This hypothesis is particularly attractive for chondrules in ordinary chondrites in which the least melted chondrules (presumed to be the least transformed and hence the closest to their precursors [55]) are aggregates of fine crystals with occasional coarser relicts, rather than looking as if they coalesced from microdroplets, as in carbonaceous chondrites [55–57]. Ordinary chondrites could thus contain mostly chondrules made from recycled precursors whereas carbonaceous chondrites would contain more chondrules of the earliest generations.

4.5. Was the system open or closed?

Chondrule compositions are a key issue as they range from refractory-rich to volatile-rich. This variety of compositions could either be inherited from the precursors or result from the chondrule-forming event, depending on whether the system was closed or open to the surrounding gas, or derive from a combination of both. In the canonical nebular models, where condensates formed dustballs as temperature fell, the system is closed: volatile-poor chondrules were made from volatile-poor material, while volatile-rich chondrules require a flash heating mechanism to preserve their volatile content.

The experiments of Hewins et al. [58] and Yu et al. [21] showed, however, that type II chondrule bulk compositions are impossible to explain with melting in a canonical nebula, even with flash heating, because retaining volatiles such as S, Na or K requires total pressures and oxygen fugacities, or partial pressures of these elements, several orders of magnitude above the nebular values. These partial pressures are not known, so the time scale for the formation of natural chondrules cannot be specified, but they ought to have been high enough to allow recondensation of volatiles into chondrules. This process may have contributed to the apparent lack of isotopic fractionation induced by evaporation during chondrule formation (e.g., [9,10]).

Comparing chondrule chemistry with their textures also points to an open system behavior; Hewins et al. [55] found an inverse correlation between the grain size of ordinary chondrite type I chondrules and their volatile content. The finest (least melted) chondrules have Mg-normalized bulk compositions close to that of CI chondrite and contain no metal but abundant Ni-bearing Fe-sulfide [56]. All the intermediates exist between these fine-grained chondrules and the typical volatile-depleted and metal-bearing coarser-grained ones. This indicates that type I chondrules may have started with unaltered CI compositions and evolved to their present compositions during melting, as was experimentally achieved by Cohen [23]. In such a case, most chondritic metal might be a product from chondrules (made primarily by S loss [56], although some carbon-reduction might also occur in chondrule formation). This appears consistent with the present mineralogy of chondrule-free CI chondrites which are metal free (but contain Ni-bearing Fe-sulfide and magnetite instead) and may have done so even before they were altered on their parent bodies.

4.6. Geochemical consequences of open system behavior

By producing metallic melts which have the ability to physically separate from silicate melts (e.g., [59]) and by losing volatiles which will only partly recondense (e.g., [37]), chondrules thus have the ability to generate both the metal/silicate and the refractory/volatile fractionations that characterize the chondrite classes (Figs. 7 and 8). In that framework, the chemical complementarity between matrix and chondrules [35–37] finds a natural explanation as chondrules and matrix would both have evolved from unaltered CI type material, one gaining part of the material (volatiles and metal) lost by the other. There is also some evidence of condensation into chondrule rim material.

The present oxygen isotopic signatures of chondrules are still poorly understood and those of their precursors are not known. Luck et al. [41] discuss in detail the origin of the correlations between bulk ^{63}Cu and ^{16}O excesses in carbonaceous and in ordinary chondrites together with Al/Mn ratios. These authors suggest mixing an ^{16}O enriched and a copper-poor but ^{63}Cu enriched refractory component together with another component close to CI in ^{16}O and ^{63}Cu or

slightly more depleted, possibly the nebular gas. Such interaction could clearly have taken place during chondrule formation as Yu et al. [20] experimentally demonstrated that an exchange with the ambient gas could modify the oxygen signatures of chondrules, as Varley et al. [60] found a mild correlation between $\Delta^{17}O$ and the extent of chondrule melting in CR chondrites and as FeO-rich chondrules in CRs [61] and CMs [62] appear more depleted in ^{16}O than their FeO-poor neighbours. This model might explain the oxygen isotopic signatures of ordinary chondrite chondrules which would have derived from unaltered CI type precursor material. Carbonaceous chondrite chondrules however require the addition of ^{16}O and ^{63}Cu, presumably in an enriched refractory component as proposed by Luck et al. [41].

Alternative explanations for the mass independent fractionation of oxygen in meteorites, involving symmetry-dependent chemical reactions (e.g., [63]) or CO self shielding [64] or linked to Galactic chemical evolution [65], might allow us to derive chondrules from both ordinary and carbonaceous chondrites from unaltered CI material. It is however unclear that they would also provide an explanation for the correlations observed between bulk ^{63}Cu and bulk ^{16}O excesses in carbonaceous and ordinary chondrites and between ^{63}Cu excesses and Al/Mn ratios. In addition, the presence of a refractory component in the precursors of chondrules from CO and CV chondrites but not from ordinary chondrites, (as shown by [7] based on REE signatures) yields support to the previous model.

4.7. The relationship between chondrules and CAIs

This raises the issue of the relationship between chondrules and CAIs, the most likely candidates for the refractory component of chondrule precursors in carbonaceous chondrites, which happen to be very abundant (~ 10–13%) in CV and CO chondrites and very rare (< 1%?) in ordinary chondrites (e.g., [66]). CAIs are the oldest objects of the solar system and appear to have been formed 2.5 ± 1.2 Ma before chondrules [12]. Many of them have also experienced melting by a process that is currently unknown and that it is tempting to relate to that which later formed chondrules. It is however unlikely that there was a gradual change with time from CAI formation to chondrule formation as the bulk of the two populations appears distinct both in age and in composition. A few intermediate objects however do exist such as Al-rich or forsterite–fassaite [30] chondrules. These might have been formed from precursors containing excess CAI material. The presence of CAI material within chondrule precursors poses another significant problem discussed in the next section: how could these objects have been prevented for more than 2 million years from drifting into the Sun [6]?

5. Where, how and when?

The wide variety of proposed chondrule-forming mechanisms (e.g., Fig. 1) fall into two broad categories depending on whether they are assumed to have operated in the vicinity of newly formed planets or within the protoplanetary nebula. Following the publication 30 years ago of a very influential paper [67], a consensus seemed to have been reached by most scientists in the field (mostly petrologists at the time) that chondrules could only have been formed in a nebular setting, although a few dissenting voices always existed (e.g., [47]). The pendulum recently started swinging the other way, mostly under the influence of isotope geochemists who inferred, based on extinct radioactive isotopes such as ^{26}Al [68,69] and ^{53}Mn [70], that the formation of chondrules lasted from < 1 up to 4 million years after CAIs.

5.1. Forming chondrules in a planetary environment

Assuming these ages are crystallization (rather than metamorphic) ages, forming chondrules over such an extended period raises the problem of preventing the oldest ones and the CAIs from drifting into the growing Sun so that they are able to mix with younger ones. While the chronological interpretation of extinct radionuclides alone could be questioned (as they might have been initially heterogeneously distributed or secondarily disturbed by parent-body processes), it was recently beautifully confirmed by the Pb–Pb data of Amelin et al. [12]. According to this interpretation, the youngest chondrules [68] have roughly the same age as the oldest phosphate age for a metamorphosed chondrite, 4563 Ma [48].

If the environment in which CAIs and chondrules formed did not change significantly over time, then protoplanets such as those which eventually became differentiated already existed soon after the oldest CAIs were melted and they provide one easy way to store these objects and prevent them from drifting into the sun.

Additional benefits of planetary models are that they explain the presence of lithic clasts within chondrites [47] and the textures of chondrites as seen in Fig. 3, and especially they provide an atmosphere for melting volatile-rich chondrules much more suitable than the open solar nebula. They do not, however, allow us to generate the chemical fractionation between chondrite classes easily in the chondrule-forming event, as seems to be suggested from the comparison between CIs and other chondrites.

The possibilities for generating chondrules in a planetary environment are numerous. Most of them involve collisions and will not be reviewed here as their physics needs to be investigated in detail, as was recently done for nebular models. A recent hybrid planetary-nebular model [71] produces chondrules by heating nebular particles or debris of first generation planetesimals by bow-shocks around eccentric planetesimals driven by Jovian resonances.

5.2. Forming chondrules in a nebular environment

The recent revisiting of planetary models seems timely as our knowledge of the time scales and processes at work in a disk containing colliding planetary embryos is evolving. It however does not appear fully justified based only on the problem of preventing older generations of CAIs and chondrules from drifting into the Sun, because many other solutions to this problem exist. One involves a first step of accretion immediately following chondrule formation to form bodies which are later disrupted and reassembled, mixing CAIs and chondrules of different ages. In a purely nebular context, outward radial diffusion in a weakly turbulent nebula will preserve CAIs [72]. The density of the gas may also be low enough that the residence time of CAIs and chondrules reaches several million years and, even if the bulk of the CAIs and chondrules spiral into the Sun, the tail of the distribution will still diffuse upstream [73].

Aside from condensation of a liquid from the nebular gas (e.g., [52]), models to form chondrules in a nebular setting are even more varied than for planetary environments. The present review concentrates on the two that were developed or significantly improved and received much attention since the summary sketched in Fig. 1.

The model of Shu et al. [2] involves the inner edge of the protoplanetary disk where matter is partly accreted into the Sun and partly ejected by bipolar outflows. Heating by the Sun and solar flares melts dustballs brought to the inner edge of the disk makes CAIs and chondrules. Some of these are subsequently thrown back by an "X-wind" onto the disk in the asteroidal belt region. This model explains the narrow size distribution of chondritic components in each chondrite class (Fig. 5) by sorting and also allows generating some of the short-lived radioactive nuclides recently detected in chondrules [15] and in CAIs (e.g., [74,75]) by irradiation. However, it fails to explain the cooling rates observed for chondrules, the generation of matrix with a different composition in different locations (to explain the apparent matrix–chondrule complementarity) and the provision of a mechanism for storing CAIs and chondrules over extended periods (to explain their Pb–Pb age difference).

The nebular shock model of Desch and Connolly [19] and of Ciesla and Hood [76] does better in some of these respects. It involves melting chondrules (and possibly also CAIs) as the result of the propagation of shock waves within the nebula and, by taking into account the radiation absorbed by the chondrules both from the heated gas and from the neighboring chondrules, generates cooling rates that match those estimated for chondrules [19]. Transient atmospheres are generated in clumps of evaporating dust and chondrules, allowing formation of volatile-rich chondrules. The matrix is formed from unevaporated dust and some volatiles lost from chondrules, so the complementarity between it and chondrules falls out of the model. Size sorting (see Fig. 5) of chondrites is achieved by concentrating chondrules between eddies in a turbulent nebula (after Cuzzi et al. [77]) and accreting separately chondrules with different aerodynamic properties (which depend on their sizes). Finally, this model [19] predicts a correlation between chondrule density in the source region, cooling

rate and temperature that generates relative abundances of the different chondrule textural types corresponding to those observed in chondrites. This represents a milestone in the present author's opinion as it is the first model that closely matches some of the significant properties of chondrites. Like other nebular models, however, it needs to appeal to a mechanism such as [72] to prevent CAIs and chondrules from being dragged by the nebular gas into the growing Sun.

5.3. The possible consequences of chondrule formation in a nebular setting

The stopping time of chondrules due to gas drag depends on their radius which correlates with their FeO content, so type I and II chondrules are expected to separate [78] and so are metallic particles which are smaller even than type I chondrules. By allowing such sorting, chondrule formation in a nebular setting thus has the ability to generate the various chondrite classes. This possibility is especially attractive for ordinary chondrites in which average chondrule diameters range from 300 μm (in H) to 700 μm (in L) to 900 μm (in LL) [66] while the vol.% of chondrule material present as type I decreases from 57% (in H) to 40% (in L) to 25% (in LL) [79] and the abundance of metal (which has aerodynamic properties closer to those of type I chondrules [80]) decreases in parallel.

This model even holds a potential explanation for the oxygen isotopic signatures of the three classes, as there is no difference between randomly chosen individual chondrules of each class (which span the whole range of values of the bulk samples from the three classes) but two reasons make it likely that volatile-rich type II (concentrated in LLs) are on average slightly more depleted in ^{16}O than volatile-poor type I: (1) Clayton et al. [81] showed that in ordinary chondrites the larger chondrules tend to be more depleted in ^{16}O; (2) type II chondrules in carbonaceous chondrites tend to be more depleted in ^{16}O than their type I counterparts [61,62]. Note that this model might even be extended to the case of the Earth if it was made mostly from type I chondrules as suggested by Hewins and Herzberg [1], as a higher content of type I chondrules than in H chondrites would naturally place it at the intercept of the $Y+R$ and of the terrestrial fractionation line in Fig. 9.

Chondrule formation in a nebular setting may also have played a role in the earliest stages of accretion by sticking together particles while they are still partly melted or above the glass transition temperature as the textures in Fig. 3 suggest. Much of the earliest accretion may however be due to electrostatic attraction [50,51].

6. Future directions

Despite the significant recent advances described above (such as obtaining the first absolute ages of chondrules), major problems still plague all the possible chondrule-forming mechanisms that have been proposed up to now and we remain unable to unequivocally decide between the two types of possible environments in which chondrules were formed. Progress may come from several directions:

6.1. More detailed observations of chondrules

Precise dating of individual chondrules has only just begun and is sure to contribute significantly to our knowledge of the chondrule forming mechanism. Another new direction will involve a much better understanding of the details of the physical mechanisms involved within a melting and cooling chondrule. This will be achieved with the help of in situ isotopic and trace measurements relevant to exchanges between the solid, liquid and gaseous phases involved, coupled with experimental simulations.

6.2. More detailed observations of chondrites

We are still unable to fully disentangle the effects of the chondrule-forming event from those of aqueous alteration and thermal metamorphism. These secondary effects may distort our present vision of chondrule precursors and of the chondrule-forming mechanism. Systematic studies of chondrule properties (such as their age, chemistry, textures, diameters) in a given meteorite and within a chondrite class will also provide tighter constraints. It is possible that all chondrules do not share the same type of precursors and/or formation mechanism as

seems to be suggested by the differences between carbonaceous and ordinary chondrites discussed throughout this paper and shown in Fig. 3. The study of new chondrite classes or (in the distant future) of samples returned from chondritic asteroids might help in that respect.

6.3. More detailed observation and modeling of young planetary systems trying to better reproduce chondrule/chondrite properties

Until fairly recently, the solar nebula has been treated as a "black box" in which any poorly understood effect could have happened and there has in the past been too much distance between meteoriticists and astrophysicists. Chondrule formation however needs to be linked to realistic astrophysical constraints. The 1994 "Chondrule and the Protoplanetary Disk Conference" followed by a book [82], started bridging that gap. It will hopefully be narrowed further with new chondrule formation models, which, after [19], will focus on making detailed petrographic predictions.

Acknowledgements

Like at least one previous Frontiers author [83], I want to thank Alex Halliday for "early encouragement and recent patience". I also want to thank for their insight my colleagues from MNHN: M. Bourot-Denise, C. Perron and F. Robert and from Rutgers University: B. Cohen and Y. Yu. P. Cassen and S. Desch made helpful suggestions concerning the astrophysical aspects, D. Ben Othman and J.-M. Luck concerning Cu isotopes and H. Palme kindly let me copy Fig. 7 from one of his manuscripts. He, S. Desch and C. M.O'D Alexander are thanked for reviews which improved this manuscript. Last but not least, I thank R. Hewins for scientific and material support throughout this work. *[AH]*

References

[1] R.H. Hewins, C.T. Herzberg, Nebular turbulence, chondrule formation, and the composition of the earth, Earth Planet. Sci. Lett. 144 (1996) 1–7.

[2] F. Shu, H. Shang, A.E. Glassgold, T. Lee, X-rays and fluctuating x-winds from protostars, Science 277 (1997) 1475–1479.

[3] J.A. Wood, Chondrites: Tight-Lipped Witnesses to the Beginning, Harold Masursky Plenary Lecture, LPSC, Houston, 2000.

[4] R.A. Kerr, A meteoriticist speaks out, his rocks remain mute, Science 293 (2001) 1581–1584.

[5] R.H. Hewins, Chondrules and the protoplanetary disk: an overview, in: R.H. Hewins, R.H. Jones, E.R.D. Scott (Eds.), Chondrules and the Protoplanetary Disk, Cambridge University Press, Cambridge, 1996, pp. 3–9.

[6] J.A. Wood, Unresolved issues in the formation of chondrules and chondrites, in: R.H. Hewins, R.H. Jones, E.R.D. Scott (Eds.), Chondrules and the Protoplanetary Disk, Cambridge University Press, Cambridge, 1996, pp. 55–69.

[7] K. Misawa, N. Nakamura, Origin of refractory precursor components of chondrules from carbonaceous chondrites, in: R.H. Hewins, R.H. Jones, E.R.D. Scott (Eds.), Chondrules and the Protoplanetary Disk, Cambridge University Press, Cambridge, 1996, pp. 99–105.

[8] H. Yurimoto, J.T. Wasson, Extremely rapid cooling of a carbonaceous-chondrite chondrule containing very ^{16}O-rich olivine and a ^{26}Mg excess, Geochim. Cosmochim. Acta 66 (2002) 4355–4363.

[9] C.M.O'D. Alexander, J.N. Grossman, J. Wang, B. Zanda, M. Bourot-Denise, R.H. Hewins, The lack of potassium isotopic fractionation in Bishunpur chondrules, Meteorit. Planet. Sci. 35 (2000) 859–868.

[10] C.M.O'D. Alexander, J. Wang, Iron isotopes in chondrules: implications for the role of evaporation during chondrule formation, Meteorit. Planet. Sci. 36 (2001) 419–428.

[11] T.D. Swindle, A.M. Davis, C.M. Hohenberg, G.J. MacPherson, L.E. Nyquist, Formation times of chondrules and Ca–Al-rich inclusions: constraints from short-lived radionuclides, in: R.H. Hewins, R.H. Jones, E.R.D. Scott (Eds.), Chondrules and the Protoplanetary Disk, Cambridge University Press, Cambridge, 1996, pp. 77–86.

[12] Y. Amelin, A.N. Krot, I.D. Hutcheon, A.A. Ulyanov, Lead isotopic ages of chondrules and calcium–aluminum-rich inclusions, Science 297 (2002) 1678–1683.

[13] A.N. Halliday, The Origin and Earliest History of the Earth, in: Treatise of Geochemistry, vol. 1, Elsevier, Amsterdam, 2004.

[14] S.J. Kortenkamp, E. Kokubo, S.J. Weidenschilling, Formation of planetary embryos, in: R.M. Canup, K. Righter (Eds.), Origin of the Earth and Moon, University of Arizona Press, Tucson, 2000, pp. 85–100.

[15] F. Robert, M. Chaussidon, Boron and lithium isotopic composition in chondrules from the Mokoïa meteorite, Lunar Planet. Sci. XXXIV (2003) 1344, (pdf.).

[16] S.J. Desch, H.C. Connolly, G. Srinivasan, An interstellar origin for the beryllium 10 in calcium-rich, aluminum-rich inclusions, Astrophys. J. 602 (2004) 528–542.

[17] P.M. Radomsky, R.H. Hewins, Formation conditions of pyroxene–olivine and magnesian olivine chondrules, Geochim. Cosmochim. Acta 54 (1990) 3475–3490.

[18] Y. Yu, R.H. Hewins, Transient heating and chondrule formation-evidence from Na loss in flash heating simulation experiments, Geochim. Cosmochim. Acta 62 (1998) 159–172.

[19] S.J. Desch, H.C. Connolly Jr., A model for the thermal processing of particles in solar nebula shocks: application to cooling rates of chondrules, Meteorit. Planet. Sci. 37 2002, pp. 183–208.

[20] Y. Yu, R.H. Hewins, R.N. Clayton, T.K. Mayeda, Experimental study of high temperature oxygen isotope exchange during chondrule formation, Geochim. Cosmochim. Acta 59 (1995) 2095–2104.

[21] Y. Yu, R.H. Hewins, C.M.O'D. Alexander, J. Wang, Experimental study of evaporation and isotopic mass fractionation of potassium in silicate melts, Geochim. Cosmochim. Acta 67 (2003) 773–786.

[22] B.A. Cohen, R.H. Hewins, Y. Yu, Evaporation in the young solar nebula as the origin of "just-right" melting of chondrules, Nature 406 (2000) 600–602.

[23] B.A. Cohen, Chondrule formation by open system melting of nebular condensates, PhD thesis, Rutgers University (2002).

[24] C.M.O'D. Alexander, Chemical equilibrium and kinetic constraints for chondrule and CAI formation, Geochim. Cosmochim. Acta (in press).

[25] A.J. Brearley, Nature of matrix in unequilibrated chondrites and its possible relationship to chondrules, in: R.H. Hewins, R.H. Jones, E.R.D. Scott (Eds.), Chondrules and the Protoplanetary Disk, Cambridge University Press, Cambridge, 1996, pp. 137–151.

[26] D.W.G. Sears, R.T. Dodd, Overview and classification of meteorites, in: J.F. Kerridge, M.S. Matthews (Eds.), Meteorites and the Early Solar System, University of Arizona Press, Tucson, 1988, pp. 3–31.

[27] M. Rotaru, J.-L. Birck, C.J. Allègre, Clues to early solar-system history from chromium isotopes in carbonaceous chondrites, Nature 358 (6386) (1992) 465–470.

[28] N. Dauphas, B. Marty, L. Reisberg, Molybdenum nucleosynthetic dichotomy revealed in primitive meteorites, Astrophys. J. 569 (2002) L139–L142.

[29] L.A. Leshin, A.E. Rubin, K.D. McKeegan, The oxygen isotopic composition of olivine and pyroxene from CI chondrites, Geochim. Cosmochim. Acta 61 (1997) 835–845.

[30] B. Zanda, G. Libourel, Ph. Blanc, Source chondrules for refractory forsterites in primitive chondrites, revision for, Meteorit. Planet. Sci. (2004), in press.

[31] F. Robert, D. Gautier, B. Dubrulle, The Solar system D/H ratio: observations and theories, ISSI: Dust to Terrestrial Planets, Berne, Kluwer, Dordrecht, 2000, pp. 201–224.

[32] E. Deloule, F. Robert, J.C. Doukhan, Interstellar hydroxyl in meteoritic chondrules: implications of water in the inner solar system, Geochim. Cosmochim. Acta 62 (1998) 3367–3378.

[33] H. Palme, Chemical and isotopic heterogeneity in protosolar matter, Philos. Trans. R. Soc. London Ser., A 359 (2001) 2061–2075.

[34] J.W. Larimer, E. Anders, Chemical fractionation in meteorites: II. Abundance patterns and their interpretation, Geochim. Cosmochim. Acta 31 (1967) 1239–1270.

[35] J.A. Wood, Meteoritic constraints on processes in the solar nebula, in: D.C. Black, M.S. Matthews (Eds.), Protostars and Planets II, University of Arizona Press, Tucson, 1985, pp. 687–702.

[36] H. Palme, S. Klerner, Formation of chondrules and matrix in carbonaceous chondrites, Meteorit. Planet. Sci. 35 (2000) A124.

[37] B. Zanda, M. Humayun, R.H. Hewins, M. Bourot-Denise, A.J. Campbell, The Relationship Between Volatile Element Patterns and Chondrule Textures in CRs and OCs, Goldschmidt Conference Abtracts, 2002.

[38] R. N. Clayton, N. Onuma, L. Grossman, T.K. Mayeda, Distribution of the pre-solar component in Allende and other carbonaceous chondrites, Earth Planet. Sci. Lett. 34 (1977) 209–224.

[39] R.N. Clayton, T.K. Mayeda, The oxygen isotope record in Murchison and other carbonaceous chondrites, Earth Planet. Sci. Lett. 67 (1984) 151–161.

[40] E.D. Young, S.S. Russell, Oxygen reservoirs in the early solar nebula inferred from an Allende CAI, Science 282 (1998) 452–455.

[41] J.-M. Luck, D. Ben Othman, J.-A. Barrat, F. Albarède, Coupled ^{63}Cu and ^{16}O excesses in chondrites, Geochim. Cosmochim. Acta 67 (2003) 143–151.

[42] R.N. Clayton, T.K. Mayeda, Oxygen isotope classification of carbonaceous chondrites, Lunar Planet. Sci. XX (1989) 169–170.

[43] R.N. Clayton, Oxygen isotopes in meteorites, Annu. Rev. Earth Planet. Sci. 21 (1993) 115–149.

[44] G.W. Kallemeyn, A.E. Rubin, J.T. Wasson, The compositional classification of chondrites: VII. The R chondrite group, Geochim. Cosmochim. Acta 60 (1996) 2243–2256.

[45] Ph. Bland, personal communication (2004).

[46] D. Ben Othman, J.-M. Luck, personal communication (2004).

[47] R. Hutchison, Chondrules and their associates in ordinary chondrites: a planetary connection? in: R.H. Hewins, R.H. Jones, E.R.D. Scott (Eds.), Chondrules and the Protoplanetary Disk, 1996, pp. 311–318, Cambridge.

[48] C. Göpel, G. Manhès, C.J. Allègre, U–Pb systematics of phosphates from equilibrated ordinary chondrites, Earth Planet. Sci. Lett. 121 (1994) 153–171.

[49] K. Misawa, A. Yamaguchi, Zircons in eucrites: occurrence, possible origin and U–Pb isotopic systematics, Lunar Planet. Sci. XXXII (2001) 1676.

[50] N. Haghighipour, Growth of dust particles and accumulation of centimeter-sized objects in the vicinity of a pressure-enhanced region of a solar nebula, Lunar Planet. Sci. XXXV (2004) 2001.

[51] S.G. Love, D.R. Pettit, Fast, repeatable clumping of solid particles in microgravity, Lunar Planet. Sci. XXXV (2004) 1119.

[52] D.S. Ebel, L. Grossman, Condensation in dust-enriched systems, Geochim. Cosmochim. Acta 64 (2000) 339–366.

[53] A.N. Krot, et al, Ferrous silicate spherules with euhedral iron–nickel metal grains from CH carbonaceous chondrites: evidence for supercooling and condensation under oxidizing conditions, Meteorit. Planet. Sci. 35 (2000) 1249–1258.

[54] C.M.O'.D. Alexander, Recycling and volatile loss in chondrule formation, in: R.H. Hewins, R.H. Jones, E.R.D. Scott (Eds.), Chondrules and the Protoplanetary Disk, Cambridge University Press, Cambridge, 1996, pp. 233–241.

[55] R.H. Hewins, Y. Yu, B. Zanda, M. Bourot-Denise, Do nebular fractionations, evaporative losses, or both, influence chondrule compositions? Antarct. Meteor. Res. 10 (1997) 294–317.

[56] B. Zanda, M. Bourot-Denise, R.H. Hewins, Chondrule precursors: the nature of the S- and Ni-bearing phase(s), Lunar Planet. Sci. XXVII (1996) 1485–1486.

[57] J.W. Nettles, G.E. Lofgren, H.Y. McSween Jr., Recycled "chondroids" in LEW86018: a petrographic study of chondrule precursors, Lunar Planet. Sci. XXXIII (2002) 1752.

[58] R.H. Hewins, B. Zanda, Y. Yu, M. Bourot-Denise, Towards a new model for chondrules, Paul Pellas Symposium abstracts, Muséum National d'Histoire Naturelle, Paris, 1998, pp. 31–33.

[59] G.K. Benedix, T.J. McCoy, T.L. Dickinson, G.E. Lofgren, Partial melting of chips of the indarch (EH4) meteorite: further insights into melt migration, Meteorit. Planet. Sci. 36 (2001) A18, (Suppl.).

[60] L.R. Varley, L.A. Leshin, Y. Guan, B. Zanda, M. Bourot-Denise, Oxygen isotopic composition of Renazzo chondrule olivine and comparison with extent of chondrule melting, Lunar Planet. Sci. XXXIV (2003) 1899.

[61] H.C. Connolly Jr., M.K. Weisberg, G.R. Huss, On the nature and origins of FeO-rich chondrules in CR2 chondrites: a preliminary report, Lunar Planet. Sci. XXXIV (2003) 1770.

[62] I. Jabeen, H. Hiyagon, Oxygen isotopes in isolated and chondrule olivines of Murchison, Lunar Planet. Sci. XXXIV (2003) 1551.

[63] M.H. Thiemens, Mass-independent isotopic effects in chondrites: the role of chemical processes, in: R.H. Hewins, R.H. Jones, E.R.D. Scott (Eds.), Chondrules and the Protoplanetary Disk, 1996, pp. 107–118, Cambridge.

[64] R.N. Clayton, Solar system: self-shielding in the solar nebula, Nature 415 (2002) 860–861.

[65] D.D. Clayton, Isotopic anomalies: chemical memory of galactic evolution, Astrophys. J. 334 (1988) 191–195.

[66] A.J. Brearley, R.H. Jones, Chondritic meteorites, in: J.J. Papike (Ed.), Planetary Materials, Reviews in Mineralogy vol. 36, 1998, 398 pp.

[67] G.K. Taylor, E.R.D. Scott, K. Keil, Cosmic setting for chondrule formation, in: E.A. King (Ed.), Chondrules and Their Origins, Lunar and Planetary Institute, Houston, 1983, pp. 262–278.

[68] S.S. Russell, G.R. Huss, G.J. MacPherson, G.J. Wasserburg, Early and late chondrule formation: new constraints for solar nebula chronology from $^{26}Al/^{27}Al$ in unequilibrated ordinary chondrites, Lunar Planet. Sci. XXVIII (1997) 1209–1210.

[69] S. Mostefaoui, N.T. Kita, S. Togashi, S. Tachibana, H. Nagahara, Y. Morishita, The relative formation ages of ferromagnesian chondrules inferred from their initial aluminum-26/aluminum-27 ratios, Meteorit. Planet. Sci. 37 (2002) 421–438.

[70] G.W. Lugmair, A. Shukolyukov, Early solar system events and timescales, Meteorit. Planet. Sci. 36 (2001) 1017–1026.

[71] S.J. Weidenschilling, F. Marzari, L.L. Hood, The origin of chondrules at Jovian resonances, Science 279 (1998) 681–684.

[72] J.N. Cuzzi, S.S. Davis, A.R. Dobrovolskis, Blowing in the wind: II. Creation and redistribution of refractory inclusions in a turbulent protoplanetary nebula, Icarus 166 (2) (2003) 385–402.

[73] S. Desch, Personal communication (2003).

[74] M. Chaussidon, F. Robert, K.D. McKeegan, S. Krot, Li, Be, B distribution and isotopic composition in refractory inclusions from primitive chondrites: a record of irradiation processes in the protosolar nebula, Meteoritics 36 2001, pp. A40, (suppl.).

[75] M. Chaussidon, F. Robert, K.D. McKeegan, Incorporation of short-lived 7Be in one CAI from the allende meteorit, Lunar Planet. Sci. XXXIII (2002) 1563.

[76] F.J. Ciesla, L.L. Hood, The nebular shock wave model for chondrule formation: shock processing in a particle–gas suspension, Icarus 158, 2002, pp. 281–293.

[77] J.N. Cuzzi, R.C. Hogan, J.M. Paque, A.R. Dorbovoskis, Size-selective concentration of chondrules and other small particles in protoplanetary nebula turbulence, Astrophys. J. 546 (2001) 496–508.

[78] J.N. Cuzzi, A.R. Dobrovolskis, R.C. Hogan, Turbulence, chondrules, and planetesimals, in: R.H. Hewins, R.H. Jones, E.R.D. Scott (Eds.), Chondrules and the Protoplanetary Disk, Cambridge Univ. Press, Cambridge, 1996, pp. 35–43.

[79] M. Bourot-Denise, unpublished data.

[80] K.E. Kuebler, H.Y. McSween, W.D. Carlson, D. Hirsch, Sizes and masses of chondrules and metal-troilite grains in ordinary chondrites: possible implications for nebular sorting, Icarus 141 (1999) 96–106.

[81] R.N. Clayton, T.K. Mayeda, J.N. Goswami, E.J. Olsen, Oxygen isotopic studies of ordinary chondrites, Geochim. Cosmochim. Acta 55 (1991) 2317–2337.

[82] R.H. Hewins, R.H. Jones, E.R.D. Scott (Eds.), Chondrules and the Protoplanetary Disk, Cambridge University Press, Cambridge, 1996, 346 pp.

[83] A.D. Anbar, Iron stable isotopes: beyond biosignatures, Earth Planet. Sci. Lett. 217 (2004) 223–236.

Brigitte Zanda is currently the curator in charge of the meteorite collection at the Paris Muséum. Most of her recent work has involved petrologic studies of chondrules and chondrites, coupled with collaborations with geochemists and experimentalists. Her goal is to investigate the role of chondrule formation in the genesis of protoplanets.

Reprinted from
Earth and Planetary Science Letters 223 (2004) 241–252

www.elsevier.com/locate/epsl

Planetary accretion in the inner Solar System

John E. Chambers*

Department of Terrestrial Magnetism, Carnegie Institution of Washington, 5241 Broad Branch Road N.W., Washington, DC 20015, USA

Received 15 December 2003; received in revised form 23 April 2004; accepted 27 April 2004
Available online 4 June 2004

Abstract

Unlike gas-giant planets, we lack examples of terrestrial planets orbiting other Sun-like stars to help us understand how they formed. We can draw hints from elsewhere though. Astronomical observations of young stars; the chemical and isotopic compositions of Earth, Mars and meteorites; and the structure of the Solar System all provide clues to how the inner rocky planets formed. These data have inspired and helped to refine a widely accepted model for terrestrial planet formation—the planetesimal hypothesis. In this model, the young Sun is surrounded by a disk of gas and fine dust grains. Grains stick together to form mountain-size bodies called planetesimals. Collisions and gravitational interactions between planetesimals combine to produce a few tens of Moon-to-Mars-size planetary embryos in roughly 0.1–1 million years. Finally, the embryos collide to form the planets in 10–100 million years. One of these late collisions probably led to the formation of Earth's Moon. This basic sequence of events is clear, but a number of issues are unresolved. In particular, we do not really understand the physics of planetesimal formation, or how the planets came to have their present chemical compositions. We do not know why Mars is so much smaller than Earth, or exactly what prevented a planet from forming in the asteroid belt. Progress is being made in all of these areas, although definitive answers may have to wait for observations of Earth-like planets orbiting other stars.
© 2004 Elsevier B.V. All rights reserved.

Keywords: earth; terrestrial planets; asteroids; accretion; solar nebula

1. Introduction

We are witnessing a revolution in planetary science. The discovery of about a hundred other planetary systems has provided a wealth of new information to a field that was previously focussed on only one. However, the new planets are probably all gas giants, akin to Jupiter and Saturn, so they tell us relatively little about the nature and origin of small, rocky planets like Earth. We know of only one other system of terrestrial planets. This is in a most unfamiliar place: orbiting a *pulsar*, the extinct remnant of a supernova explosion [1]. Remarkably, the pulsar planets bear a striking resemblance to the inner planets of the Solar System in terms of their orbits and masses, although they may have originated under very different conditions. Currently, we lack a way to detect terrestrial planets in orbit around ordinary stars [2], so we have almost no notion of how common or otherwise Earth-like planets may be. As tantalizing as the new planetary discoveries are, we must look elsewhere for clues to the origin of the Sun's terrestrial planets.

Astronomical observations of newborn stars show that many are surrounded by a disk-shaped region of

---

* SETI Institute, 2035 Landings Drive, Mountain View, CA 94043, USA. Tel.: +1-650-604-5514; fax: +1-650-604-6779.

E-mail address: john@mycenae.arc.nasa.gov (J.E. Chambers).

0012-821X/$ - see front matter © 2004 Elsevier B.V. All rights reserved.
doi:10.1016/j.epsl.2004.04.031

gas clouded with fine dust [3]. The disks generally have radii at least as large as the Sun's planetary system, and contain at least as much mass as our planets. These observations, together with the planar geometry of the Solar System, and the fact that the Sun's planets all orbit in the same direction, suggest the planets formed in a similar disk environment. This is often called the *protoplanetary disk* or *protoplanetary nebula*. If this disk had the same composition as the Sun, roughly 0.5% of the mass in Earth's locale would have existed in solid grains of rock and metal. The remaining 99.5% would have been gas: hydrogen, helium and volatile materials such as water and carbon monoxide. Since Earth is made almost entirely of rock and metal, it is clear that planet formation in the inner Solar System was a sideshow compared to the evolution of the more massive protoplanetary nebula itself, although the conventional viewpoint is just the opposite.

Interestingly, most known giant planets orbit stars (the Sun included) that contain above average amounts of dust-forming elements such as iron [4]. One way to interpret this correlation is that giant planets form most readily where solid materials are abundant. The same may be true of rocky planets. It is also apparent that planet formation is an inefficient process, because even stars with dust-poor disks contain enough material to form a respectable system of planets. Stars older than a few million years (Myr) apparently lack massive gas-rich disks [5]. If massive disks are essential in order to generate planets then planet formation must begin within a few million years. Some older stars possess tenuous disks containing some dust but apparently little gas [6,7]. Dust in these systems should be ground down to small sizes and pushed out of the system on time scales of 10^4–10^6 years by the gentle but insistent pressure of light from the star itself [8]. This may be second-generation dust formed by high speed collisions between solid bodies orbiting these stars or dust evaporating from the surface of comets [8]. If so, this suggests that dust is able to accumulate into large solid bodies in a variety of protoplanetary disks.

2. Physical and cosmochemical constraints

We can make a crude estimate of the minimum mass of the Sun's protoplanetary nebula by totalling up the rock, metal and ice that resides in the planets, and adding enough hydrogen and helium to give a composition similar to the Sun. This *minimum mass nebula* contains a few percent of a solar mass—a number that fits snugly within the range of values estimated for circumstellar disks. However, this material did not quietly metamorphose into the planets we see today. Most of the gas has gone, somewhere. Some of the dust has disappeared too, or at least it has moved around a great deal. The giant planets probably contain tens of Earth masses of rocky and icy material, while the vast expanse of the asteroid belt, between 2 and 4 astronomical units (AU) from the Sun, has only enough stuff to make a planet 1% the mass of Mercury. Just as curiously, 90% of the total mass of the inner Solar System now resides in a narrow strip between 0.7 and 1 AU.

Much of what we know about the early history of the Solar System comes from studying *primitive meteorites*. These rocks come from asteroids that never became hot enough to melt. Thus, primitive meteorites and their parent bodies act as a kind of archaeological site, preserving the detritus formed in the first few million years of the Solar System. Most primitive meteorites are composed largely of *chondrules*—beads of rock typically about 1 mm in size. The composition and texture of most chondrules suggests they were once balls of dust floating in the solar nebula that were strongly heated and cooled over the space of a few hours [9]. The heating melted the chondrules but was not sufficiently protracted to allow all the more volatile elements such as sulfur to escape. Particular types of meteorite contain chondrules with distinctive sizes and compositions. This may mean that chondrules formed in small regions of the protoplanetary nebula in a series of separate events. Theories abound for the origin of chondrules [9]. Models in which dust is melted by shock waves in the nebula are currently in vogue [10,11], although the source of these shock waves is unclear. Chondrule formation may be intimately tied to other events in the Solar System. In particular, shock driven chondrule formation could require the early formation of Jupiter [11,12]. A small fraction of chondrules appear to have formed as a result of impacts on asteroids [13], which implies that large bodies had already accreted by the time these chondrules formed.

Primitive meteorites also contain refractory components, similar in size to chondrules. These *calcium aluminium rich inclusions* (CAIs) are a minor component of meteorites, and their origin is even more enigmatic than that of chondrules. They are important here because their age places timing constraints on planet formation. The relative ages of CAIs and chondrules can be estimated from the abundances of the decay products of short-lived isotopes. Particularly useful are, ^{41}Ca, ^{26}Al and ^{53}Mn, with half lives of 0.1, 0.7 and 3.7 Myr respectively [14]. The absolute ages of chondrules and CAIs can be calculated from the modern abundances of lead isotopes formed by the decay of long-lived isotopes of uranium [15].

According to the isotopes, CAIs are the oldest Solar System materials we possess. Most formed in an interval spanning only a few hundred thousand years [16] around 4.56 billion years ago [17]. Chondrules apparently formed 1–4 Myr later than this [17,18]. Thus, some CAIs survived in the nebula for millions of years before bedding down with much younger chondrules to form asteroids. It is conceivable that something similar happened in the region containing the terrestrial planets, in which case the early stages of planet formation spanned several million years at least.

Iron meteorites tell their own tale. These meteorites come from asteroids that became hot enough to melt and differentiate. The most plausible source of heat was the decay of short-lived isotopes, especially ^{26}Al. Melting must have occurred while was still abundant, which means these asteroids took something like 2 Myr to form [19,20]. Why did some asteroids melt when others did not? Presumably, different stages of planet and asteroid formation occurred concurrently in the same region of the nebula. Some objects formed earlier than others, and their subsequent thermal evolution was different as a result.

The terrestrial planets are also differentiated, with high density iron-rich cores and low density silicate-rich mantles. Earth's mantle is highly depleted in *siderophile* (iron loving) elements, when ratioed to silicon say, compared to the Sun. Presumably, these elements sank to the core along with the iron during core formation. The process of core formation is hard to disentangle from the process of accretion itself. It is likely that the two happened concurrently [21]. The time scale for core formation can be estimated using

the decay of U isotopes to Pb, and also the short-lived isotope ^{182}Hf, which decays to ^{182}W with a half-life of 9 Myr. These isotope systems are useful because the parent nuclei are *lithophile* (silicate loving) while the daughter isotopes are more siderophile. Assuming core formation happened continuously and that accretion tailed off roughly exponentially over time, the lead isotopes indicate that Earth accreted/differentiated with a mean life of 15–40 Myr [22]. Somewhat confusingly, the Hf–W isotopes provide a shorter mean life of about 11 Myr [23]. The reason why these two systems give different results is unclear and the actual time scale probably lies somewhere in between [21].

The inner planets would have been mostly molten at the time they differentiated. Unfortunately, this melting erased much information about what happened to these bodies before their cores formed. We know rather more about subsequent events. In particular, Earth's mantle is blessed with more siderophile elements (gold etc.) than one would expect to find after its core formed. This is consistent with continued growth of the Earth after core formation ceased, although this *late veneer* constitutes less than 1% of Earth's total mass [24]. The mixture of osmium isotopes we see in Earth's mantle differs from carbonaceous chondrite meteorites (probably from the outer asteroid belt) but is similar to ordinary chondrites (probably from the inner belt) [24]. This is consistent with the late veneer coming either from material in the inner asteroid belt, or from the terrestrial-planet region itself.

Collisions shaped the inner Solar System in several ways. High speed impacts onto planetary surfaces supplied enough kinetic energy to cause melting. On small bodies, melted material tended to escape to space. Bodies the size of Ceres and larger were massive enough to hang on to some molten material, and impacts onto planet-sized bodies probably caused enough melting to trigger core formation [25]. The high density of Mercury may be the result of a violent collision with another large body, which removed much of Mercury's silicate mantle [26]. The Moon is highly depleted in both iron and volatile elements. This makes sense if the Moon formed from hot mantle material thrown into orbit around Earth after the planet was hit by another differentiated body [27]. The ancient surfaces of the Moon, Mercury and Mars bear the scars of numerous smaller impacts, although on the Moon at least, these collisions happened hundreds of millions of years after the planets formed [28].

Meteorites, together with rocks from Mars, the Moon and Earth, generally contain a similar mixture of isotopes, unlike dust grains that formed outside the Solar System. This suggests that material in the inner Solar System was thoroughly mixed on very fine scales at some point [29]. (Isotopes of oxygen do not obey this rule for reasons that are hotly debated.) Earth, Mars and the parent bodies of the various meteorite groups differ substantially in their chemistry however. Each object is made up of a different mixture of the major rock forming elements [24]. The spectral characteristics of modern asteroids vary in ways that are correlated with their distance from the Sun [30], and this is widely interpreted to reflect (literally) differences in their composition, as well as their thermal evolution. Hence, although different regions of the nebula probably exchanged a good deal of material, each of the inner planets and asteroids ultimately acquired a mixture of material unique to that body.

Compared to the Sun, many primitive meteorites are depleted in moderately volatile elements—those elements that condense and evaporate at temperatures between about 650 and 1250 K. The curious aspect is that these elements are depleted in a way that roughly correlates with their condensation temperature [31]. There is more than one way to interpret this correlation. The depletion pattern may represent a distant memory of an early hot phase in the history of the nebula [32], rather as the cosmic microwave background radiation provides a glimpse of the early history of the universe. If this interpretation is right, the inner few astronomical units of the nebula must have been hot enough to vapourise rock at some point. The CI group of primitive meteorites are not depleted in moderately volatile elements, which may mean they come from bodies that formed further from the Sun where temperatures were cooler. The depletion of moderately volatile elements can be interpreted in another way. Rather than indicating a globally hot nebula, the depletions could be caused by localized events such as those that generated chondrules [31]. In either case, planet-sized bodies probably suffered further depletions as a result of energetic collisions [33].

The inner planets possess rather little in the way of highly volatile material such as water and the noble gases. This depletion can be attributed to high temperatures in the inner nebula-volatile materials simply didn't condense while the planets were forming. However, the isotopic mixture of xenon on Earth and Mars implies that these planets have lost almost all their initial allotment of noble gases, and possibly a lot of other volatile material too [34]. The origin of Earth's meagre volatile inventory is still unclear. Helium and neon leaking from the mantle hint that the planet might have captured a massive atmosphere directly from the nebula early in its history [35]. Volatile substances in the atmosphere could have entered the mantle while the planet was still molten. It is hard to explain the abundances of the other noble gases in the atmosphere today unless some volatiles came to Earth from another source, such as impacts by comets or asteroids [36,37]. The deuterium to hydrogen ratio in terrestrial seawater differs by a factor of two from the ratio measured in comets to date, so comets were probably less important in this respect than asteroids [38].

3. The planetesimal theory

The astronomical and cosmochemical data described above generally support a model for the formation of the planets known as the *planetesimal theory*. Crudely, the theory posits that dust grains in the nebula collided and stuck together to make aggregates; these collided to form bigger bodies, etc. until the largest objects were the size of planets. Objects tended to acquire most of their mass locally, so planets and asteroids forming in different regions of the nebula came to have somewhat different compositions. Once planet-sized bodies formed, it was mostly a matter of mopping up the remaining debris or removing it from the system. Something along these lines almost certainly happened in the inner Solar System, leading to the formation of Earth and the other terrestrial planets. As with all theories however, the devil is in the details. This is particularly apparent in the early stages of planetary accretion.

The planetesimal theory is often portrayed as a sequence of steps rather like acts in a play. The characters and scenery change with time, and the audience applauds at the end of each act. This division is partly an indication of how we think about complex problems, but it also reflects changes in the relative importance of physical processes at each stage of planetary accretion. Following convention, I will describe the stages in order, with the caveat that these stages probably overlapped in both time and space.

4. Formation of planetesimals

The story begins with the gas and dust of the Sun's protoplanetary disk. Gas pressure gives the disk a definite thickness in the vertical direction. Pressure decreases with distance from the Sun, which allows the gas to orbit the Sun slightly more slowly than a solid body moving on a circular orbit would. Dust grains feel little pressure support, so they tend to settle towards the midplane of the disk, sweeping up other grains en route to form loosely bound aggregates (see Fig. 1). In the absence of turbulence in the gas, a typical dust grain will reach the midplane in about 10^4 years [39].

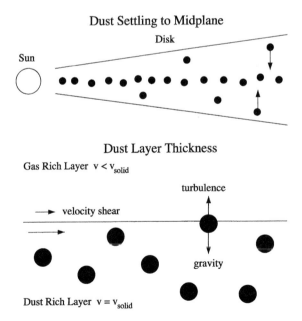

Fig. 1. Dust grains slowly settle to the midplane of the nebula due to the vertical component of the Sun's gravity, forming a solid-rich layer. This layer orbits the Sun slightly faster than the gas-rich layers above and below. The resulting wind shear generates turbulence, even if other sources of turbulence are absent. Thus, the solid-rich layer has a finite thickness.

As the dust becomes concentrated towards the midplane, the solid to gas ratio increases. The dust-rich layer begins to orbit the Sun with the speed of a solid body rather than the slightly slower speed of the gas. Gas in the dusty layer is herded along by the solid particles, moving faster than it would like to, while gas above and below the midplane moves more slowly as before. This differential velocity generates turbulence, which acts to puff up the dust-rich layer, even if no other sources of turbulence are present. A compromise is reached between gravity and turbulence, and this determines the thickness of the dust-rich layer [40].

At this point, the script for our play becomes hard to read. If the solid-rich layer becomes dense enough, the densest portions will be unable to resist the Narcissus-like attraction of their own gravity, becoming ever smaller and denser. Once this *gravitational instability* gets going, collapse can continue until solid bodies a few km in size are generated. Such bodies are dubbed *planetesimals*. Whether gravitational instability (GI) ever gets going is the subject of much debate [41–44]. It seems that GI can only occur if the ratio of solids to gas in a column of nebula material exceeds a critical value. Recent calculations suggest this critical value is several times the solid to gas ratio for material with solar composition, even when volatiles such as water and carbon monoxide have condensed [45]. In addition, bodies may have to grow to metre size or larger before conditions become right for GI [40].

Unabashed, theorists have thought of several ways to increase the solid to gas ratio and give GI a helping hand. This is done by either collecting solids in one place or removing some of the gas. Small solids have a tendency to move radially within the nebula and pile up at locations where there is a local maximum in the gas pressure [46], or where the concentration of solids is higher than average [47]. The local solid to gas ratio can also increase over time as small solids migrate inwards [42], or as gas escapes from the Solar System due to *photoevaporation* by ultraviolet light from the Sun. It remains to be seen whether these mechanisms operate with sufficient effect to permit GI.

In the absence of gravitational instability, large bodies presumably form by the gradual aggregation of dust grains and small solids such as chondrules. Experiments show that irregular dust grains can stick together if they collide at speeds of up to a few tens of metres per seconds [48]. High collision speeds are more likely to cause grains to rebound or break apart, while low collision speeds are more likely to lead to sticking. Small solids are probably strongly coupled to the motion of the gas, so they typically undergo gentle collisions leading to accretion. Charge exchange between grains and the generation of electric dipoles also aids accretion, leading to the rapid formation of dust aggregates many centimetres in size [49].

Collisional accretion becomes more challenging when bodies reach 0.1–10 m in size. These objects are too large to be swept along at the same speed as the gas, but too small to be unaffected by it entirely. Because gas orbits the Sun more slowly than a solid body, boulder-sized objects feel a headwind. If solids in the dust-rich layer are effective at dragging the gas along with them, this headwind will be quite small [40]. Otherwise, the headwind will be around 50 m s^{-1}, similar to the wind speed in a hurricane [50].

The headwind affects boulder-size objects in two ways. Dust grains entrained in the gas strike large bodies with the same speed as the headwind. In principle, this increases the amount of material that can be swept up by large bodies. However, if the dust grains hit at high speeds they are more likely to cause erosion akin to sand blasting. Second, the headwind gradually robs large bodies of their orbital angular momentum, causing them to drift towards the Sun. This drifting due to *gas drag* can be extremely rapid for metre-sized objects—as fast as 1 AU in 500 years [40]. The ultimate fate of drifting bodies depends on the thermal structure of the nebula. If the inner nebula is hot, objects will evaporate when the temperature becomes high enough; otherwise, they fall into the Sun. Radial transport of solid material by gas drag may lead to substantial variations in both the solid to gas ratio and the chemical composition in different regions of the nebula [51].

The existence of gas drag would imply that solid bodies must grow rapidly until they are many metres in size if they are to survive. It seems reasonable that boulder-sized objects will only actually stick together during rare, low-velocity collisions. These objects may gain most of their mass by accreting smaller solids and dust grains. The nebula headwind might aid accretion in some cases by blowing small fragments from erosive collisions back onto metre-size bodies [52].

If the nebular gas is turbulent, small solids will not simply accumulate in a thin layer at the midplane. However, solids will become highly concentrated in stagnant regions. These solid-rich regions could evolve rapidly into planetesimals [53]. The efficiency of this *turbulent concentration* depends on the size and compactness of the solids. Chondrule-like particles seem particularly well suited in this respect [53], so it may be no accident that they form the major component of most primitive meteorites, while larger solid particles are not seen.

Despite substantial progress in understanding the earliest stage of planet formation, the origin of planetesimals must still be regarded as an unsolved problem. The audience watching our play could be forgiven for having serious misgivings at this stage in the performance. Fortunately, things proceed more smoothly in the next act.

5. Formation of planetary embryos

The second stage of planet formation begins when much of the solid material has formed into planetesimals a few kilometres in diameter. How these bodies formed is rather less important than how big they are. For the second stage to proceed, bodies must be large enough to gravitationally perturb their neighbours during close approaches. This stage of accretion has been examined extensively using theoretical models for two reasons: (i) the evolution depends on a small number of processes that are fairly well understood, and (ii) the number of planetesimals is huge. This means their evolution can be studied in a statistical sense, just as a gas composed of trillions of molecules can be described using kinetic theory.

The weakest link in the theory is understanding the outcome of collisions. Laboratory experiments have studied impacts involving planetary materials at a wide range of collision speeds, but these experiments are limited to bodies less than a metre in size. Most of what we know about planet-forming collisions comes from numerical simulations instead. To date, these simulations provide a rather sparse coverage of collisional phase space. That said, the results suggest most collisions lead to net accretion, unless the impact speed is subtantially higher than the target's gravitational escape velocity or the impact is at a grazing angle [54,55].

A planetesimal accretes its smaller brethren at a rate that depends on the number of objects per unit volume and the planetesimal's velocity v_{rel} relative to other objects (see Fig. 2). If v_{rel} is large, a planetesimal collides only with objects that pass directly in front of it. If v_{rel} is small, a planetesimal's gravity will pull in material from further away. This *gravitational focussing* increases the frequency of collisions. More often than not, planetesimals approach each other without actually colliding, but their trajectories are altered by their gravitational interaction. The cumulative effect of many close encounters determines a planetesimal's velocity relative to other bodies in the same region of the nebula. Large bodies tend to acquire small relative velocities and vice versa, a state of affairs referred to as *dynamical friction*. All the while, gas drag is striving to make the orbits of the planetesimals circular and coplanar, effectively reducing v_{rel}.

Despite this apparent complexity, accretion is likely to proceed in one of only a few ways [56–58]. Initially, the largest planetesimals feed voraciously on smaller objects, while the collective gravitational effects of the small objects keeps v_{rel} low. This makes gravitational focussing highly effective. The largest bodies, termed *planetary embryos*, quickly outgrow all the others, a process known as *runaway growth*.

The days of unfettered growth are numbered however. Runaway growth slows when planetary embryos become about 100 times more massive than a typical planetesimal. Now it is the gravitational perturbations of the embryos that determine (v_{rel}) rather than perturbations from the more numerous planetesimals [59]. Accretion enters a new self-regulated regime called *oligarchic growth* [55]. Planetary embryos continue to outgrow smaller planetesimals, but embryos in neighbouring regions of the disk are forced to grow at similar rates. Whenever one embryo gets too greedy, events conspire to allow nearby embryos to catch up. The more massive an embryo is, the more strongly it perturbs nearby planetesimals, thereby increasing v_{rel}. Thus, gravitational focussing is reduced, and the embryo grows more slowly than a smaller embryo would.

As in any good oligarchy, each embryo stakes out a region of influence, or *feeding zone* in the disk. A typical feeding zone in the inner Solar System is a roughly annular region of order 0.01 AU in width. A combination of dynamical friction and occasional gravitational interactions between neighbouring embryos acts to keep these bodies on widely spaced orbits. Each embryo accretes most of its mass from its own feeding zone, giving the embryos distinct chemical compositions.

The oligarchic growth stage lasts for 0.1–1 Myr from the time when planetesimals first appear in large numbers [58,60]. Oligarchic growth ends when the number of planetesimals dwindles so much that they can no longer restrain the actions of the planetary embryos. Our play has reached a moment of crisis. With the demise of the planetesimals, dynamical friction shuts down. The embryos stray beyond their feeding zones and the previous order collapses as the large bodies begin to interact strongly and collide with one another. Accretion of the planets now enters a prolonged terminal phase.

6. From embryos to planets

The final stage of planetary accretion involves a few dozen embryos with masses comparable to the Moon or Mars (0.01–0.1 Earth masses). Gravitational perturbations between embryos increase their relative

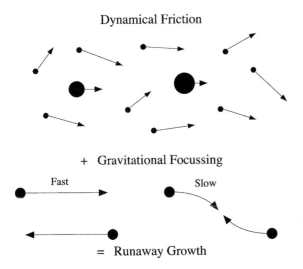

Fig. 2. The mechanics of runaway growth. Large bodies tend to have lower relative velocities than small objects as a result of numerous gravitational encounters. When large bodies pass close to each other, their trajectories are focussed by their gravitational attraction. Small bodies fly past each other too quickly to be significantly affected by their mutual attraction. Thus, large bodies grow faster than small ones.

velocities. Gravitational focussing becomes weak, and the accretion rate slows dramatically.

Over time, the embryos scatter one another inwards or outwards, and the radial ordering established during oligarchic growth becomes scrambled. Any primordial chemical and isotopic gradients are blurred as a result. The final planets are a mixture of material from a broad region of the inner Solar System, although each planet accretes more material from its own locale than elsewhere, so the mixture is different for each planet [61]. Earth and Venus are composites formed from a dozen or more embryos. Mars and Mercury contain material from only a few embryos, possibly as few as one in each case. Thus, these planets probably sampled rather less of the nebula than their larger siblings. The final stage of accretion is highly *chaotic*. That is, it depends sensitively on the outcome of individual events such as whether a close encounter between two embryos results in a collision or a near miss. To illustrate this, Fig. 3 shows the results of four numerical simulations of this stage of accretion, each beginning with the same total mass and number of embryos [62]. The planetary systems produced in each case are quite different.

The time scale for the final stage of accretion depends on whether nebula gas is still present. In the presence of a minimum-mass gas nebula, Earth may have formed in as little as 5 Myr [63], although this is hard to reconcile with the time scales derived from the U–Pb and Hf–W isotopes described above. In the absence of significant amounts of gas, numerical simulations suggest Earth took roughly 100 Myr to form, with the accretion rate declining approximately exponentially over time [62,64]. Small amounts of nebula gas can have a significant effect on late-stage accretion. In particular, the lingering presence of roughly 0.1% of a minimum-mass nebula may have helped to circularize the orbits of the inner planets as they neared completion [65,66].

The inner planets probably each accreted some planetesimals from the region that now contains the asteroid belt. These planetesimals have a different chemical composition, and are probably richer in volatile materials, than planetesimals that formed closer to the Sun. The asteroid belt may have been an important source of the water and other volatile substances that now exist on Earth [37]. Some primitive meteorites contain up to 10% water by mass, and

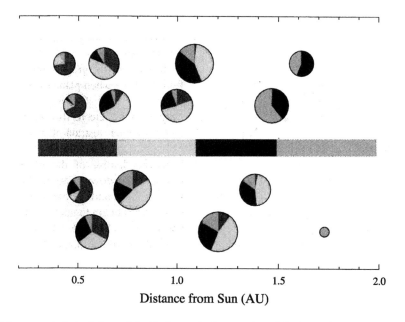

Fig. 3. The results of four numerical simulations of the final stage in the accretion of the inner planets. Each row of symbols shows one simulation, with symbol radius proportional to the radius of the planet. The segments in each pie chart show the fraction of material originating from each of the four zones of the nebula indicated by the shaded rectangles. In each simulation, the largest planet has a mass similar to Earth (results taken from Ref. [62]).

this water has a deuterium to hydrogen ratio similar to water on Earth. However, the bulk of this asteroidal matter must have arrived on Earth, or its precursors, before core formation was complete. These meteorites are rich in siderophile elements, so they would dominate the Earth's late veneer if they were added late, yet their Os isotopic and trace element compositions are distinct [24].

Late stage accretion is not a wholly efficient business. Some embryos fall into the Sun after straying into the asteroid belt, a region that contains unstable orbital resonances associated with Jupiter and Saturn. Up to 1/3 of the embryos that form within 2 AU of the Sun are likely to suffer this fiery fate [62]. High speed collisions between embryos can lead to fragmentation. Collision speeds are highest close to the Sun, which made Mercury especially vulnerable to disruptive impacts. This may explain why the innermost planet remained so small. The low mass of Mars compared to Earth and Venus is harder to fathom, and current theories have little to say on the subject. Chance events may have conspired to prevent Mars from accreting additional planetary embryos, but this explanation seems unsatisfactory. A significant fraction of collisions between embryos are likely to eject a substantial amount of material into orbit around the newly formed body [64]. It is likely that the Moon formed from such debris following the impact of a Mars-sized body onto Earth [27].

7. Formation of the asteroid belt

The story of planetary accretion 1 AU from the Sun has a happy ending but something clearly went wrong in the asteroid belt. Either planets never formed in this region or they survived only briefly. The imprint of short-lived isotopes seen in many meteorites implies the asteroids formed in a few million years. However, bodies the size of Ceres and Vesta can only have accreted this rapidly if the asteroid belt initially contained at least 100 times as much solid material as it does today [67], so the current low mass of the asteroid belt requires an explanation.

It is possible that the growth of large bodies in the asteroid belt was frustrated by the early formation of Jupiter. Gravitational perturbations from Jupiter would have increased the relative velocities of planetesimals

in general, especially for bodies that are substantially different in size. This would delay the onset of runaway growth until bodies were larger than the largest asteroids that exist today [68]. Collisional erosion may have been an important process in the asteroid region, but this alone cannot explain the low mass of the asteroid belt today. Vesta currently sports a basaltic crust that formed in the first few million years of the Solar System. It is doubtful that this crust would have survived until now if >99% of the asteroid belt has been pummelled into dust [69]. Instead, the asteroid belt probably lost most of its bulk in another way.

Two other models are currently in the running. Each makes use of the fact that the asteroid belt is crisscrossed by a number of unstable orbital resonances associated with Jupiter and Saturn. Today, an asteroid entering a resonance is quickly forced onto a highly eccentric (elliptical) orbit, such that it typically falls into the Sun or is ejected from the Solar System in about 1 Myr [70]. While the nebula is still present, objects moving on eccentric orbits experience substantial gas drag. As the nebula disperses, some of the unstable resonances sweep across the asteroid belt, possibly more than once [71]. The combination of *resonance sweeping* and gas drag causes many bodies smaller than about 100 km to migrate inwards, leaching mass from the asteroid belt, and depositing it in the region where the terrestrial planets are forming [72].

Gas drag has its limits however. Bodies the size of the Moon or Mars are too massive to drift significantly. Left to their own devices, planetesimals in the asteroid region should eventually form planetary embryos, unless the giant planets form quickly. However, even planetary embryos as massive as Earth become vulnerable once Jupiter and Saturn form. Gravitational encounters between embryos cause their orbits to wander slowly through asteroid belt. Sooner or late, each embryo enters an unstable resonance where it is likely to be removed from the asteroid belt before another close encounter scatters it out of the resonance again. Numerical simulations show that all embryos in the asteroid region are likely to be lost in this way, along with the great majority of smaller asteroids, once the giant planets form [73]. All that remain are a few small objects too puny to continue the process any further.

There appears to be a fine line between forming a system of terrestrial planets and generating an asteroid belt. The outcome depends mainly on proximity to the giant planets and their unstable resonances [74,75]. When terrestrial planets do form, their characteristics are hard to predict ahead of time since the final stage of accretion is dominated by chance events involving a small number of planetary embryos. It is probably a matter of chance that the Solar System ended up with precisely four terrestrial planets, and that one of these now resides in a pleasantly habitable location [62]. A minor change at any stage in the formation of the planets could have produced a very different, perhaps equally fascinating, outcome.

8. Looking ahead

The coming years should see progress in a number of areas that will help our understanding of the origin of planets and asteroids. Astronomical observations of protoplanetary disks by NASA's Spitzer Space telescope and other programmes will provide better models for the structure and evolution of protoplanetary disks. Continuing searches for extrasolar giant planets will soon establish whether giants with orbits similar to Jupiter are rare or commonplace. There is also the exciting prospect of finding terrestrial planets orbiting Sun-like stars via NASA's upcoming Kepler mission. In cosmochemistry, we may soon see the resolution of several key questions including a clear understanding of the source(s) of short-lived isotopes in the early Solar System, and agreement on the time scale for Earth's accretion and differentiation, and the timing of the Moon-forming impact. The hugely successful ongoing programme to collect and analyse Antarctic meteorites is sure to throw up a few surprises in the years ahead. Finally, on the theoretical front, the time is ripe for breakthroughs on a number of vexing issues, including the origin of chondrules and planetesimals, an understanding of the physics of interactions between planets and protoplanetary disks, and the origin of water and other volatiles on the terrestrial planets. We have much to look forward to.

Acknowledgements

I am very grateful to Alan Boss, Lindsey Bruesch, Jeff Cuzzi, Alex Halliday, Helen Williams, Kevin Zahnle and an anonymous reviewer for providing comments that substantially improved this article and helped to avert a number of gaffs during its preparation. *[AH]*

References

[1] M. Konacki, A. Wolszczan, Masses and orbital inclinations of planets in the PSR B1257+12 system, Astrophysical Journal 591 (2003) L147–L150.

[2] S. Seager, The search for extrasolar earth-like planets, Earth and Planetary Science Letters 208 (2003) 113–124.

[3] S.V.W. Beckwith, A.I. Sargent, R.S. Chini, R. Guesten, A survey for circumstellar disks around young stellar objects, Astronomical Journal 99 (1990) 924–945.

[4] D.A. Fischer, J.A. Valenti, Metallicities of stars with extrasolar planets, in: D. Deming, S. Seager (Eds.), Scientific Frontiers in Research on Extrasolar Planets, ASP Conference Series, vol. 294, ASP, San Francisco, 2003, pp. 117–128.

[5] K.E. Haisch, E.A. Lada, C.J. Lada, Disk frequencies and lifetimes in young clusters, Astrophysical Journal 553 (2001) L153–L156.

[6] C. Spangler, A.I. Sargent, M.D. Silverstone, E.E. Becklin, B. Zuckerman, Dusty debris around solar-type stars: temporal disk evolution, Astrophysical Journal 555 (2001) 932–944.

[7] J.S. Greaves, I.M. Coulson, W.S. Holland, No molecular gas around nearby solar-type stars, Monthly Notices of the Royal Astronomical Society 312 (2000) L1–L3.

[8] P. Artymowicz, M. Clampin, Dust around main sequence stars: nature or nurture by the interstellar medium, Astrophysical Journal 490 (1997) 863–878.

[9] R.H. Jones, T. Lee, H.C. Connolly, S.G. Love, H. Shang, Formation of chondrules and CAIs: theory vs. observation, in: V. Mannings, A.P. Boss, S.S. Russell (Eds.), Protostars and Planets vol. IV, University of Arizona Press, Tuscon, AZ, 2000, pp. 927–961.

[10] S.J. Desch, H.C. Connolly, A model of the thermal processing of particles in solar nebula shocks: application to the cooling rates of chondrules, Meteoritics 37 (2002) 183–207.

[11] S.J. Weidenschilling, F. Marzari, L.L. Hood, The origin of chondrules at jovian resonances, Science 279 (1998) 681–684.

[12] A.P. Boss, Shock-wave heating and clump formation in a minimum mass solar nebula, 31st Lunar Planetary Science Conference, Houston, Texas, 2000, abstract 1084.

[13] A.N. Krot, A.E. Rubin, Chromite-rich mafic silicate chondrules in ordinary chondrites: formation by impact melting, 24th Lunar Planetary Science Conference, Houston, Texas, 1993, pp. 827–828.

[14] J.N. Goswami, H.A.T. Vanhala, Extinct radionuclides and the origin of the Solar System, in: V. Mannings, A.P. Boss, S.S. Russell (Eds.), Protostars and Planets IV, University of Arizona Press, Tucson, 2000, p. 963.

[15] C.J. Allegre, G. Manhes, C. Gopel, The age of the Earth, Geochimica et Cosmochimica Acta 59 (1995) 1445–1456.

[16] M. Wadhwa, S.S. Russell, Timescales of accretion and differ-

entiation in the early solar system: the meteoritic evidence, in: V. Mannings, A.P. Boss, S.S. Russell (Eds.), Protostars and Planets IV, University of Arizona Press, Tucson, 2000, p. 995.

[17] Y. Amelin, A.N. Krot, I.D. Hutcheon, A.A. Ulyanov, Lead isotopic ages of chondrules and calcium–aluminum-rich inclusions, Science 297 (2002) 1678–1683.

[18] G.R. Huss, G.J. MacPherson, G.J. Wasserburg, S.S. Russell, G. Srinivasan, 26Al in CAIs and chondrules from unequilibrated ordinary chondrites, Meteoritics 36 (2001) 975–997.

[19] D.S. Woolum, P. Cassen, Astronomical constraints on nebular temperatures: implications for planetesimal formation, Meteoritics 34 (1999) 897–907.

[20] N. Sugiura, H. Hoshino, Mn–Cr chronology of five IIIAB iron meteorites, Meteoritics 38 (2003) 117–143.

[21] A.N. Halliday, Mixing, volatile loss and compositional change during impact-driven accretion of the Earth, Nature 427 (2004) 505–509.

[22] A.N. Halliday, Terrestrial accretion rates and the origin of the Moon, Earth and Planetary Science Letters 176 (2000) 17–30.

[23] Q. Yin, S.B. Jacobsen, K. Yamashita, J. Blichert-Toft, P. Telouk, A short timescale for terrestrial planet formation from Hf–W chronometry of meteorites, Nature 418 (2002) 949–952.

[24] M.J. Drake, K. Righter, Determining the composition of the Earth, Nature 416 (2002) 39–44.

[25] W.B. Tonks, H.J. Melosh, Core formation by giant impacts, Icarus 100 (1992) 326–346.

[26] W. Benz, W.L. Slattery, A.G.W. Cameron, Collisional stripping of Mercury's mantle, Icarus 74 (1988) 516–528.

[27] R.M. Canup, E. Asphaug, Origin of the Moon in a giant impact near the end of Earth's formation, Nature 412 (2001) 708–712.

[28] W.K. Hartmann, G. Ryder, L. Dones, D. Grinspoon, The time-dependent intense bombardment of the primordial Earth/Moon system, in: R.M. Canup, K. Righter (Eds.), Origin of the Earth and Moon, University of Arizona Press, Tucson, 2000, pp. 493–512.

[29] H. Becker, R.J. Walker, Efficient mixing of the solar nebula from uniform Mo isotopic composition of meteorites, Nature 425 (2003) 152–155.

[30] J. Gradie, E. Tedesco, Compositional structure of the asteroid belt, Science 216 (1982) 1405–1407.

[31] C.M.O'D. Alexander, A.P. Boss, R.W. Carlson, The early evolution of the inner solar system: a meteoritic perspective, Science 293 (2001) 64–68.

[32] P. Cassen, Nebular thermal evolution and the properties of primitive planetary materials, Meteoritics 36 (2001) 671–700.

[33] A.N. Halliday, D. Porcelli, In search of lost planets—the paleocosmochemistry of the inner Solar System, Earth and Planetary Science Letters 192 (2001) 545–559.

[34] K. Zahnle, Origins of atmospheres, in: C.E. Woodward, J.M. Shull, H.A. Thronson Jr. (Eds.), Proceedings of the International Conference at Estes Park, Colorado, 19–23 May, 1997.

[35] D. Porcelli, D. Woolum, P. Cassen, Deep Earth rare gases: initial inventories, capture from the solar nebula, and losses during Moon formation, Earth and Planetary Science Letters 237 (2001) 237–251.

[36] N. Dauphas, The dual origin of the terrestrial atmosphere, Icarus 165 (2003) 326–339.

[37] A. Morbidelli, J. Chambers, J.I. Lunine, J.M. Petit, F. Robert, G.B. Valsecchi, K.E. Cyr, Source regions and time scales for the delivery of water to Earth, Meteoritics 35 (2000) 1309–1320.

[38] F. Robert, The origin of water on Earth, Science 293 (2001) 1056–1058.

[39] S.J. Weidenschilling, Dust to planetesimals, Icarus 44 (1980) 172–189.

[40] J.N. Cuzzi, A.R. Dobrovolskis, J.M. Champney, Particle gas dynamics in the midplane of a protoplanetary nebula, Icarus 106 (1993) 102–134.

[41] S.J. Weidenschilling, J.N. Cuzzi, Formation of planetesimals in the solar nebula, in: E.H. Levy, J.I. Lunine (Eds.), Protostars and Planets III, University of Arizona, Tucson, 1993, pp. 1031–1060.

[42] A.N. Youdin, F.H. Shu, Planetesimal formation by gravitational instability, Astrophysical Journal 580 (2002) 494–505.

[43] W.R. Ward, On planetesimal formation: the role of collective particle behaviour, in: R.M. Canup, K. Righter (Eds.), Origin of the Earth and Moon, University of Arizona, Tucson, 2000, pp. 75–84.

[44] S.J. Weidenschilling, Radial drift of particles in the solar nebula: implications for planetesimal formation, Icarus 165 (2003) 438–442.

[45] M. Sekiya, Quasi-equilibrium density distributions of small dust aggregations in the solar nebula, Icarus 133 (1998) 298–309.

[46] N. Haghighipour, A.P. Boss, On pressure gradients and rapid migration of solids in a nonuniform solar nebula, Astrophysical Journal 583 (2003) 996–1003.

[47] J. Goodman, B. Pindor, Secular instability and planetesimal formation in the dust layer, Icarus 148 (2000) 537–549.

[48] T. Poppe, J. Blum, T. Henning, Analogous experiments on the stickiness of micron sized preplanetary dust, Astrophysical Journal 533 (2000) 454–471.

[49] J. Marshall, J. Cuzzi, Electrostatic enhancement of coagulation in protoplanetary nebulae, 32nd Lunar Planetary Science Conference, Houston, Texas, 2001, abstract 1262.

[50] S.J. Weidenschilling, Aerodynamics of solid bodies in the solar nebula, Monthly Notices of the Royal Astronomical Society 180 (1977) 57–70.

[51] T.F. Stepinski, P. Valageas, Global evolution of solid matter in turbulent protoplanetary disks, Astronomy and Astrophysics 319 (1997) 1007–1019.

[52] G. Wurm, J. Blum, J.E. Colwell, A new mechanism relevant to the formation of planetesimals in the solar nebula, Icarus 151 (2001) 318–321.

[53] J.N. Cuzzi, R.C. Hogan, J.M. Paque, A.R. Dobrovolskis, Size-selective concentration of chondrules and other small particles in protoplanetary nebula turbulence, Astrophysical Journal 546 (2001) 496–508.

[54] W. Benz, E. Asphaug, Catastrophic disruptions revisited, Icarus 142 (1999) 5–20.

[55] Z.M. Leinhardt, D.C. Richardson, T. Quinn, Direct N-body simulations of rubble pile collisions, Icarus 146 (2000) 133–151.

[56] E. Kokubo, S. Ida, Oligarchic growth of protoplanets, Icarus 131 (1998) 171–178.

[57] R.R. Rafikov, The growth of planetary embryos: orderly, runaway or oligarchic? Astronomical Journal 125 (2003) 942–961.

[58] E.W. Thommes, M.J. Duncan, H.F. Levison, Oligarchic growth of giant planets, Icarus 161 (2003) 431–455.

[59] S. Ida, J. Makino, Scattering of planetesimals by a protoplanet—slowing down of runaway growth, Icarus 106 (1993) 210–227.

[60] S.J. Weidenschilling, D. Spaute, D.R. Davis, F. Marzari, K. Ohtsuki, Accretional evolution of a planetesimal swarm, Icarus 128 (1997) 429–455.

[61] J.E. Chambers, P. Cassen, The effects of nebula surface density profile and giant-planet eccentricities on planetary accretion in the inner solar system, Meteoritics 37 (2002) 1523–1540.

[62] J.E. Chambers, Making more terrestrial planets, Icarus 152 (2001) 205–224.

[63] C. Hayashi, K. Nakazama, Y. Nakagawa, Formation of the Solar System, in: Protostars and Planets II, University of Arizona, Tucson, 1985, pp. 1100–1153.

[64] C.B. Agnor, R.M. Canup, H.F. Levison, On the character and consequences of large impacts in the late stage of terrestrial planet formation, Icarus 142 (1999) 219–237.

[65] J. Kominami, S. Ida, The effect of tidal interaction with a gas disk on formation of terrestrial planets, Icarus 157 (2002) 43–56.

[66] C.B. Agnor, W.R. Ward, Damping of terrestrial-planet eccentricities by density-wave interactions with a remnant gas disk, Astrophysical Journal 567 (2002) 579–586.

[67] G.W. Wetherill, An alternative model for the formation of the asteroids, Icarus 100 (1992) 307–325.

[68] S.J. Kortenkamp, G.W. Wetherill, Runaway growth of planetary embryos facilitated by massive bodies in a protoplanetary disk, Science 293 (2001) 1127–1129.

[69] D.R. Davis, E.V. Ryan, P. Farinella, Asteroid collisional evolution: results from current scaling algorithms, Planetary Science 42 (1994) 599–610.

[70] B.J. Gladman, F. Migliorini, A. Morbidelli, V. Zappala, P. Michel, A. Cellino, C. Froeschle, H.F. Levison, M. Bailey, M. Duncan, Dynamical lifetimes of objects injected into asteroid belt resonances, Science 277 (1997) 197–201.

[71] M. Nagasawa, H. Tanaka, S. Ida, Orbital evolution of asteroids during depletion of the solar nebula, Astronomical Journal 119 (2000) 1480–1497.

[72] F. Franklin, M. Lecar, On the transport of bodies within and from the asteroid belt, Meteoritics 35 (2000) 331–340.

[73] J.E. Chambers, G.W. Wetherill, Planets in the asteroid belt, Meteoritics 36 (2001) 381–399.

[74] G. Laughlin, J. Chambers, D. Fischer, A dynamical analysis of the 47 ursae majoris planetary system, Astrophysical Journal 579 (2002) 455–467.

[75] H.F. Levison, C.B. Agnor, The role of giant planets in terrestrial planet formation, Astronomical Journal 125 (2003) 2692–2713.

 **John E. Chambers** is a research scientist at the SETI Institute, located in Mountain View, California. His work involves theoretical studies of the origin of planetary systems and habitable planets using computer modeling. Chambers received a Ph.D. in physics from the University of Manchester in the United Kingdom in 1995. He has since worked at the Carnegie Institution of Washington in Washington DC, Armagh Observatory in Northern Ireland, and NASA Ames Research Center in Moffett Field, California. He has been with the SETI Institute since 2002.

Reprinted from
Earth and Planetary Science Letters 223 (2004) 1–16

www.elsevier.com/locate/epsl

The chemistry of subduction-zone fluids

Craig E. Manning*

Department of Earth and Space Sciences, University of California at Los Angeles, Los Angeles, CA 90095-1567, USA

Received 24 April 2004; accepted 27 April 2004

Abstract

Subduction zones generate voluminous magma and mediate global element cycling. Fluids are essential to this activity, yet their behavior is perhaps the most poorly understood aspect of the subduction process. Though many volatile components are subducted, H_2O is the most abundant, is preferentially fractionated into the fluid phase, and, among terrestrial volatiles, is by far the most effective solvent. H_2O therefore controls the chemical properties of subduction-zone fluids. Rising pressure (P) and temperature (T) along subduction paths yield increased H_2O ionization, which enhances dissolved solute concentrations. Under appropriate conditions, silicate solubilities may become so high that there is complete miscibility between hydrous melts and dilute aqueous solutions. Miscible fluids of intermediate composition (e.g., 50% silicate, 50% H_2O) are commonly invoked as material-transport agents in subduction zones; however, phase relations pose problems for their existence over significant length scales in the mantle. Nevertheless, this behavior provides a key clue pointing to the importance of polymerization of alkali aluminosilicate components in deep fluids. Aqueous aluminosilicate polymers may enhance solubility of important elements even in H_2O-rich fluids. Subduction-zone fluids may be surprisingly dilute, having only two to three times the total dissolved solids (TDS) of seawater. Silica and alkalis are the dominant solutes, with significant Al and Ca and low Mg and Fe, consistent with a role for aqueous aluminosilicate polymers. Trace-element patterns of fluids carrying only dissolved silicate components are similar to those of primitive island-arc basalts, implying that reactive flow of H_2O-rich, Cl-poor, alkali-aluminosilicate-bearing fluid is fundamental to element transport in the mantle wedge. Better understanding of the interaction of this fluid with the mantle wedge requires quantitative reaction-flow modeling, but further studies are required to achieve this goal.
© 2004 Elsevier B.V. All rights reserved.

Keywords: subduction zones; subduction-zone fluids; mantle wedge; metasomatism

1. Introduction

The segments of subduction zones extending from trenches to beneath volcanic arcs are sites of profound chemical change. Incoming lithosphere is stripped of elements [1–4] which are transferred to the overlying mantle [5–9] in a process that ultimately generates volcanic arcs. The chemical work done in this "subduction factory" is fundamental to the Earth's evolution, because it leads to prolific volcanism and degassing, it mediates the global cycling of elements, and over time it produced the continental crust.

The transfer of material in subduction zones occurs in steps, and the agents of transfer vary. It is generally thought that H_2O-rich fluid (Table 1) is initially

---

* Tel.: +1-310-206-3290; fax: +1-310-825-2779.
  E-mail address: manning@ess.ucla.edu (C.E. Manning).

0012-821X/$ - see front matter © 2004 Elsevier B.V. All rights reserved.
doi:10.1016/j.epsl.2004.04.030

Table 1
Glossary of terms

Compatible element	An element that is preferentially partitioned into the solid in a solid–fluid mixture
Critical end point	Point defined by the intersection of a critical curve and a solubility curve
Critical curve	Curve linking critical points of end members in a two-component system. It is the boundary between stability fields of a supercritical fluid and a two phase mixture of liquid + vapor
Critical point	The termination of the liquid + vapor field in a one component system
Fluid	A disordered, non-crystalline phase consisting of particles in motion and possessing unspecified composition
Incompatible element	An element that is preferentially partitioned into the fluid in a solid–fluid mixture
Liquid	High-density, subcritical fluid
Metasomatism	The process by which rock is compositionally modified
Miscibility	The property enabling discrete phases to mix completely to form a single phase
Polymerization	In the context of this paper, the linking of cations by shared (bridging) oxygens
Solubility curve	In a simple two component system, a curve linking the eutectic with the triple point of an end member. In the system A-H_2O, the high-T solubility curve is the line along which solid A may coexist with liquid and vapor
Solidus	Reaction boundary denoting the first appearance of melt with increasing T in a multicomponent system
Supercritical	State of a system in which a single fluid phase is stable
Supercritical fluid	A fluid stable at P and T greater than the critical point or curve in the system of interest
Total dissolved solids	The sum of masses of all solutes, in g/kg H_2O
Vapor	Low-density, subcritical fluid. Synonymous with "gas"

responsible for leaching elements from the slab [8,10–13]. The fluid is liberated by metamorphic devolatilization in subducting lithosphere and carries solutes as it migrates into the overlying wedge of mantle, resulting in chemical modification, or "metasomatism," of slab and wedge. As the process continues, the fluid triggers melting, yielding voluminous magma that is chiefly basaltic in composition. These magmas serve as the second agents of mass transfer,

bringing slab- and mantle-derived components toward the surface as they rise.

Although melting in subduction-zone settings can be quite complex [8,9,12,14], study of the sources and evolution of the magmas is well advanced. In contrast, we know little about the fluid that begins the process of material transfer in subduction zones. In general, understanding the compositional evolution of a moving fluid and its host rocks requires techniques that couple fluid flow with chemical reaction. Reactive flow characterizes many terrestrial environments— e.g., ore deposits, crustal metamorphic systems, sedimentary basins, mid-ocean ridges— and modeling element transport in such systems has met with success. The same cannot be said of subduction zones, primarily because we lack basic information on fluid composition and how it is controlled. No direct, pristine fluid sample can be collected from this environment and working backwards from evolved magmatic products yields insight into only a part of the flow system. In addition, experimental study of fluids at the requisite high pressure (P) and temperature (T) has proven to be a singular challenge. As a result, fundamental questions remain: are the fluids dilute solutions or silicate-rich mixtures intermediate between H_2O and melt? What is the role of ligands such as Cl? How does mineral solubility and element partitioning change along the flow path? Answering these questions requires a better understanding of the chemical behavior of the fluid phase at great depth.

In this paper, I highlight recent advances that offer preliminary insights into the chemical behavior of subduction-zone fluids. I first review the physical controls that operate along the paths of fluid flow. This is followed by discussion of the chemical properties of solutions at high P, which gives context to a summary of constraints on composition. The new results are an initial step toward a chemical foundation for investigating one of Earth's most important fluids.

2. Physical controls on subduction-zone fluids

As summarized in Fig. 1, rising P and T during subduction drive mineral reactions that yield a discrete fluid phase. The fluid is probably rich in H_2O relative to other volatiles (e.g., CO_2) because of greater abundance, favorable partitioning and low thermal stability of hydrated silicates [15–20]. Experiments, simulations, and isotope geochemistry indicate that volatile liberation from subducted lithologies is more or less continuous [13,21–28], although instances of episodic fluid production may be indicated by slab seismicity [18,29].

The slab's capacity to produce fluid diminishes with depth, as minerals progressively transform to less volatile-rich assemblages (Fig. 1a). Although much of the slab H_2O is lost, a fraction is retained and recycled into the deep Earth in residual phases [13,30–37]. The liberated fluid is buoyant, and upon formation begins ascending toward the surface. Some fluid may move upward within the slab [38], but most evidently migrates into the overlying mantle wedge (Fig. 1a).

Movement into the mantle wedge is an important step for subduction-zone fluids. Not only is there a sudden shift in host-rock composition, but in addition mantle minerals are strongly undersaturated in volatiles. At equilibrium, free H_2O cannot exist in the mantle until formation of a fully hydrated mineral assemblage (serpentine, chlorite, talc, and amphibole) or hydrous mineral stability is exceeded. Fig. 1b illustrates that near the slab, the mantle H_2O content can increase to >5 wt.%, but this changes with position. The consumption of fluid by the mantle begins at the shallowest levels in the forearc [18,39,40], and continues to at least sub-arc depths. During subduction, the slab and overlying mantle become mechanically coupled, causing mantle material to be dragged downward (Fig. 1). This provides a mechanism to continually supply fresh, volatile-poor mantle for the uptake of fluid. Despite the efficiency of this process, some slab-derived fluid may travel great distances before being consumed, and locally may even reach the surface (Fig. 1a). A natural example of the surficial venting of slab-derived fluids may occur at Kinki, Japan [41].

The down-going mantle cuts across mineral stability boundaries, causing volatile-bearing minerals to regenerate the fluid they earlier absorbed (Fig. 1b). Several such events may occur, until stable minerals become nominally anhydrous. From this point on, the mantle can no longer consume the fluid phase, leaving it free to flow upwards to trigger arc-magma production. Hence, a given fluid "particle" in subduction zones experiences a stepwise history, with discrete flow events interspersed between periods of movement while dissolved in solids.

Subduction-zone fluids migrate along paths of increasing temperature with decreasing pressure (Fig. 1c). The fluid path illustrated in Fig. 1 involves a T increase from ~ 500 °C at 100-km depth (~ 3.2 GPa), to ~ 1150 °C, at 80-km depth (~ 2.4 GPa). The increase of 650 °C over ~ 20 km (32.5 °C/km) probably represents a maximum gradient because the model system on which the figure is based represents an old, cold subduction system [42]. Nevertheless, increasing T during decompression is unusual among terrestrial fluid-flow systems, and it has important consequences for chemical behavior.

Fluid flow in the mantle wedge may involve porous flow, channeled flow, or a combination [20,43–45]. During porous flow, fluids intimately interact with the rock matrix, leading to a strong potential for continuous equilibration with the host material along the flow path, such that the rock matrix controls the composition of the fluid. By contrast, channelized flow produces zones in which the fluid-to-rock ratio is much higher. The fluid interacts with less rock per unit volume and retains more of its initial source composition. Porous and channelized flow will have significantly different velocities of several m/year vs. hundreds of m/year, respectively. U–Th–Ra disequilibria in arc magmas [46,47] and distances implied by Fig. 1 suggest 2.5 to ~ 100 m/year, which is consistent with both channelized and porous flow. It is quite possible that the velocity variations are real, reflecting a spectrum of hydrologic settings.

3. Chemical controls on the composition of subduction-zone fluids

The key step in identifying the chemical controls on fluid composition in the slab and mantle wedge is the direct measurement of mineral solubility at subduction-zone P and T. Recent advances have provided the first systematic observations of this kind [48–57]. Results demonstrate large solubility increases with increasing pressure. What governs this behavior? Important factors include the properties of solvent H_2O, the association/dissociation behavior of dissolved solutes, ligand concentrations, and the extent of formation of dense, solute-rich fluids that are intermediate between melts and H_2O.

3.1. H₂O at subduction-zone conditions

Because of its volumetric dominance, H_2O controls the properties of subduction-zone fluids. Water's polar character and greater tendency to dissociate make it a substantially more powerful solvent than other volatiles. The solvent properties of H_2O depend on the density, ordering, hydrogen bonding, and dissociation of H_2O molecules. H_2O density is $1.2–1.4$ g/cm^3 at sub-arc depths (Fig. 1)

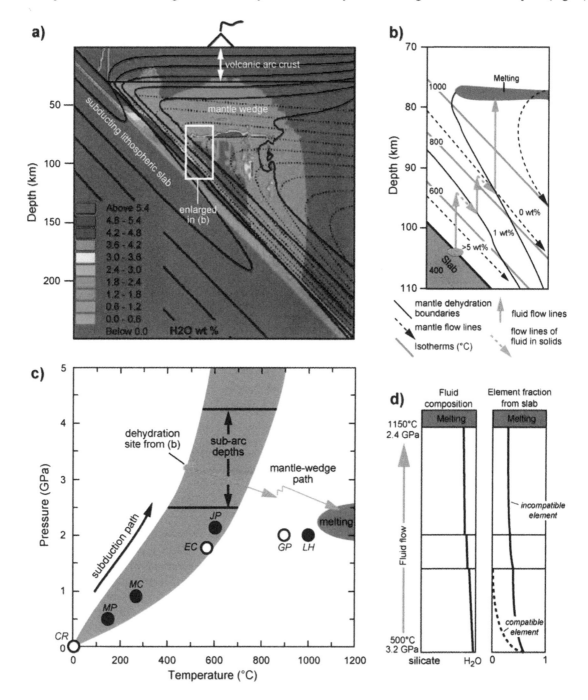

and $1.0–1.1$ g/cm^3 at the thermal maximum in the mantle wedge [58–62]. With increasing temperature, the short-range-ordered, tetrahedral packing of H_2O molecules begins to break down. Supercritical H_2O is largely disordered [63,64], and the hydrogen-bond network is disrupted. Although H_2O remains a predominantly molecular solvent up to at least 10 GPa and 1000 °C [65,66], its extent of dissociation increases significantly during subduction (Fig. 2). At sub-arc depths, pure H_2O has between 0.01 and 0.1% H^+ (neutral pH is 3–4). This is significant because, other things being equal, it will raise the concentration of solutes due to an increase in ionic strength.

3.2. Dissociation/association and ion hydration

Pressure and temperature have opposite effects on dissociation of salts in H_2O. At constant P, increasing T causes ions to associate as the dielectric constant of H_2O increases and ion-hydration shells become less stable. At constant T, increasing P causes dissociation by enhancing electrostriction of H_2O around ions, which reduces the volume of solvation. Fig. 2 shows that changes in P and T along subduction paths promote association in dilute solutions; however, dissociation increases dramatically with small increases in concentration. This further increases the capacity to carry dissolved solutes, and magnifies the effects of additional ligands. For

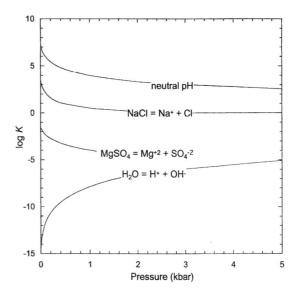

Fig. 2. Extent of dissociation (as measured by equilibrium constant, K) of H_2O, NaCl, and $MgSO_4$ as a function of pressure. H_2O and NaCl dissociation are computed along the high-T part of the envelope of subduction paths in Fig. 1c. See [92] for data sources and methods. $MgSO_4$ is taken from experiments [120] at H_2O density of 1.0 g/cm^3.

example, high Cl in arc lavas [67,68] implies that slab-derived fluids carry chloride. Increasing chlorinity favors solution of metals because of metal-chloride complexing. Calcite and anhydrite solubilities in H_2O–NaCl are dramatically enhanced with increasing NaCl concentration at high pressure [50,69], and

Fig. 1. (a) Results of a numerical simulation of H_2O production and migration in a model subduction zone [42]. The model includes solid mantle flow, assumes mineral-fluid-melt equilibrium in the slab (MORB–H_2O) and mantle wedge (peridotite–H_2O) system, and approximates aqueous fluid migration by porous flow along a pressure gradient. Subduction rate is 6 cm/year and the slab is 130 million years old. Colors represent rock H_2O content, solid lines denote isotherms (200 °C intervals, surface is 0 °C), and dashed lines indicate mantle flow paths. The location of the volcanic front is schematic. Water content decreases with depth in the slab due to H_2O loss from hydrous minerals. Large portions of the mantle in the model are at least partly hydrated by this fluid; some fluid crosses the moho and locally reaches the surface. High H_2O contents in the high-T region of the mantle beneath the volcanic arc signify sites where melting can be expected. This conceptual framework is consistent with high-precision studies of trace elements and isotopic variations across and along volcanic arcs [109–118]. (b) Enlargement of region in (a) showing schematic path of a slab-derived fluid, with mantle H_2O contents (wt.%). The fluid migrates into the mantle wedge (solid orange arrows), where it is absorbed through formation of hydrous minerals. Downward flow of solid mantle (dashed arrows) causes dehydration. After multiple hydration–dehydration steps, the fluid enters a region where it is stable with anhydrous minerals, which allows greater travel distances. (c) Pressure–temperature diagram showing representative conditions at the slab–mantle interface (shaded) in northwest and southeast Japan [119], the coolest of which (left side) corresponds to results in (a). The fluid-flow trajectory represents the schematic path from (a) and (b). Also shown are P and T of fluid compositions discussed in the text. Open symbols designate fluids equilibrated with crustal mineral assemblages; filled symbols designate fluids equilibrated with mantle (Table 1). Abbreviations: CR, Costa Rica [83]; MP, Marianas (Pacman Seamount) [82,85]; MC, Marianas (Conical Seamount), [85,108]; EC, MORB eclogite [92]; JP, jadeite peridotite [94]; GP, garnet–orthopyroxene [95]; LH, Lherzolite [96]. (d) Schematic changes in fluid composition along the flow path in (c). Concentration of silicate components derived from the rock matrix increase downstream because of rising temperature, leading to dilution of the slab component. Slab-derived compatible elements are lost close to the slab, but incompatible elements may be transported to the magma source region. In detail patterns will be more complicated owing to changes in mineral stability and fluid flux.

minor-element patterns may be strongly affected [70].

3.3. Critical curves and second critical end points

Interpretation of subduction-zone fluids is made difficult by uncertainty about melt-vapor miscibility and the existence of critical end-points in rock–H_2O systems. It is commonly assumed that fluids in subduction zones may attain properties intermediate between hydrous silicate liquid and H_2O. The expected enhancement of transport has led to the frequent indictment of intermediate fluids as metasomatic agents. However, such fluids exist only under quite restricted conditions, and their direct relevance to subduction zones is problematic. Part of the uncertainty derives from complex relationships whose depiction may seem inscrutable to all but the most ardent fan of phase diagrams. A brief overview is offered here.

3.3.1. Background

One-component systems, such as H_2O or mineral A (Fig. 3), possess a critical point marking termination of the distinction between the liquid and the vapor phase. At T and P above this point, there is only one phase, a supercritical fluid. In a system with both A and H_2O, the two critical points are linked together by a "critical curve" that marks the boundary between the stability region of a single supercritical fluid, and that of two fluids, a denser liquid and a less-dense vapor. Another curve extends from the point at which ice and mineral A coexist with liquid and vapor (the A-ice eutectic). This is the "solubility curve," which marks the stable coexistence of mineral A with liquid and vapor (Fig. 3A). Compositions of liquid, gas, and supercritical fluid vary along critical and solubility curves. For example, with increasing T along the solubility curve, the liquid composition changes continuously from nearly pure H_2O to pure A. At high T, where the liquid is rich in A, the solubility curve is equivalent to the hydrous melting curve (the "solidus" in the simple system shown). Two-component systems of geologic interest exhibit a range of behaviors. In some, the critical curve and solubility curve remain separated over their entire lengths (Fig. 3A); in others, they intersect (Fig. 3B) to yield two "critical end-points." The lower (first) critical end-point typically lies near

the critical point of H_2O. The upper, or second, critical end-point lies at high P and T, potentially near subduction paths.

3.3.2. Critical behavior in albite–H_2O

The curious relationships indicated by critical behavior, in particular of the second type (Fig. 3B), have led to speculation about a role for supercritical fluids in subduction zones. This can be explored using the system albite–H_2O for illustration. The critical curve of albite–H_2O lies at relatively low P (Fig. 4a) [71], below that of melt generation in the mantle wedge. The albite–H_2O critical curve probably intersects the solubility curve at ~ 670 °C, 1.6 GPa (Fig. 4a) [72] to generate a second critical end-point. Two schematic subduction paths illustrate the effects. Along Path 1 (Fig. 4a,b), albite solubility in H_2O increases until, at the solubility curve (H_2O-saturated melting, 685 °C, 1.5 GPa), it is ~ 10 wt.%. Above this point, only hydrous, albite-rich liquid can coexist with H_2O-rich vapor (Fig. 4b). Liquid + vapor become fully miscible at ~ 700 °C, 1.55 GPa (Fig. 4a), which signifies passage of the crest of the liquid + vapor field (Fig. 4b). Along Path 2 (Fig. 4a,c), C2 produces a narrow PT region (shaded) in which albite solubility increases continuously in a coexisting fluid that varies from dilute solution to hydrous albite-rich liquid— there is no discrete melting point.

3.3.3. Critical behavior and subduction zones

Despite the intriguing behavior of albite–H_2O, the significance of intermediate fluids in subduction zones is questionable. Critical behavior is strongly dependent on composition. Complete miscibility occurs at low to moderate P in many Na-rich systems [73]. Fluorine, boron and excess Na shift the critical curve to dramatically lower T at a given P [74]. A natural example has been reported in a pegmatite, in which complete miscibility of F-, B- and P-rich fluid occurs at 712 °C, 21.5 wt.% H_2O [75]. However, the critical curves of potassic and mafic systems lie at much higher P [76]; e.g., 12 GPa, ~ 1150 °C in MgO–SiO_2–H_2O [77]. The system K_2O–SiO_2–H_2O remains subcritical to at least 2 GPa at 1100 °C [78]. Critical curves for mantle systems (lherzolite–CO_2–H_2O) are likely 7.5 GPa or higher at ~ 1000 °C [79]. More information

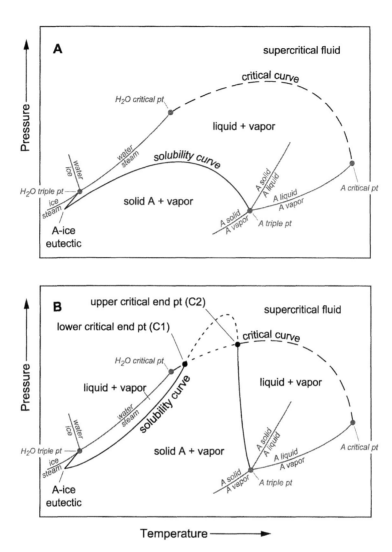

Fig. 3. Schematic $P-T$ projections of phase relations illustrating contrasting behavior of simple two-component systems. Grey lines indicate phase relations for one-component systems H_2O and hypothetical composition A; black lines represent relations for A-H_2O mixtures. Labeling of fields is for H_2O-rich systems. (A) The system A-H_2O, in which the critical curve and solubility curve do not intersect. NaCl–H_2O is an example of such a system. (B) The system A-H_2O, in which the critical curve and solubility curve intersect. This yields two critical end points (C1 and C2). Short dashed lines denote metastable portions of curves. Albite–H_2O is an example.

is needed, but it appears that critical curves for compositions more closely approximating subduction-zone rocks lie at substantially higher P than that for albite–H_2O.

Even if phase relations can be approximated by albite–H_2O, Fig. 4 reveals important restrictions on supercritical fluids. Other phases in the same chemical system may play an important role. Albite transforms to jadeite + quartz at high P, with different

solubility and a change in solidus slope (Fig. 4a). No sign of a supercritical fluid has been noted above the solidus at high P, possibly pointing to structural changes in the silicate-rich liquid that could cause the liquid + vapor field to reappear.

Another problem for intermediate fluids is that the critical curve seems to be a feature of very H_2O-rich compositions at the T of interest. Again taking albite–H_2O as a model, a closed system with several wt.%

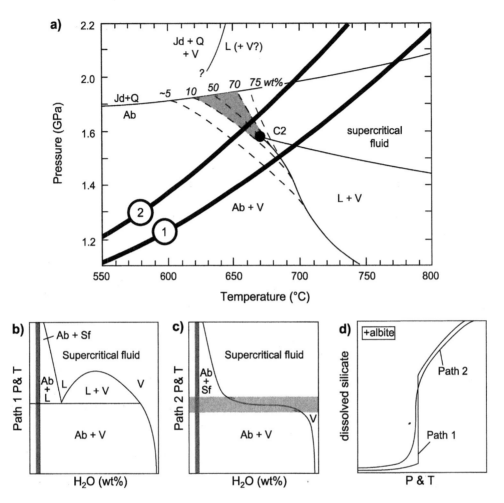

Fig. 4. (a) *PT* diagram illustrating phase relations near the second critical end point (C2) in the system albite–H₂O. Critical curve, solubility curve, and isopleths of albite solubility (dashed, in wt.%) from [71,72]. Thick solid lines represent two schematic subduction paths passing below (1) and above (2) the critical end-point. Shading indicates the region of greatest change in fluid composition. The Jd + Q solidus is queried because it is not known how it is affected by C2. Abbreviations: Ab, albite; Jd, jadeite; L, liquid; Q, quartz; Sf, supercritical fluid; V, vapor. (b) Phase relations of albite–H₂O along Path 1 as function of H₂O concentration. (c) Phase relations of albite–H₂O along Path 2 as a function of H₂O concentration. Light shading represents the region shaded in (a). In (b) and (c), the dark shaded band represents typical H₂O contents of geologic systems. (d) Comparison of the variation in dissolved silicate in the phase coexisting with albite along Paths 1 and 2.

H₂O (Fig. 4b,c) will contain albite everywhere along both paths at conditions in Fig. 4. Such a system on Path 1 will encounter neither the L + V field (which requires more than ~ 20 wt.% H₂O) nor the critical curve (~ 50 wt.% H₂O; Fig. 4b) [71,72]. On Path 2, albite coexists with an intermediate fluid only over a narrow *PT* interval (Fig. 4a,c). Comparison of the compositions of the albite-saturated fluid phase (vapor, liquid, or supercritical fluid; Fig. 4d) shows that little difference results from passage of the critical

point. Contrasts between supercritical and subcritical solubility behavior will increase with greater separation of the subduction path from C2, but only if "fanning" of iso-concentration lines increases and there are no other phases (e.g., jadeite and quartz) to change the bulk solubility of the solids (Fig. 4a).

Even if they form, intermediate fluids can only be important metasomatic agents if they separate from their source and travel significant distances. However, with movement comes a new chemical environment and different P and T. Rocks encountered along the mantle-wedge flow path will not be in equilibrium with the fluid, which will likely result in precipitation of much of the solute load over short length scales.

3.4. Polymerization of silicate components

Although I have argued that fully miscible fluids are not as important in subduction zones as has been supposed, the critical behavior holds a clue to what may be a fundamental aspect of deep fluid chemistry. A supercritical fluid can change continuously from pure H_2O to hydrous melt; i.e., solutes can evolve from hydrated ions or molecules, through small clusters, to the polymerized network of a hydrous silicate liquid. Fully miscible behavior requires that polymerization and network formation by silicate components in the aqueous phase is an important aspect of the chemistry even of dilute H_2O-rich subduction-zone fluids.

An early, direct observation of silicate polymers in dense fluids was by Mysen [78], who identified aqueous silica dimers and trimers coexisting with $K_2O–SiO_2–H_2O$ melts. These structures are marked respectively by two and three Si cations linked by bridging oxygens. Phase-equilibrium and in-situ studies confirm that silica polymerization is significant at subduction-zone P and T [49,52–54]. At 800 °C, 1.2 GPa, aqueous silica is a mix of monomers and dimers, with dimers increasing from 0 to 70% from pure H_2O to quartz saturation [51].

Its predominance in solution and ability to polymerize mean that silica plays a central role in the generation of solute polymers. However, liquid–vapor miscibility in aluminous compositions [71,73] suggests that Al solubility is enhanced because it participates in polymerization, as in silicate liquids. This is consistent with the stability of aqueous Al–Si complexes in crustal fluids [80]. Other elements may have solubility enhanced by participation in aqueous polymerization reactions (e.g., Mg [33,81]). Thus, complexing among Si, Al, and other elements via formation of polymerized solute may play an important role in controlling fluid composition at depth in subduction-zone settings. The partitioning of elements between the bulk fluid, the silicate polymer network in the fluid, and the rock matrix likely controls the overall compositional evolution of subduction-zone fluids. It is possible that concentrations

of nominally insoluble elements may be enhanced through their participation in polymers, even in fairly dilute solutions.

4. Composition of subduction-zone fluids

The discussion of physical and chemical controls on subduction-zone fluids provides a framework for examination of constraints on composition. Fig. 1 shows that fluids may be produced over a range in depth, and that their upward flow into the mantle carries them across a major compositional boundary on a path of increasing temperature with decreasing pressure. Fig. 1d illustrates the expected behavior of typical silicate components from the matrix along this path. Solubility will increase because the increase in T has a larger effect than the drop in P. For a slab-derived fluid in the mantle, this leads to dilution of the slab component as the mantle minerals progressively dissolve into the fluid. However, the relative solubility of a given element in the rock vs. the fluid will be a complex function of P, T, and composition. Those elements that are strongly partitioned into the rock relative the fluid ("compatible elements") will be lost from the fluid very near its source. The result is (1) a metasomatic zone near the slab–mantle interface that is rich in compatible elements such as Si, Ca, Al, etc., and (2) difficulty in transferring the slab signal for such elements far into the mantle. In contrast, elements that are strongly partitioned into the fluid relative to the rock ("incompatible elements") will remain in the fluid as it travels away from the slab, allowing the compositional signature of the slab to travel well into the mantle wedge. Within this framework, we can examine existing constraints on fluid composition in subduction zones (Table 2).

4.1. Major elements

Direct samples of subduction-zone fluids are available from the Costa Rica and the Izu-Bonin/Mariana convergent margins [82,83]. At Costa Rica, décollement fluids are sourced at 10–15 km, 100–150 °C [84]. They have 28 g/kg H_2O total dissolved solids (TDS), dominated by Na and Cl, with low Si and a modest load of alkali and alkaline earth metals. Chlorinity is below seawater.

Table 2
Comparison of major-element compositions of subduction-zone fluids

	Costa Rica décollement	Marianas (Pacman seamount)	Marianas (Conical seamount)	Model eclogite (~ MORB)	Jadeite peridotite	Garnet + orthopyroxene	Spinel–ilmenite–rutile lherzolite
Depth to slab (km)	0.36	~ 15	~ 25	~ 50			
P (GPa)				1.75	2.2	2.0	2.0
T (°C)				550	600	900	1000
Cl	486	404	315	400	500		
Si	0.084		0.032	356	547	905	285
Al				0.8	99	193	61
Na	411		423	509	121	28	31
K	3.3		15				
Ca	23	47	1.3	8		44	26
Mg	12.9	0	0.005	0.001	3	125	21
Fe						13	11
Ti							3
Alkalinity		1	31				
TDS	28	17	28	48	38	83	67
Source	[83]	[85]	[108]	[92]	[94]	[95]	[96]

Concentrations in millimol per kg H_2O, except alkalinity = meq/kg H_2O, total dissolved solids (TDS) = grams total solute per kg H_2O.

Pore fluids from Mariana forearc serpentine-mud volcanoes originate at the slab–mantle interface at 15–25 km [82,85,86]. Cl concentration is lower than seawater, consistent with a dehydration origin. Ca decreases and CO_2 (alkalinity) increases with depth, reflecting the onset of slab decarbonation between 15- and 25-km depth [85]. Mg and Ca are lower than in Costa Rica fluids, perhaps due to lower chlorinity. Lower Si probably reflects the serpentinite host.

Direct samples have unknown reaction-flow history prior to sampling, so conditions of last equilibration are unclear. Fluid inclusions in exhumed blueschist- and eclogite-facies oceanic mafic rocks represent an alternative sample. Fluids show 1–7 wt.% NaCl equivalent [87], or TDS = 10–75 g/kg H_2O. This spans values of shallow fluids. Increases in TDS with depth over 25–50 km are no more than about a factor of two, and low salinity and CO_2 indicate that H_2O is the dominant solvent.

It is important to note that primary fluid inclusions in eclogites associated with continental subduction may have much higher N_2, CO_2, and salinity (up to 50 wt.% NaCl equivalent) [11,88–91]. There appears to be a fundamental difference in the salinity of fluids associated with oceanic and continental subduction, as recorded by fluid inclusions. Cross-comparisons between fluids in these two environments require extreme caution.

In the absence of direct samples of unreacted fluids or fluid inclusions, we must rely on estimates from mineralogic phase relations and experiments. Theoretical analysis of deeper fluid produced at the blueschist-eclogite transition indicates yet higher TDS (48 g/kg H_2O) [92]. This is largely due to the dramatic increase in silica at similar Cl content. Na and Ca concentrations are broadly similar. Al concentration, which is so low in low-pressure fluids that it is rarely analyzed, is higher than Mg. Varying Cl does not have a large effect. The high Al suggests that deep fluids leaving the slab are Na–Ca–Al–Si rich, but low in Mg and Fe, unless significant Cl is present. This is consistent with vein minerals in blueschists and eclogites [87,93].

Insights into the compositional evolution of slab-derived fluids as they react with the mantle wedge can be gained from experimental studies. Aqueous fluid in equilibrium with jadeite–peridotite at conditions near the slab–mantle interface (Fig. 1) is Si- and Na-rich, and contains substantial Al [94]. Total dissolved solids is similar to the results for eclogite with Cl-free fluid. The higher Si reflects the strong dependence of Si concentration on temperature at high pressure [49].

In the region of arc-magma generation, a 5-molal NaCl solution in equilibrium with garnet and orthopyroxene at 900 °C, 2 GPa, has nearly twice the silica, as well as higher Mg, Ca, and Al, low Na and

Fe, TDS of 83 g/kg H_2O [95]. H_2O equilibrated with spinel lherzolite and Ti phases at 2.0 GPa, 1000 °C, shows lower Na, Al, and Si, but is otherwise similar to the garnet–orthopyroxene fluid [96]. Differences in Na between the fluids at 900–1000 °C and those at 550–660 °C are probably due to varying Na in the buffering assemblage.

Three important conclusions can be drawn from Table 2 about the major-element compositions of subduction-zone fluids. First, the deeper fluids are similar to the direct, shallow samples in one important way: total solute concentrations are low, regardless of pressure, temperature, or chlorinity. Deep fluids have TDS only two to three times that of seawater, and no more than 50% higher than shallow fluids from near the entrance to the subduction zone. TDS does not approach tens of wt.%, as would be associated with supercritical fluids of intermediate composition (Fig. 4). This is consistent with the independent arguments developed above.

The second conclusion is that TDS, though generally modest, nevertheless increases with depth. This arises from changes in solubility of rock-forming minerals due to the *PT*-enhancement of the solvent power of H_2O; additional ligands such as Cl may be present, but are not required.

Finally, there are important changes in major elements with depth. The dominant solutes in deep fluids are Si and Na. Al concentrations are higher than Ca, Fe, and usually Mg. Thus, the solutes in H_2O-rich fluids in subduction zones are dominated by alkali and aluminosilicate components. This contrasts with fluids from shallow environments, where alkali and other metals predominate and Al is virtually insoluble. It is quite likely that this change with increasing depth in subduction zones simply reflects the ability of alkali aluminosilicate components to form aqueous polymers at high *P* and *T*.

4.2. Trace elements

Trace-element patterns support the inference that the continental crust formed by island-arc magmatism (Fig. 5), but enrichments and depletions preclude simple anhydrous melting and fractionation. Instead, arc-magmas form by hydrous melting of a source metasomatized by slab-derived components. Trace-element patterns of low-Cl subduction-zone

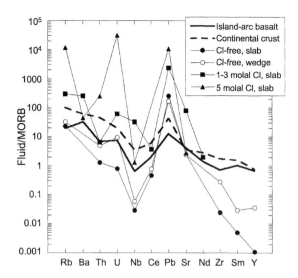

Fig. 5. Trace-element diagram normalized to MORB [121]. Elements are plotted in order of decreasing compatibility from left to right. Trace-element patterns grouped by chlorine concentration: Cl-free [97]; 1–3 molal [89]; 5 molal [70]. Island-arc basalt from [122]; continental crust from [123].

fluids are broadly similar to island-arc basalt (IAB) and continental crust, with enrichments in LILE and Pb, and depletions in high field strength elements (Fig. 5). However, >1 molal chloride changes some aspects of the patterns [70]. While Pb enrichment, Nb depletion, and a decreasing abundance with compatibility remain, the 5 molal brine shows very high Rb/Ba, Th/U ~ 0.01, enhanced Pb enrichment, and extreme differences between adjacent elements. This would appear to suggest that the trace-element signature of the IAB source is controlled by flow of low-Cl fluid [97,98]. However, the simple model compositions require more H_2O in the IAB source than is typically inferred [12], and isotopic studies support a wide range of chlorinity [99–102]. While there are likely to be real variations in Cl content in the global subduction system, there is also a need for more sophisticated reactive flow models that account for changes in mantle mineral assemblage along with evolution of fluid and rock composition.

5. Prospects for the future

This paper has attempted to highlight some of the open questions regarding the chemistry of subduction-zone fluids. H_2O is a powerful solvent at subduction zone conditions, and it can effectively mobilize many rock components. Although supercritical, intermediate fluids are commonly invoked to explain mantle-wedge metasomatism, the hypothesis is problematic in detail. Supporting this, theoretical and experimental constraints on fluids at sub-arc conditions are surprisingly dilute. It may be that the chemistry of aqueous silicate polymers, as controlled by P, T and the mineralogy of the host, plays a fundamental role in the composition of subduction-zone fluids. The effects of Cl, and by inference other ligands, may be profound, but it is unclear at this time what role is required of them.

There are excellent prospects for progress on these issues in the near future. Experimental advances, including the hydrothermal diamond anvil cell [103–106] and solid media solubility techniques [48–51], promise to provide fundamental data on mineral solubility and fluid composition. Advances in molecular dynamics and ab-initio molecular dynamics simulations are already giving unprecedented insight into the nature of H_2O and ion hydration, geometry and structure [64,107]. Finally, new geochemical tools, including light element isotope ratios (e.g., Li and B), new spectroscopic methods and ICP-MS, are being applied to fluid and melt inclusions in subducted oceanic rocks, mantle-wedge xenoliths and arc magmas. Results give new insight into the details of slabs and arc-magma sources, which constrain the initial and final conditions in the reacting system of interest here.

The new data will provide the foundation necessary to develop sophisticated reaction-flow models of this complex metasomatic environment. Once this is under way, we will have made a major advance toward unraveling the chemical workings of this part of the subduction factory.

Acknowledgements

This study was supported by NSF EAR 9909583 and 0337170. Reviews by J. Ayers, G. Bebout and an anonymous reviewer improved the manuscript. I am also indebted to R. Newton, J. Davidson, and S. Peacock for insightful critiques. *[AH]*

References

[1] G.E. Bebout, J.G. Ryan, W.P. Leeman, A.E. Bebout, Fractionation of trace elements by subduction zone metamorphism—effect of convergent margin thermal evolution, Earth Planet. Sci. Lett. 171 (1999) 63–82.

[2] D.R. Hilton, T.P. Fischer, B. Marty, Noble gases and volatile recycling at subduction zones, Rev. Mineral. Geochem. 47 (2002) 319–370.

[3] R.D. Jarrard, Subduction fluxes of water, CO_2, chlorine, and potassium, Geochim. Geophys. Geosys. 4 (2003) DOI:10.1029/2002GC000392.

[4] S.J. Sadofsky, G.E. Bebout, Record of forearc devolatilization in low-T, high P/T metasedimentary suites: significance for models of convergent margin chemical cycling, Geochem. Geophys. Geosys. 4 (4) (2003) 29 pp.

[5] J.M. Eiler, B. McInnes, J.W. Valley, C.M. Graham, E.M. Stolper, Oxygen isotope evidence for slab-derived fluids in the sub-arc mantle, Nature 393 (1998) 777–781.

[6] J.M. Eiler, A. Crawford, T. Elliott, K.A. Farley, J.W. Valley, E.M. Stolper, Oxygen isotope geochemistry of oceanic-arc lavas, J. Petrol. 41 (2000) 229–256.

[7] I.J. Parkinson, R.J. Arculus, The redox state of subduction zones: insights from arc-peridotites, Chem. Geol. 160 (1999) 409–424.

[8] R.J. Stern, Subduction zones, Rev. Geophys. 40 (4) (2002) 3-1–3-38.

[9] P.E. van Keken, The structure and dynamics of the mantle wedge, Earth Planet. Sci. Lett. 215 (2003) 323–338.

[10] B.O. Mysen, P. Ulmer, J. Konzett, M.W. Schmidt, The upper mantle near convergent plate boundaries, Rev. Mineral. 37 (1998) 97–138.

[11] M. Scambelluri, P. Phillipot, Deep fluids in subduction zones, Lithos 55 (2001) 213–227.

[12] P. Ulmer, Partial melting in the mantle wedge—the role of H_2O in the genesis of mantle-derived "arc-related" magmas, Phys. Earth Planet. Inter. 127 (2001) 215–232.

[13] S. Poli, M.W. Schmidt, Petrology of subducted slabs, Ann. Rev. Earth Planet. Sci. 30 (2002) 207–335.

[14] P.S. Hall, C. Kincaid, Diapiric flow at subduction zones: a recipe for rapid transport, Science 292 (2001) 2472–2475.

[15] J.F. Molina, S. Poli, Carbonate stability and fluid composition in subducted oceanic crust: an experimental study on H_2O-CO_2-bearing basalts, Earth Planet. Sci. Lett. 176 (2000) 295–310.

[16] D.M. Kerrick, J.A.D. Connolly, Metamorphic devolatilization of subducted oceanic metabasalts: implications for seismicity, arc magmatism and volatile recycling, Earth Planet. Sci. Lett. 189 (2001) 19–29.

[17] D.M. Kerrick, J.A.D. Connolly, Metamorphic devolatilization of subducted marine sediments and the transport of volatiles into the Earth's mantle, Nature 411 (2001) 293–296.

[18] S.M. Peacock, Are the lower planes of double seismic zones caused by serpentine dehydration in subducting oceanic mantle? Geology 29 (2001) 299–302.

[19] B.R. Hacker, G.A. Abers, S.M. Peacock, Subduction factory: 1. Theoretical mineralogy, density, seismic wave speeds, and

H₂O content, J. Geophys. Res. 108 (2003) DOI:10.1029/2001JB001127.

[20] B.R. Hacker, S.M. Peacock, G.A. Abers, S.D. Holloway, Subduction factory: 2. Are intermediate-depth earthquakes in subducting slabs linked to metamorphic dehydration reactions? J. Geophys. Res. 108 (2003) DOI:10.1029/2001JB001129.

[21] M.W. Schmidt, S. Poli, Experimentally based water budgets for dehydrating slabs and consequences for arc magma generation, Earth Planet. Sci. Lett. 163 (1998) 361–379.

[22] T. Moriguti, E. Nakamura, Across-arc variation of Li isotopes in lavas and implications for crust/mantle recycling at subduction zones, Earth Planet. Sci. Lett. 163 (1998) 167–174.

[23] S. Peacock, R.L. Hervig, Boron isotopic composition of subduction-zone metamorphic rocks, Chem. Geol. 160 (1999) 281–290.

[24] S.M. Straub, G.D. Layne, The systematics of boron isotopes in Izu arc front volcanic rocks, Earth Planet. Sci. Lett. 198 (2002) 25–39.

[25] P.E. van Keken, B. Kiefer, S.M. Peacock, High-resolution models of subduction zones: implications for mineral dehydration reactions and the transport of water into the deep mantle, Geochim. Geophys. Geosys. 3 (2002) DOI: 10.1029/2001GC000256.

[26] P.B. Tomascak, E. Widom, L.D. Benton, S.L. Goldstein, J.G. Ryan, The control of lithium budgets in island arcs, Earth Planet. Sci. Lett. 196 (2002) 227–238.

[27] G.E. Bebout, E. Nakamura, Record in metamorphic tourmalines of subduction-zone devolatilization and boron cycling, Geology 31 (2003) 407–410.

[28] T. Zack, P.B. Tomascak, R.L. Rudnick, C. Dalpé, W.F. McDonough, Extremely light Li in orogenic eclogites: the role of isotope fractionation during dehydration in subducted oceanic crust, Earth Planet. Sci. Lett. 208 (2003) 279–290.

[29] K. Obara, Nonvolcanic deep tremor associated with subduction in southwest Japan, Science 296 (2002) 1679–1681.

[30] B. Wunder, Equilibrium experiments in the system MgO–SiO₂–H₂O (MSH): stability fields of clinohumite–OH [Mg₉Si₄O₁₆(OH)₂], chondrodite [Mg₅Si₂O₈(OH)₂] and phase A [Mg₇Si₂O₈(OH)₆], Contrib. Mineral. Petrol. 132 (1998) 111–120.

[31] A. Pawley, Stability of clinohumite in the system MgO–SiO₂–H₂O, Contrib. Mineral. Petrol. 138 (2000) 284–291.

[32] A. Pawley, Chlorite stability in mantle peridotite: the reaction clinochlore plus enstatite = forsterite + pyrope + H₂O, Contrib. Mineral. Petrol. 144 (2003) 449–456.

[33] R. Stalder, P. Ulmer, Phase relations of a serpentine composition between 5 and 14 GPa: significance of clinohumite and phase E as water carriers into the transition zone, Contrib. Mineral. Petrol. 140 (2001) 670–679.

[34] Q. Williams, R.J. Hemley, Hydrogen in the deep earth, Annu. Rev. Earth Planet Sci. 29 (2001) 365–418.

[35] G.D. Bromiley, A.R. Pawley, The high-pressure stability of Mg-surassite in a model hydrous peridotite: a possible mechanism for the deep subduction of significant volumes of H₂O, Contrib. Mineral. Petrol. 142 (2002) 714–723.

[36] K. Litasov, E. Ohtani, Stability of various hydrous phases in CMAS pyrolite–H₂O system up to 25 GPa, Phys. Chem. Miner. 30 (2003) 147–156.

[37] J.F. Forneris, J.R. Holloway, Phase equilibria in subducting basaltic crust: implications for H₂O release from the slab, Earth Planet. Sci. Lett. 214 (2003) 187–201.

[38] I. Cartwright, A.C. Barnicoat, Stable isotope geochemistry of Alpine ophiolites: a window to ocean-floor hydrothermal alteration and constraints on fluid–rock interaction during high-pressure metamorphism, Int. J. Earth Sci. 88 (1999) 219–235.

[39] S.M. Peacock, R.D. Hyndman, Hydrous minerals in the mantle wedge and the maximum depth of subduction thrust earthquakes, Geophys. Res. Lett. 26 (1999) 2517–2520.

[40] M.G. Bostock, R.D. Hyndman, S. Rondenay, S.M. Peacock, An inverted continental Moho and serpentinization of the forearc mantle, Nature 417 (2002) 536–538.

[41] T. Mastsumoto, T. Kawabata, J.-I. Matsuda, K. Yamamoto, K. Mimura, ³He/⁴He ratios in well gases in the Kinki district, SW Japan: surface appearance of slab-derived fluids in a non-volcanic area, Earth Planet. Sci. Lett. 216 (2003) 221–230.

[42] H. Iwamori, Transportation of H₂O and melting in subduction zones, Earth Planet. Sci. Let. 160 (1998) 65–80.

[43] K. Mibe, T. Fujii, A. Yasuda, Control of the location of the volcanic front in island arcs by aqueous fluid connectivity in the mantle wedge, Nature 401 (1999) 259–262.

[44] G.H. Davies, The role of hydraulic fractures and intermediate-depth earthquakes in generating subduction-zone magmatism, Nature 398 (1999) 142–145.

[45] D.P. Dobson, P.G. Meredith, S.A. Boon, Simulation of subduction zone seismicity by dehydration of serpentine, Science 298 (2002) 1407–1410.

[46] S.P. Turner, On the time-scales of magmatism at island-arc volcanoes, Philos. Trans. R. Soc. Lond. Ser. A: Math. Phys. Sci. 360 (2002) 2853–2871.

[47] T. Yokoyama, E. Nakamura, K. Kobayashi, T. Kuritani, Timing and trigger of arc volcanism controlled by fluid flushing from subducting slab, Proc. Jpn. Acad., Ser. B Phys. Biol. Sci. 78 (2002) 190–195.

[48] R.C. Newton, C.E. Manning, Quartz solubility in concentrated aqueous NaCl solutions at deep crust–upper mantle metamorphic conditions: 2–15 kbar and 500–900 °C, Geochim. Cosmochim. Acta 64 (2000) 2993–3005.

[49] R.C. Newton, C.E. Manning, Solubility of enstatite + forsterite in H₂O at deep crust/upper mantle conditions: 4 to 15 kbar and 700 to 900 °C, Geochim. Cosmochim. Acta 66 (2002) 4165–4176.

[50] R.C. Newton, C.E. Manning, Experimental determination of calcite solubility in H₂O–NaCl solutions at deep crust/upper mantle pressures and temperatures: implications for metasomatic processes in shear zones, Am. Mineral. 87 (2002) 1401–1409.

[51] R.C. Newton, C.E. Manning, Activity coefficient and polymerization of aqueous silica at 800 °C, 12 kbar, from solubility measurements on SiO₂-buffering mineral assemblages, Contrib. Mineral. Petrol. 146 (2003) 135–143.

[52] Y.G. Zhang, J.D. Frantz, Enstatite–forsterite–water equilibria at elevated temperatures and pressures, Am. Mineral. 85 (2000) 918–925.

[53] N. Zotov, H. Keppler, In-situ Raman spectra of dissolved silica species in aqueous fluids to 900 °C and 14 kbar, Am. Mineral. 85 (2000) 600–603.

[54] N. Zotov, H. Keppler, Silica speciation in aqueous fluids at high pressures and high temperatures, Chem. Geol. 184 (2002) 71–82.

[55] K. Shmulovich, C. Graham, B. Yardley, Quartz, albite and diopside solubilities in H_2O–$NaCl$ and H_2O–CO_2 fluids at 0.5–0.9 GPa, Contrib. Mineral. Petrol. 141 (2001) 95–108.

[56] T. Fockenberg, M. Burchard, W.V. Maresch, Experimental determination of the solubility of natural wollastonite in pure water up to pressures of 5.0 GPa, Eos Trans. AGU 83 (47) (2002) (Fall Meet. Suppl. Abstract V72–1327).

[57] N. Caciagli, C.E. Manning, The solubility of calcite in water at 6–16 kbar and 500–800 °C, Contrib. Mineral. Petrol. 146 (2003) 275–285.

[58] S. Wiryana, L.J. Slutsky, J.M. Brown, The equation of state of water to 200 °C and 3.5 GPa: model potentials and the experimental pressure scale, Earth Planet. Sci. Lett. 163 (1998) 123–130.

[59] A.C. Withers, S.C. Kohn, R.A. Brooker, B.J. Wood, A new method for determining the P–V–T properties of high-density H_2O using NMR: results at 1.4–4.0 GPa and 700–1100 °C, Geochim. Cosmochim. Acta 64 (2000) 1051–1057.

[60] W. Wagner, A. Pruss, The IAPWS formulation (1995) for the thermodynamic properties of ordinary water substance for general and scientific use, J. Phys. Chem. Ref. Data 31 (2002) 387–535.

[61] S.V. Churakov, M. Gottschalk, Perturbation theory based equation of state for polar molecular fluid: I. Pure fluids, Geochim. Cosmochim. Acta 67 (2003) 2397–2414.

[62] E.H. Abrahmson, J.M. Brown, Equation of state of water based on speeds of sound measured in the diamond-anvil cell, Geochim. Cosmochim. Acta 68 (2004) 1827–1835.

[63] A.K. Soper, The radial distribution functions of water and ice from 220 to 673 K and at pressures up to 400 MPa, Chem. Phys. 258 (2000) 121–137.

[64] A.G. Kalinichev, Molecular simulations of liquid and supercritical water: thermodynamics, structure, hydrogen bonding, Rev. Mineral. Geochem. 42 (2001) 83–129.

[65] G.A. Lyzenga, T.J. Ahrens, W.J. Nellis, A.C. Mitchell, The temperature of shock-compressed water, J. Chem. Phys. 76 (1982) 6282–6286.

[66] R. Chau, A.C. Mitchell, R.W. Minich, W.J. Nellis, Electrical conductivity of water compressed dynamically to pressures of 70–180 GPa (0.7–1.8 Mbar), J. Chem. Phys. 114 (2001) 1361–1365.

[67] J.D. Webster, R.J. Kinzler, E.A. Mathez, Chloride and water solubility in basalt and andesite melts and implications for magmatic degassing, Geochim. Cosmochim. Acta 63 (1999) 729–738.

[68] A.J.R. Kent, D.W. Peate, S. Newman, E.M. Stolper, J.A. Pearce, Chlorine in submarine glasses from the Lau Basin: seawater contamination and constraints on the composition of slab-derived fluids, Earth Planet. Sci. Lett. 202 (2002) 361–377.

[69] R.C. Newton, C.E. Manning, Stability of anhydrite, $CaSO_4$, in $NaCl$–H_2O solutions at high pressures and temperatures: applications to fluid–rock interaction, J. Petrol. (in press).

[70] H. Keppler, Constraints from partitioning experiments on the composition of subduction-zone fluids, Nature 380 (1996) 237–240.

[71] A. Shen, H. Keppler, Direct observation of complete miscibility in the albite–H_2O system, Nature 385 (1997) 710–712.

[72] R. Stalder, P. Ulmer, A.B. Thompson, D. Günther, Experimental approach to constrain second critical end points in fluid/silicate systems: near-solidus fluids and melts in the system albite–H_2O, Am. Mineral. 85 (2000) 68–77.

[73] H. Bureau, H. Keppler, Complete miscibility between silicate melts and hydrous fluids in the upper mantle: experimental evidence and geochemical implications, Earth Planet. Sci. Lett. 165 (1999) 187–196.

[74] J.R. Sowerby, H. Keppler, The effect of fluorine, boron and excess sodium on the critical curve in the albite–H_2O system, Contrib. Mineral. Petrol. 143 (2002) 32–37.

[75] R. Thomas, J.D. Webster, W. Heinrich, Melt inclusions in pegmatite quartz: complete miscibility between silicate melts and hydrous fluids at low pressure, Contrib. Mineral. Petrol. 139 (2000) 394–401.

[76] B.O. Mysen, Solubility behavior of alkaline earth and alkali aluminosilicate components in aqueous fluids in the earth's upper mantle, Geochim. Cosmochim. Acta 66 (2002) 2421–2438.

[77] R. Stalder, P. Ulmer, A.B. Thompson, D. Günther, High pressure fluids in the system MgO–SiO_2–H_2O under upper mantle conditions, Contrib. Mineral. Petrol. 140 (2001) 607–618.

[78] B.O. Mysen, Interaction between aqueous fluid and silicate melt in the pressure and temperature regime of the earth's crust and upper mantle, N. Jb. Miner. Ab. 172 (1998) 227–244.

[79] P.J. Wyllie, I.D. Ryabchikov, Volatile components, magmas, and critical fluids in upwelling mantle, J. Petrol. 41 (2000) 1195–1206.

[80] S. Salvi, G.S. Pokrovski, J. Schott, Experimental investigation of aluminum–silica aqueous complexing at 300 °C, Chem. Geol. 151 (1998) 51–67.

[81] K. Mibe, T. Fujii, A. Yasuda, Composition of aqueous fluid coexisting with mantle minerals at high pressure and its bearing on the differentiation of the earth's mantle, Geochim. Cosmochim. Acta 66 (2002) 2273–2285.

[82] P. Fryer, J.A. Pearce, L.B. Stokking, et al, Proc. ODP, Init. Rept., 125, 1990, 1092 pp.

[83] G. Kimura, E. Silver, P. Blum, et al, Proc. ODP, Init. Rept., 170, 1997, 458 pp.

[84] E. Silver, M. Kastner, A. Fisher, J. Morris, K. McIntosh, D. Saffer, Fluid flow paths in the middle America Trench and Costa Rica margin, Geology 28 (2000) 679–682.

[85] P. Fryer, C.G. Wheat, M.J. Mottl, Mariana blueschist mud volcanism: implications for conditions within the subduction zone, Geology 27 (1999) 103–106.

[86] L.D. Benton, J.G. Ryan, F. Tera, Boron isotope systematics of slab fluids as inferred from a serpentine seamount, Mariana forearc, Earth Planet. Sci. Lett. 187 (2001) 273–282.

[87] J. Gao, R. Klemd, Primary fluids entrapped at blueschist to eclogite transition: evidence from the Tianshan meta-subduction complex in northwest China, Contrib. Mineral. Petrol. 142 (2001) 1–14.

[88] M. Scambelluri, P. Philippot, G. Pennacchioni, Salt-rich aqueous fluids formed during eclogitization of metabasites in the Alpine continental crust (Austroalpine Mt. Emilius unit, Italian western Alps), Lithos 43 (1998) 151–161.

[89] M. Scambelluri, P. Bottazzi, V. Trommsdorff, R. Vannucci, J. Hermann, M.T. Gòmez-Pugnaire, V. Lòpez-Sànchez Vizcaìno, Incompatible element-rich fluids released by antigorite breakdown in deeply subducted mantle, Earth Planet. Sci. Lett. 192 (2002) 457–470.

[90] B. Fu, J.L.R. Touret, Y.F. Zheng, Fluid inclusions in coesite-bearing eclogites and jadeite quartzite at Shuanghe, Dabie Shan (China), J. Metamorph. Geol. 19 (2001) 531–547.

[91] L. Franz, R.L. Romer, R. Klemd, R. Schmid, R. Oberhänsli, T. Wagner, S.W. Dong, Eclogite-facies quartz veins within metabasites fo the Dabie Shan (eastern China): pressure–temperature–time–deformation path, composition of the fluid phase and fluid flow during exhumation of high-pressure rocks, Contrib. Mineral. Petrol. 141 (2001) 322–346.

[92] C.E. Manning, Fluid composition at the blueschist–eclogite transition in the model system $Na_2O–MgO–Al_2O_3–SiO_2–H_2O–HCl$, Swiss Bull. Mineral. Petrol. 78 (1998) 225–242.

[93] H. Becker, K.P. Jochum, R.W. Carlson, Constraints from high-pressure veins in eclogites on the composition of hydrous fluids in subduction zones, Chem. Geol. 160 (1999) 291–308.

[94] M.E. Schneider, D.H. Eggler, Fluids in equilibrium with peridotite minerals: implications for mantle metasomatism, Geochim. Cosmochim. Acta 50 (1986) 711–724.

[95] J.M. Brenan, H.F. Shaw, F.J. Ryerson, Experimental evidence for the origin of lead enrichment in convergent-margin magmas, Nature 378 (1995) 54–56.

[96] J. Ayers, S.K. Dittmer, G.D. Layne, Partitioning of elements between peridotite and H_2O at 2.0–3.0 GPa and 900–1100 °C, and application to models of subduction zone processes, Earth Planet. Sci. Lett. 150 (1997) 381–398.

[97] J. Ayers, Trace element modeling of aqueous fluid–peridotite interaction in the mantle wedge of subduction zones, Contrib. Mineral. Petrol. 132 (1998) 390–404.

[98] R. Stalder, S.F. Foley, G.P. Brey, I. Horn, Mineral-aqueous fluid partitioning of trace elements at 900–1200 °C and 3.0–5.7 GPa: new experimental data for garnet, clinopyroxene, and rutile, and implications for mantle metasomatism, Geochim. Cosmochim. Acta 62 (1998) 1781–1801.

[99] K. Righter, J.T. Chesley, J. Ruiz, Genesis of primitive, arc-type basalt: constraints from Re, Os, and Cl on the depth of melting and role of fluids, Geology 20 (2002) 619–622.

[100] A. Laurora, M. Mazzucchelli, G. Rivalenti, R. Vannucci, A. Zanetti, M.A. Barbieri, A. Cingolani, Metasomatism and melting in carbonated peridotite xenoliths from the mantle wedge: the Gobernador Gregores case (southern Patagonia), J. Petrol. 42 (2001) 69–87.

[101] E. Widom, P. Kepezhinskas, M. Defant, The nature of metasomatism in the sub-arc mantle wedge: evidence from Re–Os isotopes in Kamchatka peridotite xenoliths, Chem. Geol. 196 (2003) 283–306.

[102] P. Kepezhinskas, M.J. Defant, E. Widom, Abundance and distribution of PGE and Au in the island-arc mantle: implications for sub-arc metasomatism, Lithos 60 (2002) 113–128.

[103] W.A. Bassett, A.J. Anderson, R.A. Mayanovic, I.M. Chou, Modified hydrothermal diamond anvil cells for XAFS analyses of elements with low energy absorption edges in aqueous solutions at sub- and supercritical conditions, Zeit. Kristallog. 215 (2000) 711–717.

[104] C. Schmidt, M.A. Ziemann, In-situ Raman spectroscopy of quartz: a pressure sensor for hydrothermal diamond-anvil cell experiments at elevated temperatures, Am. Mineral. 85 (2000) 1725–1734.

[105] C. Schmidt, K. Rickers, In-situ determination of mineral solubilities in fluids using a hydrothermal diamond-anvil cell and SR-XRF: solubility of AgCl in water, Am. Mineral. 88 (2003) 288–292.

[106] M. Burchard, A.M. Zaitsev, W.V. Maresch, Extending the pressure and temperature limits of hydrothermal diamond anvil cells, Rev. Sci. Instrum. 74 (2003) 1263–1266.

[107] D.M. Sherman, Quantum chemistry and classical simulations of metal complexing in aqueous solutions, Rev. Mineral. Geochem. 42 (2001) 273–317.

[108] M.J. Mottl, Pore waters from serpentinite seamounts in the Mariana and Izu-Bonin forearcs, Leg 125 (1992) 373–385.

[109] J.D. Woodhead, S.M. Eggins, R.W. Johnson, Magma genesis in the New Britain Island Arc: further insights into melting and mass transfer processes, J. Petrol. 39 (1998) 1641–1668.

[110] T. Ishikawa, F. Tera, Two isotopically distinct fluid components involved in the Mariana arc: evidence from Nb/B ratios and B, Sr, Nd, and Pb isotope systematics, Geology 27 (1999) 83–86.

[111] C. Class, D.M. Miller, et al, Distinguishing melt and fluid subduction components in Umnak Volcanics, Aleutian Arc, Geochem. Geophys. Geosys. 1 (2000) DOI:10.1029/1999GC000010.

[112] P.W. Reiners, P.E. Hammond, J.M. McKenna, R.A. Duncan, Young basalts of the central Washington Cascades, flux melting of the mantle, and trace element signatures of primary arc magmas, Contrib. Mineral. Petrol. 138 (2000) 249–264.

[113] A. Hochstaedter, J. Gill, R. Peters, P. Broughton, P. Holden, B. Taylor, Across-arc geochemical trends in the Izu-Bonin arc: contributions from the subducting slab, Geochim. Geophys. Geosys. 2 (2001) DOI:10.1029/2000GC000105.

[114] T. Ishikawa, F. Tera, T. Nakazawa, Boron isotope and trace element systematics of the three volcanic zones in the Kamchatka arc, Geochim. Cosmochim. Acta 65 (2001) 4523–4537.

[115] T. Churikova, F. Dorendorf, G. Wörner, Sources and fluids in the mantle wedge below Kamchatka, evidence from across-arc geochemical variation, J. Petrol. 42 (2001) 1567–1593.

[116] M.A. Elburg, M. van Bergen, J. Hoogewerff, J. Foden, P. Vroon, I. Zulkarnain, A. Nasution, Geochemical trends across an arc-continent collision zone: magma sources and slab-wedge transfer processes below the Pantar Strait volcanoes, Indonesia, Geochim. Cosmochim. Acta 66 (2002) 2771–2789.

[117] P. Cervantes, P.J. Wallace, The role of H_2O in subduction zone magmatism: new insights from melt inclusions in high-Mg basalts from Central Mexico, Geology 31 (2003) 235–238.

[118] O. Ishizuka, R.N. Taylor, J.A. Milton, R.W. Nesbitt, Fluid–mantle interaction in an intra-oceanic arc: constraints from high-precision Pb isotopes, Earth Planet. Sci. Lett. 211 (2003) 221–236.

[119] S.M. Peacock, K. Wang, Seismic consequences of warm versus cool subduction metamorphism: examples from southwest and northeast Japan, Science 286 (1999) 937–939.

[120] J.D. Frantz, J. Dubessy, B.O. Mysen, Ion-pairing in aqueous $MgSO_4$ solutions along an isochore to 500 °C and 11 kbar using Raman spectroscopy in conjunction with the diamond-anvil cell, Chem. Geol. 116 (1994) 181–188.

[121] S.-S. Sun, W.F. McDonough, Chemical and isotopic systematics of oceanic basalts: implications for mantle composition and processes, in: A.D. Saunders, M.J. Norry (Eds.), Magmatism in the Ocean Basins, Geol. Soc. London Spec. Publ., vol. 42, (1989) 313–345.

[122] M.T. McCulloch, J.A. Gamble, Geochemical and geodynamical constraints on subduction zone magmatism, Earth Planet. Sci. Lett. 102 (1991) 358–374.

[123] B.S. Kamber, A. Ewart, K.D. Collerson, M.C. Bruce, G.D. MacDonald, Fluid-mobile trace element constraints on the role of slab melting and implications for Archaean crustal growth models, Contrib. Mineral. Petrol. 144 (2002) 38–56.

Craig Manning is a Professor of Geology and Geochemistry in the Department of Earth and Space Sciences at the University of California Los Angeles. He received BA degrees in Geology and in Geography from the University of Vermont, and MS and PhD degrees in Geology from Stanford University. His current research focuses on experimental and theoretical study of mineral solubility in geologic fluids at high pressure and temperature, the role of subduction-zone fluids in geochemical cycles, Archean hydrothermal systems, and the geology and tectonics of central Asia.

Reprinted from
Earth and Planetary Science Letters 222 (2004) 697–712

www.elsevier.com/locate/epsl

Solar and solar-wind isotopic compositions

Roger C. Wiens [a,*], Peter Bochsler [b],
Donald S. Burnett [c], Robert F. Wimmer-Schweingruber [d]

[a] Space and Atmospheric Sciences, Los Alamos National Laboratory, Mail Stop D-466,
Los Alamos, NM 87544, USA
[b] Physikalisches Institut, University of Bern, Sidlerstrasse 5, CH-3012 Bern, Switzerland
[c] Department of Geology, Mail Stop 100-23, California Institute of Technology, Pasadena, CA 91125, USA
[d] Institut fuer Experimentelle und Angewandte Physik, University of Kiel, Leibnizstrasse 11, D-24118 Kiel, Germany

Received 26 November 2003; received in revised form 5 March 2004; accepted 21 March 2004

Abstract

With only a few exceptions, the solar photosphere is thought to have retained the mean isotopic composition of the original solar nebula, so that, with some corrections, the photosphere provides a baseline for comparison of all other planetary materials. There are two sources of information on the photospheric isotopic composition: optical observations, which have succeeded in determining a few isotopic ratios with large uncertainties, and the solar wind, measured either in situ by spacecraft instruments or as implanted ions into lunar or asteroidal soils or collection substrates. Gravitational settling from the outer convective zone (OCZ) into the radiative core is viewed as the only solar modification of solar-nebula isotopic compositions to affect all elements. Evidence for gravitational settling is indirect, as observations are presently less precise than the predictions of $<10‰$ effects for the isotopes of solid-forming elements. Additional solar modification has occurred for light isotopes (D, Li, Be, B) due to nuclear destruction at the base of the convection zone, and due to production by nuclear reactions of photospheric materials with high-energy particles from the corona. Isotopic fractionation of long-term average samples of solar wind has been suggested by theory. There is some evidence, though not unambiguous, indicating that interstream (slow) wind is isotopically lighter than high-speed wind from coronal holes, consistent with Coulomb drag theories. The question of fractionation has not been clearly answered because the precision of spacecraft instruments is not sufficient to clearly demonstrate the predicted fractionations, which are $<30‰$ per amu between fast and slow wind for most elements. Analysis of solar-wind noble gases extracted from lunar and asteroidal soils, when compared with the terrestrial atmospheric composition, also suggests solar-wind fractionation consistent with Coulomb drag theories. Observations of solar and solar-wind compositions are reviewed for nearly all elements from hydrogen to iron, as well as the heavy noble gases. Other than Li and the noble gases, there is presently no evidence for differences among stable isotopes between terrestrial and solar photosphere compositions. Although spacecraft observations of solar-wind isotopes have added significantly to our knowledge within the past decade, more substantial breakthroughs are likely to be seen within the next several years with the return of long-exposure solar-wind samples from the Genesis mission, which should yield much higher precision measurements than in situ spacecraft instruments.
© 2004 Elsevier B.V. All rights reserved.

Keywords: solar wind; solar abundances; solar nebula

---

* Corresponding author. Tel.: +1-505-667-3101; fax: +1-505-665-7395.
E-mail address: rwiens@lanl.gov (R.C. Wiens).

0012-821X/$ - see front matter © 2004 Elsevier B.V. All rights reserved.
doi:10.1016/j.epsl.2004.03.025

Terminology

CME	Coronal mass ejection: episodic explosive release of coronal material frequently carried within closed magnetic field loops.
Coulomb drag	An aspect of solar-wind acceleration theory referring to the acceleration of heavier species by ionized hydrogen that is escaping from the photosphere.
FIP fractionation	Elemental abundances in the solar wind are fractionated (the ratios of their abundances are modified) during ionization and acceleration according to the first ionization potential (FIP) of each element. Low FIP elements are preferentially accelerated.
High-speed streams	Fast solar wind characterized by He/H ratios of ~0.045 and by relatively low elemental fractionation, emanating from cooler, UV-dark regions of the corona, as evidenced by low charge state ions in the stream. High-speed streams are usually 600–800 km/s at 1 AU.
Interstream wind	Low-speed wind characterized by relatively high elemental fractionation, emanating from relatively hot regions of the corona. Interstream wind is always <600 km/s and often <400 km/s.
L1 point	The metastable Lagrangian point 1.5 million km sunward of the Earth.
LSCRE	*Lunar Surface Cosmic Ray Experiment:* Cosmic ray exposure experiment developed by Washington University. On Apollo 17, it included a platinum solar wind exposure foil.
OCZ	Outer convection zone: the outer third of the Sun by radius, where heat is transferred to the surface by convection. The composition of the OCZ is often referred to as "solar" composition.
Permil (‰)	Deviations in parts per thousand relative to a given standard. Example: $\delta^{15}N = -50‰$ means a 5% depletion of ^{15}N relative to $^{15}N/^{14}N$ in the standard.
Regolith	Layer of loose incoherent material at the surface of a planetary body. "Soil".
SEP component	A noble-gas and nitrogen component of lunar and meteoritic regolith grains that is more deeply buried than the solar-wind component.
Solar gravitational settling	Differential effect of gravity on elements and isotopes of different mass. It results in a depletion of heavy species in the OCZ.
Solar nebula	Disk of gas and dust from which the solar system was formed.
SWC	*Solar Wind Composition* experiments, developed by the University of Bern, consisted of specialized foils exposed to the solar wind for periods of hours to ~2 days during the Apollo lunar missions.
Wave–particle interactions	Interactions between electromagnetic waves in the solar-wind plasma and its particles which can modify the velocity distributions of the particles.

Spacecraft

ACE	*Advanced Composition Explorer,* launched August 1997 to an L1 halo orbit and currently operating.
Genesis	NASA mission returns samples in September 2004 of solar wind collected on high-purity substrates over a period of 2 years at L1.
ISEE-3	*International Sun-Earth Explorer 3,* launched in 1978, was the first space probe to use the L1 point for solar observations.
SOHO	*SOlar and Heliospheric Observer,* launched in December 1995 to an L1 halo orbit and currently operating.

Spacecraft

Ulysses	Joint NASA and European Space Agency mission launched in 1990 to study the polar regions of the Sun and the 3-D structure of the heliosphere.
Wind	NASA mission launched in 1994 to study plasma mass, momentum, and energy properties in the upstream near-Earth regions as part of the Global Geospace Science program.

1. Introduction

The isotopic composition of the sun's outer convection zone (OCZ) is of very great interest to planetary science because it is relatively unchanged in composition from the solar nebula, the starting material of the solar system. The solar nebula, and by extension, the OCZ, can be used as a baseline for comparison of present-day planetary, asteroidal, meteoritic, atmospheric, and cometary isotopic compositions. Studying compositional differences relative to the original solar nebula is one of the few ways we have of understanding the evolution of our solar system. Unfortunately, very little information on the OCZ isotopic composition is directly available from astronomical data (e.g., [1–6]). The alternative to direct solar observations is the measurement of solar-wind isotopic compositions, the main subject of this paper.

In order to use the solar-wind compositions to infer nebular isotopic compositions, important considerations must be made in two major areas. One is modification of the OCZ relative to the original solar nebula, and the other is modification of solar-wind compositions relative to the OCZ. In the first area, major modifications to the OCZ relative to its precursor nebular composition are (a) nuclear destruction and production of some light isotopes, and (b) some amount of gravitational settling from the OCZ to the radiative zone below. Both of these issues have been indirectly substantiated, but experimental confirmation of the extent of these processes via solar-wind measurements is interesting in its own right. Gravitational settling has been confirmed by the improvement of the model seismic velocities with observed helium abundances. These models predict fractionation on the order of 33‰ for ^{3}He/^{4}He, 5‰ for δ^{13}C, 4‰ for δ^{15}N, 3‰ per amu for the oxygen isotopes, dropping gradually with increasing mass number to ~ 1‰ per amu for iron [7,8]. Observational constraints indicate that nuclear burning of light isotopes during the fully convective stage of solar evolution resulted in complete loss of deuterium, a factor of ~ 140 depletion in lithium and a beryllium loss of less than a factor of 2 (e.g., [9]).

In the second area of consideration, modification of solar-wind compositions relative to the OCZ, solar-wind compositions are known to be variable over short time scales. Long-term samples of solar wind may have substantial isotopic fractionation relative to the OCZ. Significant advances have recently been made in understanding this issue, as will be discussed in the section describing solar-wind characteristics.

If a true solar nebula isotopic composition can be inferred from the solar wind or by any other means, the topics of interest are numerous, and span a number of isotopic systems. The isotopic precision desired to answer these questions spans a range, depending on the topic, but is generally from ± 10‰ to ± 0.1‰. The Genesis mission, which in September 2004 returns 2-year samples of collected solar wind, promises to enable high precision solar-wind isotopic measurements for the first time. This paper reviews the state of knowledge just prior to this important sample return. The two critical areas reviewed are: what we know about the recently emerging issue of solar-wind isotopic fractionation, and what we can infer for the solar composition from the most recent solar-wind measurements. Application of the new high-precision Genesis measurements is discussed in the final section of the paper.

2. Measurements of isotopes in the sun and solar wind

Solar and solar-wind measurements come from four sources: one is direct solar photospheric observations and three sources involve measurement of the solar wind in various ways. The photospheric observations are of D/H, He, CO and Li absorption lines, which have yielded some information on their isotopic compositions [1–6]. The three solar-wind sources consist of: (i) measurements of solar wind collected in the Solar-Wind Composition (SWC) and Lunar Surface Cosmic Ray Experiment (LSCRE) foil experiments of the University of Bern and Washington University during the Apollo missions [10]. (ii) Solar-wind measurements made on samples of planetary regoliths—both lunar soils and mineral grains from gas-rich meteorites which were exposed to solar wind and compacted and ejected from their parent body regolith, and (iii) solar-wind measurements made since the late 1970s for helium first on ISEE-3 and then Ulysses, and since the mid-1990s for heavier isotopes using time-of-flight (TOF) sensors on WIND, ACE, and SOHO spacecraft. Because of the potential

Table 1
Solar and solar-wind isotope ratios

Isotope	Ratio	δ	Permil	Wind	Source	Reference	Comments
D/H	Variable, $< 10^{-4}$	δD	< -400	–	Optical	[1]	Highest for sunspots
D/H	$< 8 \times 10^{-6}$	δD	< -950	Mean	Lunar regolith	[41]	Lowest D/H measured
D/H	$< 3 \times 10^{-6}$	δD	< -980	Mean	Lunar regolith	[47]	Extrapolated
^{4}He/^{3}He	2500^{+2500}_{-830}			–	Optical	[4]	Photospheric observation
^{4}He/^{3}He	2350 ± 120			Mean	SWC	[10]	Average of SWC experiments
^{4}He/^{3}He	2190 ± 40			Mean	Lunar regolith	[57]	71501 ilmenite
^{4}He/^{3}He	2050 ± 200			Mean	ISEE-3/ICI	[21]	3.4 year average
^{4}He/^{3}He	1900 ± 200			Fast	ISEE-3/ICI	[21]	3.4 year average
^{4}He/^{3}He	2290 ± 200			Mean	Ulysses/SWICS	[20]	500 day average
^{4}He/^{3}He	2450^{+160}_{-140}			Slow	Ulysses/SWICS	[25]	1991–1996, re-analysis
^{4}He/^{3}He	3030^{+300}_{-250}			Fast	Ulysses/SWICS	[25]	1991–1996, re-analysis
^{4}He/^{3}He	2645^{+170}_{-150}			–		[26]	Extrapolated to photosphere
^{7}Li/^{6}Li	>33			–	Optical	[6]	Photospheric observation
^{7}Li/^{6}Li	31 ± 4			Mean	Lunar regolith	[48]	Corrected for spallation
^{11}B/^{10}B	$3-5$			Mean	Lunar regolith	[48]	
		δ^{13}C	-60 ± 100	–	Optical	[3]	Photospheric observation
^{12}C/^{13}C	84 ± 5	δ^{13}C	59^{+67}_{-60}	–	Optical	[5]	Photospheric observation
		δ^{13}C	-30 to $+30$	Mean	Lunar regolith	Various	Range of lunar compositions
		δ^{13}C	$< -105 \pm 20$	Mean	Lunar regolith	[37]	Solar wind as end-member
^{14}C/^{12}C	$< 4.5 \times 10^{-10}$			Mean	Lunar regolith	[49]	
^{14}N/^{15}N		δ^{15}N	-250 to $+190$	Mean	Lunar regolith	Various	Range of lunar compositions
^{14}N/^{15}N		δ^{15}N	< -240	Mean	Lunar regolith	[41]	Correlation with D/H
^{15}N/^{14}N	$3.8 \pm 1.8 \times 10^{-3}$	δ^{15}N	30 ± 500	Mean	SOHO/MTOF	[43]	Revised value
^{16}O/^{18}O	440 ± 50	δ^{18}O	130^{+145}_{-115}	–	Optical	[5]	Photospheric observation
^{16}O/^{18}O	450 ± 130	δ^{18}O	110^{+450}_{-250}	Slow	Wind/MASS	[33]	2.5 years of data
^{16}O/^{18}O	446 ± 90	δ^{18}O	120^{+280}_{-190}	Fast	ACE/SWIMS	[34]	2 years of data
^{20}Ne/^{22}Ne	13.7 ± 0.3	δ^{20}Ne	400 ± 30	Mean	SWC	[10]	
^{20}Ne/^{22}Ne	13.8 ± 0.1	δ^{20}Ne	410 ± 10	Mean	Lunar ilmenite	[57]	
^{20}Ne/^{22}Ne	13.85 ± 0.04	δ^{20}Ne	412 ± 4	Mean	Combined regolith	[56]	Lunar, Kapoeta, and Irons
^{20}Ne/^{22}Ne	13.8 ± 0.7	δ^{20}Ne	410 ± 70	Mean	SOHO/MTOF	[65]	
^{20}Ne/^{22}Ne	13.6 ± 0.7	δ^{20}Ne	390 ± 70	Slow	WIND/MASS	[66]	2 years of data
^{22}Ne/^{21}Ne	30 ± 4	δ^{21}Ne	150^{+175}_{-135}	Mean	SWC	[10]	
^{22}Ne/^{21}Ne	29.9 ± 0.3	δ^{21}Ne	152 ± 10		Combined regolith	[56]	Lunar, Kapoeta, and Irons
^{25}Mg/^{24}Mg	0.132 ± 0.013	δ^{25}Mg	46 ± 87	Slow	Wind/MASS	[27]	17 months of data
^{25}Mg/^{24}Mg	0.128 ± 0.011	δ^{25}Mg	9 ± 103	Fast	Wind/MASS	[27]	17 months of data
^{26}Mg/^{24}Mg	0.153 ± 0.013	δ^{26}Mg	98 ± 86	Slow	Wind/MASS	[27]	17 months of data
^{26}Mg/^{24}Mg	0.138 ± 0.012	δ^{26}Mg	-10 ± 93	Fast	Wind/MASS	[27]	17 months of data
^{25}Mg/^{24}Mg	0.1408 ± 0.011	δ^{25}Mg	112 ± 87	Slow	SOHO/MTOF	[28]	7 months of data
^{25}Mg/^{24}Mg	0.1352 ± 0.013	δ^{25}Mg	68 ± 103	Fast	SOHO/MTOF	[28]	7 months of data
^{26}Mg/^{24}Mg	0.1499 ± 0.012	δ^{26}Mg	75 ± 86	Slow	SOHO/MTOF	[28]	7 months of data
^{26}Mg/^{24}Mg	0.1416 ± 0.013	δ^{26}Mg	16 ± 93	Fast	SOHO/MTOF	[28]	7 months of data
^{24}Mg/^{26}Mg	7.7 ± 0.12	δ^{26}Mg	-68 ± 14	Slow	SOHO/MTOF	[67]	
^{24}Mg/^{26}Mg	7.1 ± 0.14	δ^{26}Mg	10 ± 20	Fast	SOHO/MTOF	[67]	
^{29}Si/^{28}Si	0.0499 ± 0.0017	δ^{29}Si	-17 ± 33	Slow	Wind/MASS	[66]	2 years of data
^{30}Si/^{28}Si	0.0339 ± 0.0019	δ^{30}Si	24 ± 58	Slow	Wind/MASS	[66]	2 years of data
^{29}Si/^{28}Si	0.0447	δ^{29}Si	-118	Mean	SOHO/MTOF	[29]	1.3 years of data
^{30}Si/^{28}Si	0.0304	δ^{30}Si	-82	Mean	SOHO/MTOF	[29]	1.3 years of data
^{34}S/^{32}S	0.043 ± 0.006	δ^{34}S	-29 ± 135	Fast	ACE/SWIMS	[68]	
^{36}Ar/^{38}Ar	5.3 ± 0.3	δ^{38}Ar$_{36}$	9 ± 60	Mean	SWC	[69]	
^{36}Ar/^{38}Ar	5.48 ± 0.05	δ^{38}Ar$_{36}$	-23.6 ± 8.9	Mean	Lunar regolith	[57]	Ilmenite grains
^{36}Ar/^{38}Ar	5.80 ± 0.06	δ^{38}Ar$_{36}$	-78 ± 10	Mean	Lunar regolith	[56]	Ilmenite grains
^{36}Ar/^{38}Ar	5.5 ± 0.6	δ^{38}Ar$_{36}$	-27 ± 110	Fast, mean, slow	SOHO/MTOF	[70,71]	3.5 years of data

Table 1 (continued)

Isotope	Ratio	δ	Permil	Wind	Source	Reference	Comments
^{40}Ca/^{42}Ca	128 ± 47	δ^{42}Ca	165^{+700}_{-300}	Mean	SOHO/MTOF	[72]	1 year of data
^{40}Ca/^{44}Ca	50 ± 8	δ^{44}Ca	-70^{+180}_{-130}	Mean	SOHO/MTOF	[72]	1 year of data
		δ^{53}Cr	+0.5	Mean	Lunar plagioclase	[51]	Individual isotope anomalies independent of mass fractionation
		δ^{54}Cr	+1		Lunar plagioclase	[51]	Same as above
^{54}Fe/^{56}Fe	0.068 ± 0.004	δ^{54}Fe	68 ± 63	Mean	SOHO/MTOF	[73]	2 years of data, 380–400 km/s
^{57}Fe/^{56}Fe	0.025 ± 0.005	δ^{57}Fe	80 ± 215	Mean	SOHO/MTOF	[73]	2 years of data, 380–400 km/s

Nearly all data are 1-σ uncertainties; [10,34,56,69] are clearly 2-σ.

complications involved in interpreting solar energetic particle data, they are not included in our discussion.

The solar wind foil experiments yielded isotopic information on solar wind ^{3}He/^{4}He, ^{20,21,22}Ne, and ^{36}Ar/^{38}Ar obtained over brief periods between 1.25 and 45 h, from Apollo 11 to Apollo 17. The Apollo 17 measurement is reported here for the first time. These measurements are still more precise than can be obtained by spacecraft instruments. However, each foil exposure was for only a short duration, making these measurements subject to short-term variations in solar-wind composition, as discussed below. Historically, these measurements were a watershed in that they established that terrestrial atmospheric neon is distinctly different from solar, laying the foundation for studying the role of hydrodynamic escape in the atmospheric evolution of the terrestrial planets. Measurements of solar wind in planetary regoliths suffer from parent-body effects which must be removed to determine the true solar-wind composition. This deconvolution has become increasingly more effective, so that for noble gases the regolith analyses now provide the most accurate measurements of solar-wind composition. Another advantage of regolith measurements is their uniqueness in allowing studies of the constancy of solar-wind compositions through time. A disadvantage is their limitation mostly to measurements of noble gases.

In situ measurements by the MASS sensor on the SMS instrument on the Wind spacecraft, the MTOF sensor on the CELIAS instrument on SOHO, and the SWIMS instrument on ACE have been made on a wide variety of elements up to and including iron [11–13]. Using these instruments, measurements where the dynamic range between two isotopes is less than 100 yield results with uncertainties generally at or somewhat under 100‰. Isotope ratios with dynamic ranges between 100 and 600 (e.g. δ^{15}N, δ^{18}O) yield results with larger uncertainties, while isotope ratios with dynamic ranges >600 (e.g., δ^{17}O) have not been reported. Low abundances rule out the study of Li, Be, B, as well as elements heavier than about mass 60, e.g., heavier than iron group elements.

Table 1 is a compilation of solar and solar-wind isotopic compositions, given with 1-σ uncertainties except where noted. This table will serve as the basis for discussions below.

3. Solar-wind characteristics

To infer solar compositions from solar-wind measurements, it is crucial to understand many of the characteristics of the solar wind that result in the potential for compositional differences between it and the OCZ.

The solar wind is the continuous flow of ionized plasma streaming outward from the sun along magnetic field lines. Solar wind elemental compositions are known to be fractionated relative to the photosphere. The primary fractionation occurs in the ionization process; with the result that fractionation is a function of the elements' first ionization potential (FIP). Elemental fractionation is weakest in high-speed streams, with a factor of one to two enrichment of low FIP over high FIP elements relative to the photosphere. In low-speed (interstream) wind, elemental fractionation is greater and more variable, at factors of 2.5–5 [14,15]. A transient state of the solar wind, coronal mass ejections (CMEs), exhibits highly variable composition that, if not accounted for, makes the determination of solar composition difficult because it can contaminate measurements in the other solar-wind streams. We will therefore not consider the composition of CMEs in this work.

The FIP by itself does not affect isotopes, but there may be other, underlying mechanisms that result in isotopic fractionation. For example, the differential velocities of various species in the corona [16] have lent support to solar-wind acceleration models based on the opposing forces of gravitation and Coulomb drag, which predict isotopic fractionation. Wave–particle interactions may also contribute to fractionation effects, though if the effects are transient, long-term solar-wind samples may avoid these effects. In the Coulomb drag concept [17], the solar wind acceleration process proceeds through multiple collisions with electromagnetically accelerated protons. Inefficiencies in this process cause heavier species to be left behind. In these models, the fractionation relative to H of a species of mass number A and atomic charge q scales as $(2A - q - 1)*[(1 + A)/A]^{1/2}*q^{-2}$. These models predict significant isotopic fractionation of up to 300‰ for $^3He/^4He$ and $-50‰$ for $\delta^{18}O$, with similar fractionation/amu for other heavy ions as mentioned for oxygen. Fractionation in high-speed wind is predicted to be significantly less, at $\sim 85‰$ for $^3He/^4He$ and $-17‰$ for $\delta^{18}O$ [7,18,19]. However, as will be shown below, observational evidence for differences of this magnitude between high-speed and interstream material is still somewhat ambiguous.

3.1. Solar-wind variability

Temporal variations in solar-wind fractionation result in compositional variability. Solar wind helium isotopes measured on 1-h timescales exhibit order-of-magnitude variations, but long-term averages tend to converge. For example, the average of hourly sampling of $^4He/^3He$ from the ISEE-3 Ion Composition Instrument (ICI) over a 3.4-year period from 1978 to 1981 (2050 ± 200) is within uncertainty of the average of the Apollo SWC results (2350 ± 120) which represent about 5 days total exposure near solar maximum. Daily averages from Ulysses over 500 days in 1992–1993 yielded similar results (2290 ± 200) [20]. Here the $\pm$ represents variability of the data, independent of systematic error estimates. Thus, despite short-term variability, there appears to be a well-defined long-term average value for $^4He/^3He$, and by extension, the same

is assumed for the isotopes of heavier elements. In fact, heavier elements show significantly less isotopic variability than $^4He/^3He$ even in short-term measurements. The Apollo solar-wind foil data show >25% variation in helium isotopes, but only $\sim 5\%$ variation in $^{20}Ne/^{22}Ne$ among the six different samples [10].

3.2. Constraints on solar-wind isotopic fractionation

Are time-averaged measurements, for which short-term variability is averaged out, isotopically fractionated relative to the photosphere? The existence and magnitude of solar-wind isotopic fractionation must be determined by either (a) correlations between isotopes in different parcels of solar wind (e.g., isotope correlations for different collection periods), or by (b) comparison of isotopes between different types of solar wind which would be expected to be fractionated differently. Both techniques are discussed here. The Apollo foil experiments give relatively accurate measurements of different parcels of solar wind collected over 3 years, but specific details of the solar wind dynamics that would help determine the type of solar wind are lacking. On the other hand, in situ spacecraft have allowed comparisons of long-term average compositions for high-speed versus interstream wind.

The Apollo foil experiments yielded unsurpassed information on the light noble gas isotopes in terms of a measurement of pure solar wind—not extracted from a planetary regolith—with uncertainties of as little as $\pm 30‰$ for an individual measurement. The Apollo foils collected solar wind predominantly during periods of low-speed solar wind. Fig. 1 shows $^4He/^3He$ vs. $^{20}Ne/^{22}Ne$ data for collection periods made during each of the lunar landings, including the Apollo 17 results first published here ($^4He/^3He = 2000 \pm 200$; $^{20}Ne/^{22}Ne = 13.55 \pm 0.80$, 2-*sigma*; Bühler, personal communication). Also shown in Fig. 1, the fractionation predicted by the Coulomb drag model matches the slope of the best-fit linear correlation to the Apollo foil data, though the uncertainty for the linear correlation is quite large, allowing a complete absence of fractionation or fractionation in the other direction. Based on Fig. 1, a current best estimate of the photospheric $^{20}Ne/^{22}Ne$ ratio using both data and theory is 13.35 ± 0.80.

Using spacecraft data, He as well as heavier elements, particularly Ne, Mg, and Si, have recently been

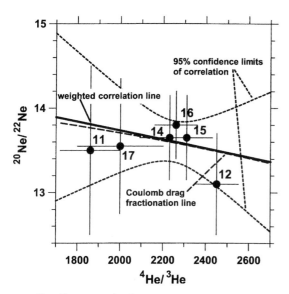

Fig. 1. $^{20}Ne/^{22}Ne$ versus $^{4}He/^{3}He$ from the SWC experiments [10] with the added LSCRE data point (17) heretofore unpublished. Data points are labeled according to the respective Apollo missions during which the foils were exposed to solar wind. A weighted correlation line is shown (solid line) with 95% confidence limits. The trend predicted by Coulomb drag fractionation theory is indicated by the dashed line.

studied for indications of isotopic mass fractionation between fast and slow wind. Fig. 2 shows various reported $^{4}He/^{3}He$ ratios, some at different wind speeds. The ISEE-3 $^{4}He/^{3}He$ ratio for v>450 km/s (1900±200) was within errors the same as the average for all speeds [21]. Ulysses has alternately sampled interstream wind and high-speed streams. Reports over the first several years of the mission (e.g., [18,20,22] suggested no

systematic variation with solar-wind speed. However, more recent analyses [23–26] have reached the opposite conclusion. The difference appears to be the result of a change in the relative weighting placed on data from different collectors [23]. With this new interpretation, Geiss and Gloeckler [23,24,26] recently used correlations of $^{4}He/^{3}He$ with the elemental ratios of He/H and Si/O in two high-speed periods and two interstream periods, chosen based on ion charge state, to derive photospheric $^{4}He/^{3}He$ ratios of 2620 and 2670, respectively, about 15% higher than the long-term average ratios measured in the solar wind (Table 1). The trend in the isotopic data of Geiss and Gloeckler confirms the expectations from the Coulomb drag models.

Data from heavier elements measured by spacecraft instruments are more ambiguous. Fig. 3 shows a three-isotope plot of Mg for a 17-month period from December 1994 to May 1996 as measured by the MASS instrument on WIND [27], and for a 7-month period as measured by MTOF on SOHO during 1996 [28]. Aside from He, Mg is the easiest test for fractionation with spacecraft instruments, as it has three relatively abundant isotopes. The data were separated into two bins by velocity, with the dividing line at 400 km/s. The high-speed stream is within uncertainty of the terrestrial ratio. However, the nominal value for the low-speed wind is fractionated in favor of the heavy isotope, opposite of that suggested by the Coulomb drag models [7,18,19], although the error bars easily permit interpretation in either direction. Other data from in situ solar wind measurements have suggested that the heavy isotopes are indeed depleted in low-speed wind

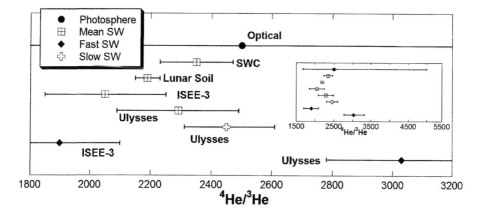

Fig. 2. $^{4}He/^{3}He$ ratios reported for the photosphere and the solar wind ("SW"), from the data in Table 1.

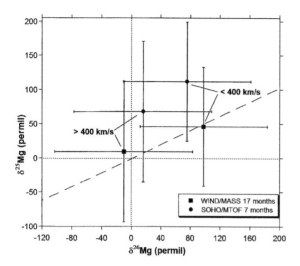

Fig. 3. δ^{25}Mg vs. δ^{26}Mg for in situ solar-wind observations made by the MASS instrument on the WIND spacecraft [27] and by SOHO/ MTOF [28]. The data are divided into two velocity bins, >400 and <400 km/s. The dashed line shows a mass-dependent-fractionation trend passing through the terrestrial composition for reference.

for silicon (e.g., [29]), but in essentially all cases the uncertainties leave room for both interpretations.

In summary, the data from the Apollo foils, to which we attribute the highest reliability, are consistent with solar-wind isotopic fractionation (Fig. 2), as are the recently published helium isotopic results. Still, in comparing all data in Table 1, due to the large uncertainties and lack of agreement between different observations, one cannot conclude that a fractionation trend has clearly been discovered, nor can we conclude the absence of a trend. A top priority of the Genesis mission will be to determine the presence or absence of solar-wind isotopic fractionation.

4. Discussion: solar and solar-wind isotopic abundances

An accurate database of the OCZ isotopic composition is useful for understanding the initial inputs to the solar nebula, providing a baseline for the planetary, asteroidal, and cometary compositions. In addition, it is important for understanding the current and ancient solar surface environment, for which issues are important such as gravitational settling out of the convective zone, deep solar mixing, and the solar radiation environment controlling production of rare isotopes at the

solar surface. Here, we review recent results for both solar and solar-wind data sets. In many cases, the precision of the solar wind measurements is not so great that the unsolved question of solar-wind isotopic fractionation causes a problem. The exceptions are the noble gases, for which He and Ne have already been discussed in the context of isotopic fractionation.

4.1. D/H and ^{4}He/^{3}He

These ratios are key parameters for pre-solar conditions, including galactic evolution and parameters for big bang nucleosynthesis. The primordial deuterium is thought to have been destroyed at the base of the convection zone, where the temperature is sufficient to cause nuclear burning. A discussion of flare-produced D is given in a later section. The early solar destruction of D produced ^{3}He directly, significantly modifying the solar ^{4}He/^{3}He. As expected, the Galileo probe ^{4}He/^{3}He ratio for Jupiter is much higher than the solar wind ratios, with a value of 6000 ± 200 [30]. This ratio is the best available estimate of the nebular ^{4}He/^{3}He, although if liquid helium rainout occurred in the interior of Jupiter, it could have produced significant ^{3}He enrichment in the atmosphere. Several factors may have worked to modify the photospheric ^{4}He/^{3}He ratio relative to primordial ^{4}He/(^{3}He$_i$ + D), where the denominator is the sum of initial ^{3}He plus deuterium. These factors include gravitational settling and the addition of ^{3}He produced via the p−p reaction at greater depths. A recent investigation [31] of He and Ne in lunar grains suggests the absence of a temporal trend, leading to a conclusion that primordial ^{4}He/(^{3}He$_i$ + D) was the same as the present OCZ ^{4}He/^{3}He. Recent estimates put the (D/H)$_i$ ratio at 19.7 ± 3.6 ppm based on solar data, and 21 ± 4 based on Jovian data [26].

4.2. Oxygen and carbon

The oxygen isotopic ratio of the OCZ has been suggested to hold important clues to the formation of the solar nebula (e.g., [32] and references therein). The solar-wind δ^{18}O values reported so far from in situ measurements are 110^{+450}_{-250}‰ for slow wind [33] and 120^{+280}_{-190}‰, 2-σ, for fast wind [34], relative to standard mean ocean water (SMOW). At present, the uncertainties of these measurements do not constrain them to be different from terrestrial. The δ^{18}O value reported for direct solar photospheric molecular measurement over sunspots, at $130^{+145}_{-115}\%o$ [5], is nearly identical to the nominal solar wind values, but with the smaller uncertainties it is slightly heavy relative to SMOW at the 1-σ level. Fig. 4 shows solar-wind $\delta^{18}O$ values in comparison with other solar-system materials. Meteoritic materials are typically displayed on a three-isotope plot, including the $\delta^{17}O$. However, there are currently no measurements of solar $\delta^{17}O$. For $\delta^{18}O$, bulk meteoritic values lie between $-5\%o$ and $+25\%o$, while oxygen in calcium–aluminum inclusions in chondritic meteorites has been measured to be as depleted in ^{17}O and ^{18}O as $-60\%o$. The single measurement of cometary water ice [35,36] is also shown.

The carbon isotopic ratio was also reported from solar CO absorption lines. The two values vary widely, at $\delta^{13}C = -60 \pm 100\%o$ [3] and $+59^{+67}_{-60}\%o$ [5] relative to the terrestrial standard, but both essentially overlap with the bulk terrestrial carbon. No solar-wind carbon isotopic measurements have been reported from spacecraft, as distinction from ions released from the carbon foils used in these instruments is difficult.

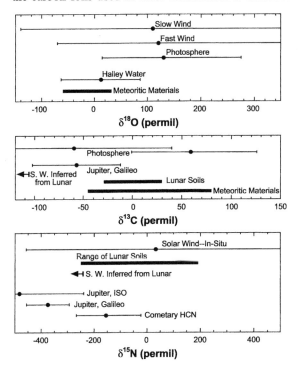

Fig. 4. Isotopic compositions measured to date for the sun, solar wind, and other relevant solar-system reservoirs. See text for explanations and references.

Surface-correlated carbon in bulk lunar soils falls in the range of $\pm 30\%o$. Carbon abundances are higher than expected relative to solar-wind noble-gas abundances in lunar soils, suggesting that the bulk of the surface-correlated C is not solar wind. A recent ion probe study found $\delta^{13}C$ values as light as $-105\%o$, leading the authors to suggest that solar-wind carbon is at least this depleted in ^{13}C [37]. For comparison, Fig. 4 shows the Jovian atmospheric carbon value of $-57 \pm 44\%o$ [38]. Cometary CN observations [39] give even lighter values (-460^{+170}_{-105}, -225^{+160}_{-115}), below the scale of Fig. 4. The context of these recent isotopically light measurements is unclear at this time.

4.3. Nitrogen

The solar-wind nitrogen isotopic ratio has long been of interest as a baseline for planetary atmospheric compositions. The lunar soils have trapped and retained nitrogen regarded initially as being implanted solar wind. However, analyses of bulk lunar soils have yielded a range of approximately 400‰ in isotopic compositions, while the abundance of N is typically enriched by up to an order of magnitude relative to noble gases when normalized to solar abundances. One interpretation was that (a) nitrogen is more efficiently retained in lunar soils and noble gases are partially lost, and along with this, (b) a secular change in the solar-wind nitrogen isotopic composition was postulated. However, the higher nitrogen abundance and the fact that it varies from grain to grain in a given sample [40] strongly suggested domination by a larger, non-solar-wind component. Hashizume et al. [41] recently reported a correlation between $\delta^{15}N$ and D/H in ion probe analyses, which they used to distinguish solar contributions (e.g., trending towards an absence of deuterium) from planetary contributions, e.g., from micrometeorite bombardment. Their results showed that isotopically light nitrogen correlated with low deuterium content. As solar wind is considered to be deuterium-free, this suggests that solar-wind nitrogen is depleted in ^{15}N by at least 240‰ relative to the terrestrial atmosphere.

Nitrogen is at the limits of what can be reasonably measured by in situ spacecraft instruments due to the large difference in abundances between ^{15}N

and ^{14}N. An early report [42] of isotopically heavy solar-wind nitrogen has recently been revised to $30 \pm 500\%o$ [43], consistent with all other N measurements. Besides the photosphere, another potential reservoir of primordial nitrogen gas is the atmosphere of Jupiter. Fig. 4 shows two Jovian measurements, one made telescopically [44] and one from the Galileo probe mass spectrometer [45,46]. Both measurements give an isotopically light composition relative to terrestrial, suggesting that the solar ratio is more like that measured in lunar soils than the nominal value of the relatively uncertain in situ solar-wind measurement. Cometary observations indicate that nitrogen in CN is isotopically heavy, off the plot in Fig. 4, but HCN is isotopically light, as shown in Fig. 4 ([39] and references therein). Measurement of the solar-wind nitrogen by Genesis is eagerly anticipated, as it could confirm or reject the agreement of the solar composition with most of these other solar-system nitrogen reservoirs, and decide whether the Earth's atmospheric composition is enriched in ^{15}N relative to the solar nebula.

4.4. Solar surface isotope production

D, ^{6}Li, ^{10}Be, and ^{14}C. Abundances of several rare isotopes, including radioactive isotopes, have been inferred from observations of the solar surface and measurements in the solar wind. These species can potentially constrain the extents and timescales of mixing in the upper portion of the solar convective zone. At present, these include D, Li, ^{10}Be, and ^{14}C. Although deuterium is destroyed at the base of the OCZ, it is produced by radiative capture of neutrons which are a by-product of energetic particle events, and in the most energetic events D can also be produced directly by proton–proton collisions. Constraints on the D/H ratio by lunar samples limit the minimum solar-wind D/H ratio to less than ~ 7 ppm, or around $-950\%o$ [41] based on individual measurements, and to <3 ppm ($-980\%o$) based on correlations [47]. Attempts to spectroscopically detect deuterium in the solar atmosphere have had mixed success, as reviewed recently in [2]. Early observations over sunspots [1] produced an estimate of D/H ~ 100 ppm, while other observations over the quiet solar surface failed to detect D. Mullan and Linsky [2] noted that solar D has been observed in solar ener-

getic particles, but so far not directly in the solar wind.

Lithium is depleted by a factor of ~ 140 in the photosphere relative to C1 meteoritic abundances. This depletion was postulated to be due to nuclear destruction of lithium at the base of the OCZ. In the absence of any other inputs, this process would be expected to leave the remaining lithium strongly enriched in ^{7}Li relative to the original $^{7}Li/^{6}Li$ ratio of 12, the chondritic and terrestrial value. Optical observations of lithium absorption lines have been made in regions above sunspots, where lower ionization facilitates their observation. Reported values are consistent with the enhancement of this ratio, constraining it to be >33 [6]. The observation appeared to measure some ^{6}Li, with a most probable $^{7}Li/^{6}Li$ ratio of ~ 50, well below that expected if the factor of 140 depletion in elemental abundance were due solely to nuclear destruction without any additional inputs [6]. An attempt was recently made to measure the $^{7}Li/^{6}Li$ ratio of the solar wind concentrated in the outer layers of lunar grains. Chaussidon and Robert [48] reported an extrapolated ratio of 31 ± 4 after correcting for contamination and spallation reactions induced by energetic particles in the lunar soil. They explained this ratio as the result of a combination of nuclear destruction at the base of the OCZ with addition of lithium with a $^{7}Li/^{6}Li$ ratio of ~ 2 produced by spallation reactions involving solar energetic particles and carbon and oxygen in the solar atmosphere. The reported $^{7}Li/^{6}Li$ ratio suggests that 6–19% of solar lithium is produced in this way.

This spallation production of lithium appears consistent with estimates of solar-wind ^{14}C and ^{10}Be abundances, also based on lunar measurements. Using a leaching technique, the solar-wind-derived $^{14}C/H$ mixing ratio was estimated to be $2.2-3.5 \times 10^{-14}$, assuming a constant ratio over the last several thousand years [49]. In a similar experiment, ^{10}Be was recovered corresponding to a $^{10}Be/H$ mixing ratio of 1×10^{-14} [50]. Based on these radioactive isotope abundances, relative to the inferred production rates near the solar surface, essentially all of these species must be ejected in the solar wind and energetic-particle events rather than being mixed into the bulk of the OCZ. The mean residence time for spallation-produced isotopes near the surface of the sun appears, from these studies, to be at most between several hundred years and several thousand years.

4.5. Solid-forming elements from magnesium to iron

These make up the bulk of the in situ solar-wind measurements in Table 1. The results are basically all consistent with terrestrial compositions, within the roughly 60‰ to 100‰ uncertainties. Attempts to measure solar-wind compositions of these elements in lunar regolith samples have been nearly all unsuccessful because the natural abundances of these elements tend to be high in lunar rocks and soils. Minerals must be used which exclude the element of interest during formation, such as using ilmenite for solar-wind Si or plagioclase for Cr. Kitts et al. [51] recently analyzed Cr from Apollo 16 plagioclase grains in search of isotope-specific anomalies (their study was not sensitive to mass-dependent fractionation departures from isotopically normal Cr). Using successive etch steps, they found a surface-correlated chromium abundance enrichment corresponding to approximately 70 times that expected for solar wind, relative to argon abundances. Isotopically, a ^{54}Cr enrichment of approximately 1‰ (10 epsilon units), and an enrichment about half that at ^{53}Cr was reported, relative to planetary ^{50}Cr and ^{52}Cr. The significance and source of these isotopic anomalies, whether solar or otherwise, are unclear.

4.6. Noble gases

The solar noble gas composition is important in understanding the source and composition of the initial volatile inputs to the planets, and in understanding the history of degassing and atmospheric evolution of the planets. Spacecraft measurements made on helium, neon, and argon isotopes (Table 1 and previous discussions) are in agreement with the higher precision measurements made on the Apollo foils and on regolith samples. Krypton and xenon abundances were too low for measurement in the Apollo foils, along with ^{40}Ar/^{36}Ar, expected to be $<5\times10^{-4}$ in the sun. Planetary regoliths have yielded the most accurate compositions for neon and argon, and for krypton and xenon they are the only source of information on the solar composition.

Noble gases in planetary regoliths are characterized by two distinct surface-correlated isotopic components. The two are implanted at different depths, one corresponding to solar-wind energies and compositions, and the other corresponding to 10–100 keV/amu and characterized by enrichment of the heavy isotopes. The deeper component has been labeled the Solar Energetic Particle (SEP) component, based on the fact that it would take more energetic particles to be implanted at the depths at which this component is found. However, the SEP abundances are orders of magnitude higher than that expected from any present-day flux of high-energy solar particles. The SEP isotopic fractionation relative to solar wind is proportional to the square of the mass, so for example $(^{20}Ne/^{22}Ne)_{SEP}=(^{20}Ne/^{22}Ne)_{SW}*(20/22)^2=(13.7)*(20/22)^2 \cong 11.3$. The abundance ratio of SEP to solar wind is nearly constant for all noble gases, in the range of 20–30% (e.g., [52]). The SEP component is found equally in the lunar regolith and in mineral grains of solar gas-rich meteorites, so the phenomenon is not specific to one particular body. It also appears in regolith materials exposed to the sun at different points in time, so it cannot be from a particular epoch in solar history. However, SEP compositions were not measured in the Apollo SWC foils (Fig. 1), and recent interplanetary dust particle measurements [53] show a substantially reduced SEP component. As the isotopic composition can be matched simply by implanting solar-wind composition isotopes with greater energies, the suprathermal energy tail of the solar wind has been suggested as a possible source, if its intensity were substantially higher in the past [52,54]. A second recent suggestion is that the SEP component consists of interstellar pick-up ions which are ionized and accelerated in the heliosphere and subsequently implanted in regolith grains [55]. The suprathermal explanation would require some mechanism to cause much higher suprathermal fluxes in the past. The pick-up ion suggestion has an explanation for higher fluxes in that the solar system has passed through regions of significantly higher neutral particle densities in the past. However, this suggestion would constrain interstellar noble gas compositions, after fractionation by the ionization process within the heliosphere, to be isotopically and elementally very similar to solar.

Table 2

Solar-wind krypton isotope ratio and δ-values (‰) relative to terrestrial air [61]

^{78}Kr	^{80}Kr	^{82}Kr	^{83}Kr	^{84}Kr	^{86}Kr
0.6365 (34)	4.088 (14)	20.482 (54)	20.291 (26)	≡ 100	30.24 (10)
+46 (6)	+32 (3)	+13 (3)	+8 (2)	≡ 0	−9 (3)

Values are normalized to ^{84}Kr. Two-sigma in the last two digits are given in parentheses.

The fact that there are two isotopically distinct components in regolith materials makes it more difficult to determine the pure solar wind composition. Some uncertainty remains on the composition of the end-members. Solar-wind neon from regolith samples agrees with the Apollo foil ^{20}Ne/^{22}Ne values, and has much higher precision (Table 1). However for argon, there is disagreement over the solar-wind end-member composition. Palma et al. [56] report a value of ^{36}Ar/^{38}Ar = 5.80 ± 0.06, while Benkert et al. [57] and Becker et al. [58] obtained ratios of ^{36}Ar/^{38}Ar = 5.48 ± 0.05 and 5.58 ± 0.03, respectively. Which ratio is correct has significant implications for planetary atmospheric evolution models. An important point is that, for the lower range of regolith values, application of solar-wind isotopic fractionation factors based on the Coulomb drag model (e.g., [7]) can yield a photospheric composition identical to the terrestrial atmosphere (5.35) [59].

Data for solar-wind krypton and xenon isotopes are given in Tables 2 and 3. For krypton, the data in Table 2 are an average of 13 samples, mostly lunar ilmenite grains. Krypton isotopes show a straight fractionation trend relative to the terrestrial atmosphere of ~ 8‰/amu. Plausible Coulomb drag fractionation factors can reduce the difference between photosphere and terrestrial atmosphere to between 1‰/amu and 2‰/amu, but does not seem to remove the difference entirely. Stronger fractionation in the past, during the bulk of the lunar implantation, could possibly account for solar-wind-terrestrial differences for both Kr and Ar. The solar-wind xenon isotopic pattern is more complicated, and solar-wind fractionation cannot account for the bulk of the difference between terrestrial and solar wind Xe.

The xenon data in Table 3 are based on the initial two etch steps of a lunar ilmenite separate [60], with some modification, e.g., using data from additional samples to reduce the uncertainty for ^{124}Xe [61]. Solar-wind Xe has been interpreted as a combination of a primordial component called U-Xe, a proposed building block for the Xe found in most meteorites, the Sun, and the terrestrial atmosphere [62,63], and a heavy-isotope component contributing substantially to ^{134,136}Xe, and possibly also to ^{131}Xe. This heavy-isotope component could either be H*-Xe or the heavy-isotope portion of Xe-HL*. H*-Xe is a component consisting solely of ^{134,136}Xe [61]. Xe-HL* is found in presolar diamonds and is characterized by strong enrichments in both heavy and light isotopes. The problem with invoking addition of the heavy-isotope portion of Xe-HL* is that the heavy and light components have never been found separate from each other. At any rate, the inferred presence of U-Xe combined with variable amounts of heavy-isotope Xe in the Sun, meteorites, and the atmosphere of the Earth suggests a possible change between the time of accretion of material to the Sun and the formation of the meteorites and the terrestrial planets [63].

5. Prospects and final remarks

Solar and solar-nebula compositions have long been inferred from primitive meteorites in the absence of direct solar data. This link was most tenuous for isotopic data, where there was essentially no proof that solar isotopic compositions were the same as planetary or meteoritic compositions. The return of samples and solar-wind-irradiated foils from the Moon during the Apollo program significantly increased our isotopic under-

Table 3

Solar-wind xenon isotope ratios and δ values (‰) relative to terrestrial air [61], with 1-sigma uncertainties

^{124}Xe	^{126}Xe	^{128}Xe	^{129}Xe	^{130}Xe	^{131}Xe	^{132}Xe	^{134}Xe	^{136}Xe
2.948 (17)	2.549 (82)	51.02 (54)	627.3 (42)	≡100	498.0 (17)	602.0 (33)	220.68 (90)	179.71 (55)
+261 (6)	+169 (33)	+82 (11)	>−34 (7)	≡0	−45 (4)	−89 (6)	−139 (4)	−174 (3)

standing for certain key elements, such as for He, Ne, and Ar through the SWC experiments, and for Ar, Kr, Xe through the analysis of lunar soils. However, a relatively complete understanding of the isotopic systematics for other elements such as N and Li have only recently been obtained from lunar samples. Within the next several years, we should see significant additional progress. The return of the Genesis capsule to Earth following two-plus years of solar-wind exposure promises a wealth of information from analyses of implanted samples [64]. The primary goals of the Genesis mission address all of the issues mentioned here. Genesis will enable measurement of solar-wind isotopic compositions to unprecedented precisions, with a goal of $\pm 1\%o$ for some of the most important measurements.

With the significant improvement in solar-wind isotopic precision promised by Genesis, other issues will rise in importance for interpreting the results, specifically for understanding the nebular isotopic composition. Three major issues will be (a) to confirm or model accurately and with confidence the effects of gravitational separation on the composition of the OCZ, (b) to characterize any isotopic fractionation in solar-wind acceleration, and (c) to understand possible biases in the Genesis samples in light of the significant short-term variability reported for solar-wind helium isotope ratios. The Genesis mission attempts to address each of these issues. To characterize fractionation (b), separate samples were collected for each of the major regimes: interstream, high speed, and coronal mass ejection. The zeroth-order test of isotopic fractionation is the comparison of compositions from these different regimes. No difference will suggest the absence of isotopic fractionation. Differences, if found, will be important for a better understanding of the solar-wind acceleration, and should aid in modeling this process. The models can in turn be used to estimate the isotopic compositions of the OCZ, potentially satisfying the original goal of Genesis, though the de-convolution of the fraction effects and their dependence on solar conditions could be challenging.

The issue of solar-wind variability (c) is addressed in part by the long Genesis collection times. Interstream and high-speed collectors each have more than ten months of exposure, while a continuous collector has over 850 days of exposure compared with <2 full-day exposure for the SWC foils. A number of different solar-wind parameters, including ion speed, flux, temperature, charge states, and compositions similar to those given in Table 1 from spacecraft measurements are being compiled for the Genesis collection period, and will be important for understanding the context of the Genesis samples. Additional studies of long-term variability of solar-wind isotope ratios can be carried out over other time periods. For example, SOHO and Wind have nearly completed one solar cycle of measurements with which comparisons can be made between solar maximum and minimum compositions. Finally, the issue of gravitational separation in the sun (a) will be addressed by comparisons between meteoritic and solar compositions for isotopes of heavy and light elements. Assuming the other two issues can be understood sufficiently, the precision afforded by Genesis on solar material will enable us to confirm the models for gravitational separation for the first time. Clearly each of these three issues—solar gravitational settling, isotopic fractionation during solar-wind acceleration, and possible biasing of the samples—will remain important areas of study to fully understand the solar isotopic composition.

In summary, great progress is being made in understanding the solar-wind and solar isotopic compositions. For the solar composition, fruitful areas include in situ spacecraft solar-wind measurements, continuing studies of the solar wind implanted in lunar and asteroidal soil samples, solar spectroscopic studies, and now with the return of Genesis samples, the measurement of directly implanted solar wind in high purity substrates. The long-sought goal of basing solar and solar-nebula isotopic ratios on actual solar composition is finally being realized.

Acknowledgements

Constructive comments were given by R.O. Pepin, F. Buehler and J. Gosling. Helpful reviews by R. Wieler, T. Zurbuchen, and R. von Steiger, and editorial assistance by A. Halliday were greatly appreciated. Work on this manuscript at Los Alamos was supported by NASA contract W-19,272 and at

Bern by the Swiss National Science Foundation.
[AH]

References

[1] A.B. Severnyi, A spectroscopic investigation of the deuterium D_α line in active regions on the Sun, Soviet Astron. 1 (1957) 324–331.

[2] D.J. Mullan, J.L. Linsky, Nonprimordial deuterium in the interstellar medium, Astrophys. J. 511 (1998) 502–512.

[3] D.N.B. Hall, Detection of the ^{13}C, ^{17}O, and ^{18}O isotope bands of CO in the infrared solar spectrum, Astrophys. J. 182 (1973) 977–982.

[4] D.N.B. Hall, Spectroscopic detection of solar ^3He, Astrophys. J. 197 (1975) 509–512.

[5] M.J. Harris, D.L. Lambert, A. Goldman, Carbon and oxygen isotope ratios in the solar photosphere, Mon. Not. R. Astron. Soc. 224 (1987) 237–255.

[6] S. Ritzenhoff, E.H. Schröter, W. Schmidt, The lithium abundance in sunspots, Astron. Astrophys. 328 (1997) 695–701.

[7] P. Bochsler, Abundances and charge states of particles in the solar wind, Rev. Geophys. 38 (2000) 247–266.

[8] S. Turcotte, R.F. Wimmer-Schweingruber, Possible in situ tests of the evolution of elemental and isotopic abundances in the solar convection zone, J. Geophys. Res. 107 (2002) 1442 (doi: 10.1029/2002JA009418).

[9] S. Vauclair, Transport phenomena and light element abundances in the sun and solar type stars, IAU Symp. 198 (2000) 470–475.

[10] J. Geiss, F. Bühler, H. Cerutti, P. Eberhardt, C. Filleux, Solar wind composition experiment, Apollo 16 Prelim. Sci. Rep., 1972, pp. 14-1–14-10. NASA SP-315.

[11] D. Hovestadt, M. Hilchenbach, A. Bürgi, B. Klecker, P. Laeverenz, M. Scholer, H. Grünwaldt, W.I. Axford, S. Livi, E. Marsch, B. Wilken, H.P. Winterhoff, F.M. Ipavich, P. Bedini, M.A. Coplan, A.B. Galvin, G. Gloeckler, P. Bochsler, H. Balsiger, J. Fischer, J. Geiss, R. Kallenbach, P. Wurz, K.-U. Reiche, F. Gliem, D.L. Judge, H.S. Ogawa, K.C. Hsieh, E. Mobius, M.A. Lee, G.G. Managadze, M.I. Verigin, M. Neugebauer, CELIAS—charge, element and isotope analysis system for SOHO, Sol. Phys. 162 (1995) 441–481.

[12] G. Gloeckler, H. Balsiger, A. Bürgi, P. Bochsler, L.A. Fisk, A.B. Galvin, J. Geiss, F. Gliem, D.C. Hamilton, T.E. Holzer, D. Hovestadt, F.M. Ipavich, E. Kirsch, R.A. Lundgren, K.W. Ogilvie, R.B. Sheldon, B. Wilken, The solar wind and suprathermal ion composition investigation on the WIND spacecraft, Space Sci. Rev. 71 (1995) 79–124.

[13] G. Gloeckler, J. Cain, F.M. Ipavich, E.O. Tums, P. Bedini, L.A. Fisk, T. Zurbuchen, P. Bochsler, J. Fischer, R.F. Wimmer-Schweingruber, J. Geiss, R. Kallenbach, Investigation of the composition of solar and interstellar matter using solar wind and pickup ion measurements with SWICS and SWIMS on the ACE spacecraft, Space Sci. Rev. 86 (1998) 497–539.

[14] R. von Steiger, J. Geiss, G. Gloeckler, Composition of the solar wind, in: J.R. Jokipii, C.P. Sonett, M.S. Giampapa (Eds.), Cosmic Winds and the Heliosphere, U. AZ Press, Tucson, AZ, 1997, pp. 581–616.

[15] R. von Steiger, J. Geiss, L.A. Fisk, N.A. Schwadron, S. Hefti, T.H. Zurbuchen, G. Gloeckler, B. Wilken, R.F. Wimmer-Schweingruber, Composition of quasi-stationary solar wind flows from SWICS/Ulysses, J. Geophys. Res. 105 (2000) 27217–27238.

[16] J.L. Kohl, G. Noci, E. Antonucci, G. Tondello, C.E. Huber, S.R. Cranmer, L. Strachan, A.V. Panasyuk, L.D. Gardner, M. Romoli, S. Fineschi, D. Dobrzycka, J.C. Raymond, P. Nicolosi, O.H.W. Siegmund, D. Spadaro, C. Benna, A. Ciaravella, S. Giordano, S.R. Habbal, M. Karovska, X. Li, R. Martin, J.G. Michels, A. Modigliani, G. Naletto, R.H. O'Neal, C. Pernechele, G. Poletto, P.L. Smith, R.M. Suleiman, UVCS/SOHO empirical determinations of anisotropic velocity distributions in the solar corona, Astrophys. J. 501 (1998) L127–L131.

[17] J. Geiss, P. Hirt, H. Leutwyler, On acceleration and motion of ions in corona and solar wind, Sol. Phys. 12 (1970) 458–483.

[18] R. Bodmer, P. Bochsler, The helium isotopic ratio in the solar wind and ion fractionation in the corona by inefficient Coulomb drag, Astron. Astrophys. 337 (1998) 921–927.

[19] R. Bodmer, P. Bochsler, Influence of Coulomb collisions on isotopic and elemental fractionation in the solar wind acceleration process, J. Geophys. Res. 105 (2000) 47–60.

[20] R. Bodmer, P. Bochsler, J. Geiss, R. von Steiger, G. Gloeckler, Solar-wind helium isotopic composition from SWICS Ulysses, Space Sci. Rev. 72 (1995) 61–64.

[21] M.A. Coplan, K.W. Ogilvie, P. Bochsler, J. Geiss, Interpretation of ^3He abundance variations in the solar wind, Sol. Phys. 93 (1984) 415–434.

[22] R. Bodmer, The helium isotopic ratio as a test for minor ion fractionation in the solar wind acceleration process: SWICS/ULYSSES data compared with results from a multifluid model, Thesis. U. Bern, 1996.

[23] J. Geiss, G. Gloeckler, Abundances of deuterium and helium-3 in the protosolar cloud, Space Sci. Rev. 84 (1998) 239–250.

[24] G. Gloeckler, J. Geiss, Deuterium and helium-3 in the protosolar cloud, in: L. da Silva, M. Spite, J.R. de Medeiros (Eds.), The Light Elements and Their Evolution, IAU Symp., vol. 198, ASP Publ., 2000, pp. 224–233.

[25] G. Gloeckler, J. Geiss, Measurement of the abundance of helium-3 in the Sun and in the local interstellar cloud with SWICS on Ulysses, Space Sci. Rev. 84 (1998) 275–284.

[26] J. Geiss, G. Gloeckler, Isotopic composition of H, He, and Ne in the protosolar cloud, Space Sci. Rev. 106 (2003) 3–18.

[27] P. Bochsler, H. Balsiger, R. Bodmer, O. Kern, T. Zurbuchen, G. Gloeckler, D.C. Hamilton, M.R. Collier, D. Hovestadt, Limits of the efficiency of isotope fractionation processes in the solar wind derived from the magnesium isotopic composition as observed with the WIND/MASS experiment, Phys. Chem. Earth 22 (1997) 401–404.

[28] H. Kucharek, F.M. Ipavich, R. Kallenbach, P. Bochsler, D. Hovestadt, H. Grünwaldt, M. Hilchenbach, W.I. Axford, H.

Balsiger, A. Bürgi, M.A. Coplan, A.B. Galvin, J. Geiss, G. Gloeckler, K.C. Hsieh, B. Klecker, M.A. Lee, S. Livi, G.G. Managadze, E. Marsch, E. Möbius, M. Neugebauer, K.U. Reiche, M. Scholer, M.I. Verigin, B. Wilken, P. Wurz, Magnesium isotope composition in the solar wind as observed with the MTOF sensor on the CELIAS experiment on board the SOHO spacecraft, ESA SP-404, (1997) 473–476.

[29] R. Kallenbach, F.M. Ipavich, H. Kucharek, P. Bochsler, A.B. Galvin, J. Geiss, F. Gliem, G. Gloeckler, H. Grünwaldt, S. Hefti, M. Hilchenbach, D. Hovestadt, Fractionation of Si, Ne, and Mg isotopes in the solar wind as measured by SOHO/CELIAS/MTOF, Space Sci. Rev. 85 (1998) 357–370.

[30] P.R. Mahaffy, T.M. Donahue, S.K. Atreya, T.C. Owen, H.B. Niemann, Galileo probe measurements of D/H and ^{3}He/^{4}He in Jupiter's atmosphere, Space Sci. Rev. 84 (1998) 251–263.

[31] V.S. Heber, H. Baur, R. Wieler, Helium in lunar samples analyzed by high-resolution stepwise etching: implications for the temporal constancy of solar wind isotopic composition, Astrophys. J. 597 (2003) 602–614.

[32] R.C. Wiens, G.R. Huss, D.S. Burnett, The solar oxygen isotopic composition: predictions and implications for solar nebula processes, Meteorit. Planet. Sci. 34 (1999) 99–108.

[33] M.R. Collier, D.C. Hamilton, G. Gloeckler, G. Ho, P. Bochsler, R. Bodmer, R. Sheldon, Oxygen-16 to oxygen-18 abundance ratio in the solar wind observed by Wind/MASS, J. Geophys. Res. 103 (1998) 7–13.

[34] R.F. Wimmer-Schweingruber, P. Bochsler, G. Gloeckler, The isotopic composition of oxygen in the fast solar wind: ACE/SWIMS, Geophys. Res. Lett. 28 (2001) 2763–2766.

[35] P. Eberhardt, M. Reber, D. Krankowsky, R.R. Hodges, The D/H and O-18/O-16 ratios in water from comet P/Halley, Astron. Astrophys. 302 (1995) 301–316.

[36] H. Balsiger, K. Altwegg, J. Geiss, D/H and O-18/O-16 ratio in the hydronium ion and in neutral water from in-situ ion measurements in comet Halley, J. Geophys. Res. 100 (1995) 5827–5834.

[37] K. Hashizume, M. Chaussidon, B. Marty, K. Terada, Protosolar carbon isotopic composition: implications for the origin of meteoritic organics, Astrophys. J. 600 (2004) 480–484.

[38] H.B. Niemann, S.K. Atreya, G.R. Carignan, T.M. Donahue, J.A. Haberman, D.N. Harpold, R.E. Hartle, D.M. Hunten, W.T. Kasprzak, P.R. Mahaffy, T.C. Owen, N.W. Spencer, S.H. Way, The Galileo probe mass spectrometer: composition of Jupiter's atmosphere, Science 272 (1996) 846–849.

[39] C. Arpigny, E. Jehin, J. Manfroid, D. Hutsemekers, R. Schulz, J.A. Stüwe, J.-M. Zucconi, I. Ilyin, Anomalous nitrogen isotope ratio in comets, Science 301 (2003) 1522–1524.

[40] R. Wieler, F. Humbert, B. Marty, Evidence for a predominantly non-solar origin of nitrogen in the lunar regolith revealed by single grain analyses, Earth Planet. Sci. Lett. 167 (1999) 47–60.

[41] K. Hashizume, M. Chaussidon, B. Marty, F. Robert, Solar wind record on the moon: deciphering presolar from planetary nitrogen, Science 290 (2000) 1142–1145.

[42] R. Kallenbach, J. Geiss, F.M. Ipavich, G. Gloeckler, P. Bochsler, et al, Isotopic composition of solar wind nitrogen: first in-situ determination by CELIAS/MTOF on Board SOHO, Astrophys. J. 507 (1998) L185–L188.

[43] R. Kallenbach, Isotopic fractionation by plasma processes, Space Sci. Rev. 106 (2003) 305–316.

[44] T. Fouchet, E. Lellouch, B. Bezard, T. Encrenaz, P. Drossart, H. Feuchtgruber, T. de Graauw, ISO-SWS observations of Jupiter: measurement of the ammonia tropospheric profile and of the N-15/N-14 isotopic ratio, Icarus 143 (2000) 223–243.

[45] S.K. Atreya, P.R. Mahaffy, H.B. Niemann, M.H. Wong, T.C. Owen, Composition and origin of the atmosphere of Jupiter—an update, and implications for the extrasolar giant planets, Planet. Space Sci. 51 (2003) 105–112.

[46] T. Owen, P.R. Mahaffy, H.B. Niemann, S. Atreya, M. Wong, Protosolar nitrogen, Astrophys. J. 553 (2001) L77–L79.

[47] S. Epstein, H.P. Taylor Jr., The isotopic composition and concentration of water, hydrogen, and carbon in some Apollo 15 and 16 soils and in the Apollo 17 orange soil, Proc. 4th Lunar Sci. Conf., Supplement 4, Geochim. Cosmochim. Acta, 1973, pp. 1559–1575.

[48] M. Chaussidon, F. Robert, Lithium nucleosynthesis in the Sun inferred from the solar-wind Li-7/Li-6 ratio, Nature 402 (1999) 270–273.

[49] A.J.T. Jull, D. Lal, D.J. Donahue, Evidence for a non-cosmogenic implanted ^{14}C component in lunar samples, Earth Planet. Sci. Lett. 136 (1995) 693–702.

[50] K. Nishiizumi, M. Caffee, Beryllium-10 from the Sun, Science 294 (2001) 352–354.

[51] B.K. Kitts, F.A. Podosek, R.H. Nichols Jr., J.C. Brannon, J. Ramezani, R.L. Korotev, B.L. Jolliff, Isotopic composition of surface-correlated chromium in Apollo 16 lunar soils, Geochim. Cosmochim. Acta 67 (2003) 4881–4893.

[52] R. Wieler, The solar noble gas record in lunar samples and meteorites, Space Sci. Rev. 85 (1998) 303–314.

[53] R.O. Pepin, R.L. Palma, D.J. Schlutter, Noble gases in interplanetary dust particles: I. The excess helium-3 problem and estimates of the relative fluxes of solar wind and solar energetic particles in interplanetary space, Meteorit. Planet. Sci. 35 (2000) 495–504.

[54] R.A. Mewaldt, R.C. Ogliore, G. Gloeckler, G.M. Mason, A new look at neon-C and SEP-neon, in: R.F. Wimmer-Schweingruber (Ed.), Solar and Galactic Composition, AIP Conf. Proc., Melville, NY, vol. 598, 2001, pp. 393–398.

[55] R.F. Wimmer-Schweingruber, P. Bochsler, Lunar soils: a long-term archive for the galactic environment of the heliosphere? in: R.F. Wimmer-Schweingruber (Ed.), Solar and Galactic Composition, AIP Conf. Proc., Melville, NY, vol. 598, 2001, pp. 399–404.

[56] R.L. Palma, R.H. Becker, R.O. Pepin, D.J. Schlutter, Irradiation records in regolith materials: II. Solar wind and solar energetic particle components in helium, neon, and argon extracted from single lunar mineral grains and from the Kapoeta howardite by stepwise pulse heating, Geochim. Cosmochim. Acta 66 (2002) 2929–2958.

[57] J.-P. Benkert, H. Baur, P. Signer, R. Wieler, He, Ne, and Ar from the solar wind and solar energetic particles in lunar ilmenites and pyroxenes, J. Geophys. Res. 98 (1993) 13147–13162.

[58] R.H. Becker, D.J. Schlutter, P.E. Rider, R.O. Pepin, An acid-

etch study of the Kapoeta achondrite: implications for the argon-36/argon-38 ratio in the solar wind, Meteorit. Planet. Sci. 33 (1998) 109–113.

[59] R. Wieler, Noble gases in the solar system, Rev. Mineral. Geochem. 47 (2002) 21–70.

[60] R. Wieler, H. Baur, Krypton and xenon from the solar wind and solar energetic particles in two lunar ilmenites of different antiquity, Meteoritics 29 (1994) 570–580.

[61] R.O. Pepin, R.H. Becker, P.E. Rider, Xenon and krypton isotopes in extraterrestrial regolith soils and in the solar wind, Geochim. Cosmochim. Acta 59 (1995) 4997–5022.

[62] R.O. Pepin, On the isotopic composition of primordial xenon in terrestrial planet atmospheres, Space Sci. Rev. 92 (2000) 371–395.

[63] R.O. Pepin, On noble gas processing in the solar accretion disk, Space Sci. Rev. 106 (2003) 211–230.

[64] D.S. Burnett, B.L. Barraclough, R. Bennett, M. Neugebauer, L.P. Oldham, C.N. Sasaki, D. Sevilla, N. Smith, E. Stansbery, D. Sweetnam, R.C. Wiens, The genesis discovery mission: return of solar matter to earth, Space Sci. Rev. 105 (2003) 509–534.

[65] R. Kallenbach, F.M. Ipavich, P. Bochsler, S. Hefti, D. Hovestadt, H. Grünwaldt, M. Hilchenbach, W.I. Axford, H. Balsiger, A. Bürgi, M.A. Coplan, A.B. Galvin, J. Geiss, F. Gliem, G. Gloeckler, K.C. Hsieh, B. Klecker, M.A. Lee, S. Livi, G.G. Managadze, E. Marsch, E. Möbius, M. Neugebauer, K.-U. Reiche, M. Scholer, M.I. Verigin, B. Wilken, P. Wurz, Isotopic composition of solar wind neon measured by CELIAS/MTOF on board SOHO, J. Geophys. Res. 102 (1997) 26895–26904.

[66] R.F. Wimmer-Schweingruber, P. Bochsler, O. Kern, G. Gloeckler, D.C. Hamilton, First determination of the silicon isotopic composition of the solar wind: WIND/MASS results, J. Geophys. Res. 103 (1998) 20621–20630.

[67] H. Kucharek, B. Klecker, F.M. Ipavich, R. Kallenbach, H. Grünwaldt, M.R. Aellig, P. Bochsler, Isotopic fractionation in slow and coronal hole associated solar wind, in: P. Brekke, B. Fleck, J.B. Gurman (Eds.), Recent Insights into the Physics of the Sun and Heliosphere: Highlights From SOHO and Other Space Missions, IAU Symp., vol. 203, 2001, pp. 562–564.

[68] R.F. Wimmer-Schweingruber, The composition of the solar wind, Adv. Space Res. 30 (2002) 23–32.

[69] H. Cerutti, Die Bestimmung des Argons im Sonnenwind aus Messungen an den Apollo-SWC-Folien, PhD thesis (1974), Univ. of Bern, Switzerland.

[70] J. Weygand, F. Ipavich, P. Wurz, J. Paquette, P. Bochsler, Determination of the argon isotopic ratio of the solar wind using SOHO/CELIAS/MTOF, ESA SP-446, (1999) 22–25.

[71] J.M. Weygand, F.M. Ipavich, P. Wurz, J.A. Paquette, P. Bochsler, Determination of the $^{36}Ar/^{38}Ar$ isotopic abundance ratio of the solar wind using SOHO/CELIAS/MTOF, Geochim. Cosmochim. Acta 65 (2001) 4589–4596.

[72] R. Kallenbach, F.M. Ipavich, P. Bochsler, S. Hefti, P. Wurz, M.R. Aellig, A.B. Galvin, J. Geiss, F. Gliem, G. Gloeckler, H. Grünwaldt, M. Hilchenbach, D. Hovestadt, B. Klecker, Isotopic composition of solar wind calcium: first in situ

measurement by CELIAS/MTOF on board SOHO, Astrophys. J. 498 (1998) L75–L78.

[73] F.M. Ipavich, J.A. Paquette, P. Bochsler, S.E. Lasley, P. Wurz, Solar wind iron isotopic abundances: Results from SOHO/CELIAS/MTOF, in: R.F. Wimmer-Schweingruber (Ed.), Solar and Galactic Composition, AIP Conf. Proc., Melville, NY, 2001, pp. 121–126.

Roger Wiens began working on the Genesis mission with Don Burnett in 1990 after completing his PhD on volatiles in the Mars meteorites and in the Mars atmosphere. After 7 years of working on the Genesis mission at Caltech, he moved to Los Alamos, where he led the Genesis instrument development effort there. His interests continue to include both Mars geochemistry and instrumentation, and the solar abundances that are the focus of the Genesis mission.

Peter Bochsler is a professor and co-director of the Physikalisches Institut at the University of Bern. He has recently served as Dean of the Faculty of Sciences. His research interests have focused on space physics, especially the acceleration and the composition of the solar wind. Professor Bochsler has been a co-investigator on a number of missions, including ISEE-3, SOHO, Wind, ACE, and Genesis.

Donald Burnett is a professor of Geochemistry at CalTech and the Principal Investigator of the Genesis Discovery Mission. His interests encompass a broad range of problems in meteoritics and planetary science. Outside of Genesis, one of his main research areas is on meteoritic Ca–Al-rich inclusions.

Robert Wimmer-Schweingruber earned his PhD at the University of Bern, Switzerland. He is currently professor at the University of Kiel, Germany, and co-director of the Institute for Experimental and Applied Physics. His interests include solar wind, suprathermal and energetic particles in the heliosphere, their composition and its relation to solar-system bodies, as well as the evolution and history of the heliosphere and solar system.

Reprinted from
Earth and Planetary Science Letters 222 (2004) 333–348

www.elsevier.com/locate/epsl

The importance of ocean temperature to global biogeochemistry

David Archer[a,*], Pamela Martin[a], Bruce Buffett[a], Victor Brovkin[b],
Stefan Rahmstorf[b], Andrey Ganopolski[b]

[a] *Department of the Geophysical Sciences, University of Chicago, Chicago, IL 60637, USA*
[b] *Potsdam Institute for Climate Impact Research, P.O. Box 601203, 14412 Potsdam, Germany*

Received 3 December 2003; received in revised form 5 March 2004; accepted 10 March 2004

Abstract

Variations in the mean temperature of the ocean, on time scales from millennial to millions of years, in the past and projected for the future, are large enough to impact the geochemistry of the carbon, oxygen, and methane geochemical systems. In each system, the time scale of the temperature perturbation is key. On time frames of $1-100$ ky, atmospheric CO_2 is controlled by the ocean. CO_2 temperature-dependent solubility and greenhouse forcing combine to create an amplifying feedback with ocean temperature; the $CaCO_3$ cycle increases this effect somewhat on time scales longer than $\sim 5-10$ ky. The CO_2/T feedback can be seen in the climate record from Vostok, and a model including the temperature feedback predicts that 10% of the fossil fuel CO_2 will reside in the atmosphere for longer than 100 ky. Timing is important for oxygen, as well; the atmosphere controls the ocean on short time scales, but ocean anoxia controls atmospheric pO_2 on million-year time scales and longer. Warming the ocean to Cretaceous temperatures might eventually increase pO_2 by approximately 25%, in the absence of other perturbations. The response of methane clathrate to climate change in the coming century will probably be small, but on longer time scales of $1-10$ ky, there may be a positive feedback with ocean temperature, amplifying the long-term climate impact of anthropogenic CO_2 release.

Keywords: carbon cycle; climate; ocean temperature

1. Introduction

The mean temperature of the ocean has varied by more than 10°C over the past 65 My. The mean ocean temperature in a coupled climate model is roughly as sensitive to pCO_2 as is the mean surface temperature of the earth [1]. Such large changes in temperature must impact biogeochemical cycling in the ocean and atmosphere. In this paper, we consider the implications of the temperature sensitivity of dissolved gases O_2 and CO_2 and methane in clathrate deposits below the seafloor within the context of Cenozoic climate change. The mechanism and extent of the interaction between these systems and temperature depends strongly on the duration of the change in ocean temperature. The largest temperature changes inferred from the geologic record took place on the 10-My time scale, which is long enough to allow weathering feedbacks to imprint their signatures. The glacial cycles are faster than silicate or

* Corresponding author. Tel.: +1-773-702-0823; fax: +1-773-702-9505.
E-mail address: d-archer@uchicago.edu (D. Archer).

0012-821X/$ - see front matter © 2004 Elsevier B.V. All rights reserved.
doi:10.1016/j.epsl.2004.03.011

organic carbon weathering time scales but do interact with the $CaCO_3$ cycle. The response time for changing deep-ocean temperature is about a millennium, so we will not be concerned with time scales shorter than that.

Terminology

ΔT_{2x}: The temperature change resulting from a doubling of the CO_2 concentration of the atmosphere. This term is used in an atypical way in this paper, to describe the mean ocean temperature, rather than the mean surface temperature.

AABW: Antarctic bottom water. A source of surface water to the deep, originating in the Southern Ocean.

Clathrate: Water frozen into a cage structure trapping a gas molecule. Ocean margin sediments contain huge amounts of methane trapped in clathrate deposits. These are also known as hydrates.

Foraminifera: Single-celled $CaCO_3$-secreting heterotrophic protista. The chemistry of their shells provides information about the chemical and physical conditions in which they grew. Planktontic foramifera lived near the sea surface; benthic foraminifera lived on the seafloor.

Gton: 10^9 metric tons, used here exclusively to measure carbon (not CO_2).

LGM: Last Glacial Maximum, 21–18 ky.

NADW: North Atlantic deep water. A source of water carrying the chemical imprint of the surface ocean into the deep.

PETM: Paleocene Eocene Thermal Maximum, an excursion of $\delta^{13}C$ and $\delta^{18}O$ that is generally interpreted as an abrupt warming and release of isotopically light methane from ocean clathrate deposits.

Radiative equilibrium: The balance between influx and output of energy from the planet. With an increase in greenhouse trapping of outgoing infrared light from the surface, the surface temperature must increase to maintain radiative equilibrium.

---

2. The record of deep-ocean temperature

2.1. Methods

2.1.1. $\delta^{18}O$

The primary tool for reconstructing deep ocean temperature is the stable oxygen isotopic composition ($\delta^{18}O$) of the calcite shells of foraminifera [2–4]. The temperature dependence varies somewhat for different species and across the full range of ocean temperature, but for benthic foraminifera, the response is $\sim 0.25‰$ per °C, relative to a precision in measurement of $\sim \pm 0.02‰$ and a change in $\delta^{18}O$ of $\sim 5.5‰$ over the past 65 My. Reconstruction of temperatures from $\delta^{18}O$ is complicated by a correlation between local salinity and $\delta^{18}O$ in seawater, and by whole-ocean shifts in $\delta^{18}O$ reflecting storage in isotopically light continental ice sheets. The earth was probably ice-free before the late Eocene, 40 My, so this issue affects the more recent part of the record. However, the older part of the record is more affected by calcite recrystallization, which biases the reconstructed temperature toward the colder pore waters [5,6].

2.1.2. Mg/Ca ratio in $CaCO_3$

The Mg/Ca ratio of biogenic calcites increases exponentially with temperature. While the relationship probably has thermodynamic underpinnings, biogenic calcite carries a species-dependent overprint [7,8] which must be determined by empirical calibration [7,9]. Most low-Mg foraminifera used in paleoceanographic studies have a Mg/Ca response of $\sim 9–10\%$ per °C, relative to an analytic precision of better than 2% and variation in Mg/Ca of more than 50% over the Cenozoic [10,11].

The concentrations of Mg and Ca ought to have been relatively stable over several million years, but on the 40-million-year time scale of the Cenozoic, ocean chemistry is less certain, and the calibration of extinct species of foraminifera becomes more difficult [10–12]. Calcite solubility increases with Mg content, so that partial dissolution tends to deplete Mg, biasing the reconstructed temperature cold [13]. In spite of these difficulties, Mg paleothermometry is useful for separating ice volume effect in the $\delta^{18}O$ record.

2.2. My time scale changes

The temperature of the deep sea has changed considerably over the last 65 million years (see Fig. 1). The lack of evidence for ice sheets prior to the end of the Eocene suggests that $\delta^{18}O$ primarily reflects temperature during that time. Deep-ocean temperature appears to have been around 8°C at the start of the Paleocene, warming to a Cenozoic maximum of nearly 12°C in the early Eocene (~ 50 Ma). Both Mg/Ca and $\delta^{18}O$ indicate a steady cooling throughout the Eocene of ~ 5–7°C, culminating in an additional abrupt cooling to ~ 4°C coincident with the development of Antarctic Ice Sheets that persisted during most of the Oligocene [10,14]. The $\delta^{18}O$ records imply a warming with the temporary waning of 'permanent' Antarctic Ice Sheets in the late Oligocene [15]. Mg/Ca, however, suggests that the cold temperatures persisted through the middle to late Miocene [10,11] and possibly until the middle Pliocene [11], with temperature fluctuations of around 1–2°C persisting on time scales of millions of years. Most of the records suggest a rapid, steady cooling to the present mean ocean temperature of 1.5°C over the last 5–10 million years. Both $\delta^{18}O$ and Mg/Ca show a ~ million-year warming event around 10 Ma, the Mid-Miocene Climatic Optimum. There are also several ~ 100,000-year spikes in the Cenozoic deep temperature record, the most well-defined spike, a warming event called the Paleocene/Eocene Thermal Maximum, is often attributed to the release of methane from hydrates.

2.3. Millennial to glacial time scale changes

Temperature cycles are also documented on shorter, glacial/interglacial time scales (Fig. 2). Interpretation of $\delta^{18}O$ is complicated by large changes of ice volume; but the emerging consensus is that half of the ~ 2‰ $\delta^{18}O$ change from full glacial to interglacial conditions during the Pleistocene is due to ice, and the other half reflects a temperature change of at least 3°C with deep-ocean temperature dropping below − 1°C. This is consistent with evidence from benthic Mg/Ca records for the Quaternary [13,16] and deep core pore water reconstructions for the Last Glacial Maximum [17,18]. Comparison of ice core records and deep-sea geochemical records suggests that there is a strong 100,000 component of deep-sea temperature change, but the

deep-sea records also show fluctuations on the order of 2°C associated with the higher frequency Milankovitch cycles (41 and 23 ky) [13,19]. At an even higher resolution, Mg/Ca and $\delta^{18}O$ records from moderate sedimentation rate deep-sea cores imply short warming events of 0.5–1.5°C on the order of several thousand years during the last glacial, consistent with comparison of $\delta^{18}O$ records and high-resolution records of sea level change reconstructed from coral reefs [20].

3. Mechanisms of ocean temperature change

3.1. My time scale

On the longest time scales, the mechanism by which the deep-ocean temperature achieves the warmest observed levels is something of a mystery. The central issue is whether the ocean flips from its present dominantly temperature-driven to a salinity-driven overturning circulation, or alternatively if temperature-driven circulation can generate the observed warming. Salt-driven circulation would lead naturally to warmer deep-ocean temperatures, as the sites of convection move to lower latitudes. The coupled atmosphere/ocean simulations of the Eocene from Huber and Sloan [21] were described as "quasi halothermal" circulation. Deep-water formation took place in a cool (~ 9°C) but salty subpolar North Atlantic. The deep-ocean temperature in these simulations reached 6°C above modern, at the lower end of Eocene deep temperature reconstructions. Higher atmospheric pCO_2 than their assumed 560 µatm would help to get a warmer deep ocean, but perhaps at the expense of excessive tropical warming. The simulation of Zhang et al. [22,23] achieved a true salt-driven overturning circulation, with much higher deep-ocean temperatures, but their simulation was unstable to occasional temperature-driven deep sea purges, and thus the temperature of the deep sea oscillated between a range of 10–16°C.

3.2. Millennial to glacial time scale

On shorter, glacial/interglacial time scales, the observed temperature changes are somewhat easier to explain. Deep convection continues to occur at high latitudes. On the one hand, surface temperature

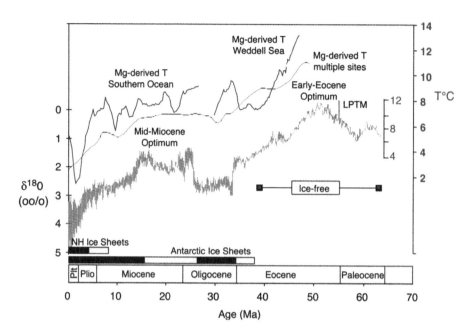

Fig. 1. Comparison of the benthic foraminiferal composite oxygen isotope record for the Cenozoic [14] (bottom gray line) and smoothed Mg/Ca derived temperature records (red and blue lines) [10,12]. The blue line is derived from Mg/Ca of multiple genera of benthic foraminifera from several, disparately located cores [10]. The red line from 0 to 26 Ma represents the temperatures derived from the Mg/Ca record from site 747 (77°E, 55°S, 1695 m) [11]. The red line from ~ 30 to 50 Ma represents the temperatures derived from the Mg/Ca record from site 689 (3°E, 65°S, 2080 m) [11]. The left axis is scaled for the oxygen isotope record. The right axis is scaled to temperature based on the temperature equations of Lear et al. [10] and Martin et al. [13]. Prior to ~ 40 Ma, the $\delta^{18}O$ record is thought to primarily reflect changes in deep-water temperature. A temperature scale is inset to estimate the cooling implied by the oxygen-isotoped data based on the temperature equation of Shackleton [4]. Banding on the lower left represents ephemeral (gray) and permanent (black) ice sheets in Antarctica (bottom) and the Northern Hemisphere (modeled after Zachos et al. [14]).

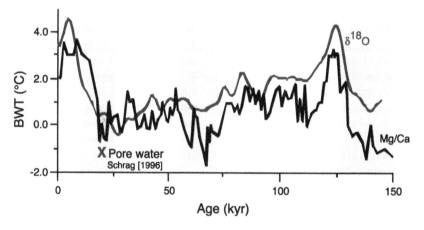

Fig. 2. Bottom water temperature changes over the last two glacial cycles for the deep tropical Atlantic. The red line is the temperature record from M12392 (25°N, 17°W, 2573 m) derived from regional comparison of $\delta^{18}O$ records [59]. The black line is the Mg-derived bottom water temperature from M16772 (1°S, 12°W, 3912 m) [13]. The 'X' is the temperature at the last glacial maximum derived from pore water $\delta^{18}O$ measurements. Figure adapted from Martin et al. [13].

changes there are larger than at low latitudes because of the ice albedo feedback. On the other hand, deep convection tends to occur near the sea ice edge and can shift in location with the ice edge, which tends to stabilize the temperature at which deep water forms.

The mechanism for glacial cooling of the deep ocean beyond its already cool modern mean of $1.5\,^\circ$C probably derives from the North Atlantic, where North Atlantic Deep Water (NADW) currently warms the deep ocean with a steady stream of $4\,^\circ$C water. During glacial time, Glacial NADW (GNADW) was colder than at present and formed further to the south [24,25]. Because deep water masses of southern origin are close to the freezing point and could not cool much farther, the increased density of colder Glacial North Atlantic Deep Water (GNADW) would by itself push the boundary between northern and southern component waters in the Atlantic deeper, in contrast to proxy observations [26]. However, a salinity increase in Antarctic Bottom Water (AABW) and a decrease in GNADW [18,26], driven perhaps by an increase in sea ice export from the Antarctic [27], may compensate for the cooler GNADW in the glacial ocean [27].

The cooling of the glacial climate was the combined result of lowered CO_2 and higher planetary albedo due to large ice sheets. For the effect of CO_2 alone on deep-sea temperature, we can turn to the coupled modeling results of Stouffer and Manabe [1]. They found that deep-sea temperature increased by about $3\,^\circ$C for doubling CO_2, in a range of 0.5–4 times a 300 µatm reference level. That is to say, deep-ocean temperature changes roughly follow changes in the mean temperature of the surface of the earth. The time constant for changing deep-ocean temperature in these simulations is about 1000 years for cooling, and about twice as long for a warming. The maximum predicted increase in deep-ocean temperature was about $6.5\,^\circ$C under a $4 \times CO_2$ atmosphere, and cooling by $3\,^\circ$C under $0.5 \times CO_2$.

4. Chemical impacts of ocean temperature

4.1. CO_2

4.1.1. Millenial time scale

The ocean is a larger reservoir for CO_2 than is the atmosphere, so on time scales of 1–100 ky, the relative variability of the CO_2 concentration of the ocean is smaller than that of the atmosphere. In addition to ocean temperature, atmospheric pCO_2 on this time scale is affected by biological redistribution of carbon in the ocean (the biological pump) and other factors. The relationship between ocean temperature and atmospheric pCO_2 is interesting, though, because CO_2 is affected by T and T is affected by CO_2 simultaneously [16].

The two relationships between temperature and CO_2 are radiative equilibrium and CO_2 solubility. We approximate these relationships as linear sensitivities about the present-day condition, although in reality, the solubility of CO_2 is exponential in temperature, and radiative equilibrium temperature goes as the log of the CO_2. Radiative equilibrium is also amplified by the ice albedo feedback, which becomes larger as it gets colder. We will facilitate direct comparisons of these relations by expressing both in a common metric of $\Delta CO_2/\Delta T$.

The radiative equilibrium relationship of concern to us here is the impact of CO_2 as a greenhouse gas on the mean temperature of the ocean. As described above, model simulations [1,28] and data from glacial time [16,18] indicate that changes in the deep-sea temperature roughly parallel those of the mean earth's surface. Based on an assumed value of the climate sensitivity to doubling CO_2, ΔT_{2x}, of $3\,^\circ$C, and assuming that mean ocean temperature follows mean surface temperature, we would expect CO_2/T to covary with a slope of about 50–70 µatm/$^\circ$C, depending on the magnitude of the pCO_2 change (because we are linearizing an exponential). In the direction of cooling, the magnitude of the slope decreases; the transition from glacial to interglacial climate was driven by ice albedo forcing as well as changes in CO_2, resulting in a slope of covariation of approximately 30 µatm/$^\circ$C.

The other relationship is the solubility of CO_2. Simple thermodynamics of a homogeneous seawater sample with no circulation or biological pump yields a pCO_2 sensitivity of 4.23% pCO_2/$^\circ$C, which translates to ~ 10 µatm/$^\circ$C. All of the box models and GCMs tested by Martin [16] exhibited similar or slightly smaller slopes of covariation than this, as do new model results presented below.

If we depict the radiative (R) and solubility (S) relations as lines on a pCO_2/T diagram (Fig. 3a), the ocean temperature and atmospheric pCO_2 find the intersection of these two lines, satisfying the radiative equilibrium and CO_2 solubility constraints simultaneously. This takes place on time scales of the ventilation of the deep sea, order of 1 ky. A source of CO_2 to the system, or an external forcing of temperature, manifests itself as a change in the position of one of the two lines, driving the system to find the new intersection.

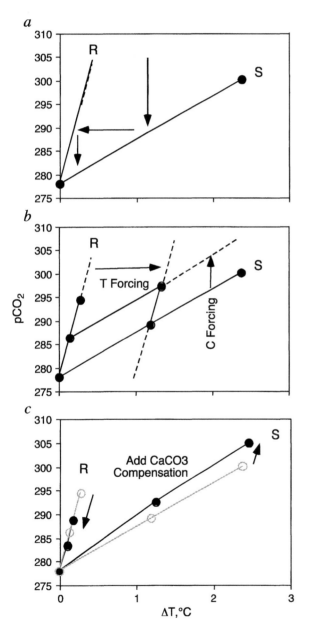

We demonstrate this idea using the Hamocc2 ocean carbon cycle coupled to a simple $CaCO_3$ sediment model [29–31]. To this model, we imposed a uniform whole-ocean temperature sensitivity to atmospheric pCO_2 of the form

$$\Delta T_{ocean} = 3\,^{\circ}C/\ln(2) \cdot \ln(pCO_2/278) \qquad (1)$$

where $3\,^{\circ}C$ is from [1] and is a typical value for ΔT_{2x}, the climate sensitivity (the range of uncertainty given for this value in IPCC 2001 is $1.5–4.5\,^{\circ}C$). Hamocc2 is an offline tracer advection model, so that changes in temperature affect the solubility of CO_2 but not the circulation.

The results in Fig. 3b arise from two types of perturbation experiments, representing externally driven changes of either temperature or CO_2. We perturb CO_2 by adding 100 or 200 Gtons C as CO_2 to the atmosphere, and running to equilibrium after 6000 years. The added CO_2 partitions itself between the atmosphere and the ocean, with its solubility modified by the change in temperature due to increased atmospheric pCO_2. Because we are adding CO_2 by means other than changing the ocean temperature, this perturbation has the effect of moving the solubility relation vertically on Fig. 3b, in a direction of increasing CO_2 at unchanging temperature. The CO_2/T system then finds the intersection of the radiative and the modified solubility relationship. The solubility line moves with CO_2 addition, but the radiative line does not. The results of several CO_2 addition experiments trace out the location and slope

Fig. 3. (a) Temperature and CO_2 forcing experiments in the Hamocc2 ocean carbon cycle model are used to determine the interaction between the radiative equilibrium and solubility relationships between deep-ocean temperature and atmospheric pCO_2. Temperature as a function of CO_2, or radiative equilibrium, denoted by "R" in the figure, is hard wired into the ocean model, as a uniform temperature offset from a present-day annual mean model result according to ΔT_{2x} of $3\,^{\circ}C$. (b) The "R" relationship is found by adding 100 or 200 Gtons of CO_2 to the atmosphere and running to equilibrium. The CO_2 solubility or "S" relationship is found by perturbing the ocean temperature by 1 or $2\,^{\circ}C$ and allowing CO_2 to equilibrate. The R and S relationships combine into a positive feedback which amplifies any external perturbation of T or CO_2 by $\sim 19\%$. (c) When $CaCO_3$ compensation is added to the ocean carbon cycle model, the R and S relationships are altered somewhat, and the amplification increases to $\sim 24\%$.

of the radiative relation. For small pCO_2 changes, order of 10 µatm, the slope of this relation is about 60 µatm/°C.

We perturb the temperature by adding a uniform offset to the temperature field when calculating the solubility of CO_2. The change in temperature causes CO_2 to degas, increasing the ocean temperature a bit further according to Eq. (1). Because we are altering temperature by means other than changing atmospheric pCO_2, the T forcing perturbation offsets the radiative relation horizontally on Fig. 3b, in the direction of increasing temperature at constant pCO_2. The new intersection of the two relations therefore traces out the trajectory of the solubility relation. The slope of the solubility relation is predicted to be about 9–10 µatm/°C.

Either type of perturbation provokes a positive feedback; for example, externally forced warming drives a CO_2 degassing which warms the ocean a bit further. The magnitude of the feedback depends on the relative slopes of the two relations. We linearize the radiative relation as

$$C = \alpha_R T(C) + r$$

where C is the atmospheric pCO_2, $T(C)$ is temperature as a variable dependent on CO_2, r is some offset, and α_R is the radiative slope, the inverse of the climate sensitivity, estimated above to be 60 µatm/°C above. The linearized solubility relation is

$$C(T) = \alpha_S T + s$$

where $C(T)$ is CO_2 but now dependent on T, s is an offset, and α_S is estimated above to be 10 µatm/°C. Then the simultaneous solution (the intersection of the two relations) is

$$T = (s - r)/(\alpha_R - \alpha_S).$$

This solution is stable for $\alpha_R > \alpha_S$; in the reverse case, the feedback is unstable and a runaway CO_2 degassing results. If we make an initial temperature perturbation ΔT, this is equivalent to offsetting the radiative line horizontally by a distance ΔT. This requires a change in r given by

$$\Delta r = \alpha_R \Delta T.$$

The final change in temperature, after the feedback, is given by

$$\Delta T_{final} = \Delta r \delta T / \delta r = \alpha_R \Delta T_{forcing} / (\alpha_R - \alpha_S)$$

and rearranging,

$$\Delta T_{final} = \Delta T_{forcing}(1 + \alpha_S / (\alpha_R - \alpha_S))$$

where the extent of amplification factor is represented by the term $\alpha_S/(\alpha_R - \alpha_S)$. For the values of α_R and α_S estimated above, this positive feedback comes to about 18–20%. For example, an initial temperature forcing of 1°C will generate an ultimate temperature change, after the CO_2 feedback, of 1.2°C.

4.1.2. Glacial time scale

On time scales of the glacial cycles (from 5 to 200 ky) the alkalinity and $CaCO_3$ cycles regulate the pH of the ocean, affecting the pCO_2 of the atmosphere. The fundamental constraint in the ocean is a balance between the influx of dissolved $CaCO_3$ and its removal by burial in sediments, called $CaCO_3$ compensation. Externally driven fluctuations in pCO_2, such as by changes in the biological pump or fossil fuel combustion, are damped by $CaCO_3$ compensation. However, $CaCO_3$ compensation adds a new sensitivity to the carbon cycle, the weathering and production of $CaCO_3$ [32].

The concentration of Ca^{2+} exceeds that of $CO_3^=$ by several orders of magnitude, and its residence time is longer despite a slight buffering of $CO_3^=$ by HCO_3^- and CO_2. The ocean therefore uses $CO_3^=$ as the regulator of $CaCO_3$ burial. If $CaCO_3$ burial is slower than weathering, for example, dissolved $CaCO_3$ builds up in the ocean, increasing $CO_3^=$, until burial equals weathering.

$CaCO_3$ compensation has a small but noticeable effect on the CO_2 and T feedbacks (Fig. 3c) by buffering the CO_2 concentration of the atmosphere against external sources and sinks. A 100-Gton CO_2 addition initially increases pCO_2 by 8 µatm, but $CaCO_3$ compensation ameliorates the atmospheric response to 5 µatm ("R" response, Fig. 3c). In addition, the solubility relation sensitivity is increased by $CaCO_3$ compensation, from about 9–10 µatm/°C to about 11–12 µatm/°C. This can be understood from the following chain of events. An increase in ocean T causes CO_2 to degas, decreasing ocean CO_2 and therefore shifting the pH of the ocean toward the basic. As a result, ocean $CO_3^=$ increases, roughly proportionally to the CO_2 decrease. $CaCO_3$ compensation, by insisting on equilibrium with $CaCO_3$, restores $CO_3^=$ toward its original value, and as $CO_3^=$ falls, the pH of the ocean shifts back toward the acidic and atmospheric pCO_2 rises still a bit further. Using the $CaCO_3$ compensation value for α_S, we calculate a feedback amplitude of about $23-25\%$ for the CO_2/T relation, on time scales of $5-10$ ky. This same feedback amplitude applies to perturbations of CO_2 or T.

4.1.3. Impacts of the CO_2/T feedback

The effects of the CO_2/T relation can be read in the tea leaves of the Vostok ice-core record [33]. The large CO_2 transition at the deglaciation is difficult to take apart, because everything is happening at once. Sea level is rising, dust fluxes are dropping, and temperatures are rising. The onset of glaciation is simpler. The Vostok record implies that the Laurentide ice sheet nucleated at the end of MIS 5e (~ 120 ky before present), although atmospheric pCO_2 was high at that time. In ongoing simulations with CLIMBER-2, an Earth system model of intermediate complexity (see Ganopolski et al. [24] for model description), the small ice sheet has a minor effect on climate except for the Northern high latitudes. That is, sea level drop is minor, and yet atmospheric pCO_2 begins to drop (Ganopolski, work in progress). We hypothesize that the North Atlantic is a plausible route to cooling the deep ocean, explaining the initial CO_2 drawdown as a response to ocean cooling. The Mg-derived temperature record shows a deep ocean decrease of $2-3\,^\circ C$ over $10-20$ ky on the 5e-4 transition, explaining potentially $30-45$ µatm of CO_2 drawdown. After this initial drawdown, changes in sea level, iron fluxes, and planktonic functional groups can be invoked to explain the rest of the drawdown to LGM levels [34].

The CO_2/T relation can also be seen in correlated spikes in CO_2 and T during stage 3. Martin [16] estimated that the slope of the CO_2/T relation during these times was ~ 10 µatm/$^\circ C$, similar to the model solubility relation.

The ocean temperature feedback will affect the atmospheric residence time and eventual fate of fossil fuel CO_2 in the future as well. This effect was not considered by [30,31], who used the same model as we are using here to forecast the dynamics of fossil fuel neutralization over a time scale of tens of thousands of years. A comparison of their model with and without the temperature feedback is shown in Fig. 4 and summarized in Table 1. The temperature feedback in this case is applied with a 1000-year relaxation time to a target temperature

$$\Delta T_{target} = 3\,^\circ C/\ln(2) \cdot \ln(pCO_2/278)$$

using

$$\delta\Delta T_{ocean}/\delta t = 10^{-3}\ \text{year}^{-1}(\Delta T_{target} - \Delta T_{ocean}).$$

The impact of the temperature feedback is to increase the long-term fraction of fossil carbon in the atmosphere. Dissolution in the ocean removes most of the fossil fuel CO_2 on an e-folding time scale of ~ 300 years, but the temperature feedback increases the fraction of the carbon that resides in the atmosphere for longer than that, by about $20-21\%$, consistent with the feedback analysis above. Some fraction of the CO_2 remains in the atmosphere even after neutralization with $CaCO_3$; and this fraction also increases when the temperature feedback is included, from $\sim 7\%$ to $\sim 9\%$, an increase of $23-29\%$. The mean atmospheric residence time of fossil carbon increases by $20-30\%$ over the case with no temperature feedback, to $40-50$ ky.

4.1.4. My time scale

On time scales of 400 ky and longer, atmospheric pCO_2 is controlled by the silicate weathering cycle. The fundamental balance is between degassing of CO_2 from the earth and the uptake of CO_2 by reaction with igneous rocks, mainly the CaO component [35]. This balance between reaction rates is achieved by modulating the pCO_2 of the atmosphere, which mainly affects weathering rates indirectly through variations in the hydrological cycle. It takes a million years or less to achieve the silicate weathering balance [36–38]. On this time scale, the pCO_2 of the atmosphere controls the carbon concentration of the ocean. Most of the interesting changes in

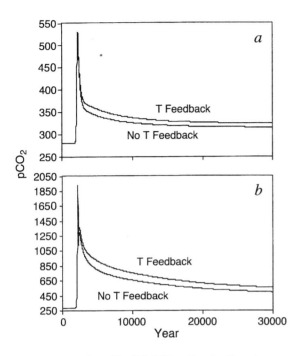

Fig. 4. Long-term fate of fossil fuel CO_2 with and without the ocean temperature feedback. (a) 1000 Gtons C addition experiment, and (b) 5000 Gtons C addition experiment.

carbon chemistry are due to factors other than ocean temperature.

4.2. Oxygen

4.2.1. Millenial to glacial time scale

The direct effect of warming the ocean to a maximum of $12\,°C$ in the early Eocene would be to decrease the saturation O_2 concentration by about 25%. The O_2 content of the atmosphere is approximately 3.6×10^{19} moles, while the O_2 capacity of the ocean, at $4\,°C$, is two orders of magnitude smaller than this, 3.6×10^{17} moles. Because the atmosphere contains more oxygen than the ocean does, the short-term effect of changing the ocean temperature T would be to decrease O_2 in the ocean without changing the concentration in the atmosphere very much. For example, an increase in ocean temperature by $1\,°C$ would decrease the solubility by about 2.6%. O_2 concentrations in the deep ocean are lower than saturation because of biological uptake, but assuming that the biological uptake remains constant, warming the ocean by $1\,°C$ would decrease the ocean inventory by 9.4×10^{15} moles, increasing O_2 in the atmosphere by only 0.026%.

During the Last Glacial Maximum, cooler deep-ocean temperatures would have increased the solubility of oxygen by approximately 10%. The significance of this to the glacial research community is that one candidate for decreasing atmospheric pCO_2 is an increase in the biological pump, which would sequester CO_2 in the deep ocean, along with a corresponding decrease in O_2. As best we can tell, however, the O_2 concentration increased in the present-day oxygen minimum zone [39–41]. Increased solubility of oxygen, aided perhaps by a change in the pattern of intermediate water ventilation, may help the research community explain this observation.

4.2.2. My time scale

On time scales of millions of years, the presence of O_2 in the ocean may be the switch in a chemostat which controls pO_2 in atmosphere. Anoxic sediments are more efficient than oxic sediments at preserving and burying organic carbon, rather than allowing it to respire. Oxygen gas, produced during photosynthesis,

Table 1
Long-term fate of fossil fuel CO_2 with and without the temperature/CO_2 feedback

	Time scale (years)	% of fossil fuel CO_2 left in the atmosphere		Amplification factor (%)
		No T feedback	T feedback	
1000 Gtons				
Ocean invasion	300	15.2	17.8	17.2
After $CaCO_3$ neutralization	5000–8000	7.4	9.0	22.3
5000 Gtons				
Dissolution	300	26.5	31.0	17.2
$CaCO_3$ neutralization	5000–8000	8.2	10.5	28.2

is left behind when organic carbon is buried, resulting in a net source of O_2 to the atmosphere. Oxygen levels may also affect the ocean inventories of the nutrients PO_4^{3-}, NO_3^-, and Fe, limiting nutrients which serve to pace the biological cycle in the ocean, further feeding back to O_2. The extent of anoxia in the ocean, driven by ocean temperature, therefore has the potential to affect many aspects of the biosphere.

Walker [42] makes the observation that the ocean contains just enough PO_4^{3-} to bring the deep sea, on average, to the brink of anoxia. He proposes an analogy to a thermostat, which ought to have its switch contact close to closing, if the thermostat is regulating the temperature closely. The deep ocean is close to anoxic because if it went anoxic, it could easily generate more O_2 for the atmosphere, thereby pulling itself back to oxygenation.

Walker's story is complicated by the behavior of PO_4^{3-} in the ocean. O_2 in the deep ocean is consumed biologically, but the drawdown is limited by the availability of PO_4^{3-} to fuel photosynthesis in the surface ocean. (We assume that the ocean inventory of biologically available nitrogen is controlled, on geologic time scales, by the availability of PO_4^{3-}). Walker's thought experiment assumes that the surface ocean is saturated in O_2 and depleted in PO_4^{3-}. The PO_4^{3-} concentration of the deep ocean is close to the ocean mean (because it is a large reservoir). These assumptions yield

$$O_2(\text{deep ocean}) = O_2(\text{sat})$$
$$- (O/P)_{\text{plankton}} PO_4^{3-}(\text{mean})$$

where the term $(O/P)_{\text{plankton}}$ represents the oxygen demand for phytoplankton degradation, analogous to the Redfield ratio for N/P in plankton. The saturated O_2 concentration, $O_2(\text{sat})$, is defined by Henry's law

$$O_2(\text{sat}) = K_H(T)*pO_2(\text{atm})$$

where the equilibrium constant K_H decreases by about 25% with an increase in ocean temperature from 2 to 15 °C. Assuming hypothetically that the deep ocean approaches anoxia in the limit of a fully efficient biological pump, Walker derived the implication of these assumptions

$$pO_2(\text{atm}) = K_H(O/P)_{\text{plankton}} PO_4^{3-}(\text{mean})$$

and points out that this relation is nearly satisfied for today's ocean. The real deep ocean is not anoxic because not all of the upwelling PO_4^{3-} is utilized by phytoplankton, but it could be. The ocean contains just enough PO_4^{3-} to take the deep ocean to anoxia, neither much more nor less.

If we postulate that an ocean anoxia chemostat exists, maintaining this relationship through time, then we might speculate about the impact of a change in the O_2 solubility K_H. An increase in ocean temperature, driving a decrease in O_2 solubility, would have to be compensated by an increase in atmospheric O_2 or a decrease in ocean PO_4^{3-}. Several intriguing papers have been published on the interplay between O_2 and PO_4^{3-} [43–45]. In all of these models, anoxia in the ocean affects the burial of P, while the ocean inventory of P and the surface ocean O_2 concentration determine the O_2 concentration in the deep sea. Surface ocean O_2 depends on atmospheric pO_2 scaled by its solubility, which depends on temperature. An increase in temperature would decrease the solubility of oxygen, bringing anoxia that much closer to within reach of the biological pump. Oxygen solubility during the Cretaceous was about 25% lower than today. The effect of higher temperature, by itself, might have been to increase atmospheric pO_2 by approximately 25%, to offset the decrease in O_2 solubility. The physiology of Cretaceous insects [46] has lead to the speculation of higher O_2 concentrations at this time, but we have no reliable geochemical indications of atmospheric O_2 levels from this time [47].

4.3. Methane clathrates

Enormous amounts of methane exist in continental margin sediments of the ocean, frozen into cages of water ice in structures known as methane hydrates or clathrates. The total inventory of methane in the clathrate reservoir is poorly constrained, but has been estimated to exceed the inventory of extractable coal and other fossil fuels, by as much as a factor of 3 [48–51]. The distribution of methane clathrates in sediments of the ocean is limited by the temperature of the water and sediment columns (Fig. 5). At low pressure near the sea surface, the temperature of the overlying water is always too warm to support clathrate stability, except perhaps in the Arctic. Within the sediment

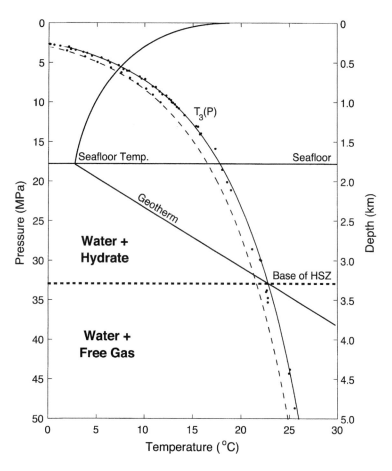

Fig. 5. Schematic illustration of temperature through the ocean and uppermost marine sediments. The temperature for clathrate stability, $T_3(P)$, increases with pressure (or depth). Experimental data for $T_3(P)$ in pure water (blue) and seawater (red) are extrapolated using a thermodynamic model [61]. A representative profile of temperature through the ocean intersects $T_3(P)$ at a depth of roughly 600 m, although clathrate is normally confined to the top few hundred meters of sediments. The base of the clathrate stability zone is defined by the intersection of the geotherm with $T_3(P)$. Adapted from Davie and Buffett [60].

column, the temperature increases with increasing depth along the geotherm. At depth in the sediment, the geotherm crosses the pressure-dependent clathrate stability temperature, demarking the lower bound for clathrate stability. The stability zone in the sediment increases with increasing water depth, but the organic carbon rain from the sea surface decreases toward the abyss. There is therefore a maximum abundance of clathrates at an intermediate depth range in the ocean, from 1 to 3 km. An increase in ocean temperature would decrease the thickness of the clathrate stability field [52], presumably decreasing the maximum inventory of clathrates in the global ocean.

Methane is a greenhouse gas, as is its oxidation product CO_2. The release of methane to the ocean or atmosphere during a transient warming has the potential to amplify the warming, creating a positive feedback analogous to that for the CO_2/T system described above. The two ingredients for the feedback are the temperature of the deep sea as driven by the radiative effect of the released carbon, and the sensitivity of the clathrate reservoir as a function of the temperature of the ocean. As for the O_2 and CO_2 systems, the time scales of the perturbations and responses play a major role in determining the behavior of the system.

4.3.1. Deep-ocean temperature response to clathrate methane release

If methane from decomposing clathrate manages to get to the atmosphere, its radiative impact will be 20 times stronger than CO_2 per molecule. However, the residence time of methane in the atmosphere, at present concentrations, is only around a decade, much shorter than the millennium time scale thermal response time of the deep sea. Therefore, the deep-ocean temperature rise associated with a short, catastrophic methane release depends on the dynamics of CO_2 rather than methane. The ocean carbon cycle model results presented above demonstrates that some fraction of any CO_2 added to the atmosphere/ocean reservoir remains in the atmosphere for order of 100 ky, limited by the time scale of the silicate weathering thermostat. A model of the PETM carbon cycle [53,54] predicted a 70-µatm rise in atmospheric pCO_2, assuming a 1000-Gton methane release spread over 10 ky. The atmospheric fraction of their clathrate carbon is 14%, broadly consistent with our model behavior, but we note that the atmospheric CO_2 signature would be higher if the methane release were larger (2000 Gtons is a more typical estimate) or faster (because $CaCO_3$ compensation neutralizes CO_2 if it has time). A 2000-Gton release in 1000 years might transiently increase pCO_2 by 600 Gtons, or about 300 µatm.

The radiative effect of the CO_2 increase depends on the initial pCO_2, because of the saturation of the infrared absorption bands of the CO_2. A doubling of the CO_2 concentration has roughly the same warming effect regardless of whether the doubling is from 100 to 200 µatm or from 1000 to 2000 µatm. The initial deep ocean temperature during the Paleocene, before the degassing event, was $\sim 8\,°C$, suggesting an atmospheric pCO_2 much higher than today. If we add 100–200 µatm to an initial pCO_2 of say 1000 µatm, the radiative signature will be much too small to explain the $\sim 5\,°C$ warming inferred from the $\delta^{18}O$ record. This suggests that the warming was driven externally, rather than simply by the clathrate decomposition event. On the other hand, the recovery from the warming parallels the recovery from the carbon isotope anomaly very closely, and the time scale for these recoveries are similar to the 100-ky time scale for silicate weathering neutralization of CO_2 and $\delta^{13}C$. Perhaps we are somehow underestimating the

atmospheric pCO_2 response to the clathrate degassing, or its radiative impact.

4.3.2. Methane clathrate inventory as a function of ocean temperature

The other half of the clathrate/temperature system is the response of the clathrate inventory to changes in ocean temperature. The longest time scale in the system is the recharging of the clathrate reservoir after it is depleted, which takes place on a million-year time scale, limited by the rate of methane production and sediment advection. The real unknown is the time scale for a methane release in response to deep-ocean warming. A change in deep-ocean temperature could propagate through the order of 100-m-thick sedimentary clathrate zone in order 10^3 years. If methane clathrate dissociates rapidly, the released gas elevates the pressure of pore water in the sediments, potentially leading to failure and slumping of the marine sediments [55,56]. The potential for these is documented by numerous pockmarks and submarine landslides on the sea floor [57]. These would probably be local events rather than global, however, so they do not suggest a mechanism by which the entire global clathrate inventory might adjust itself in a catastrophic way.

Very different consequences are expected if the dissociation of methane clathrate occurs slowly. In this case, bubbles of methane gas can remain trapped in the sediments as the clathrate dissociates. Some of this methane dissolves into the surrounding pore water and is transported toward the seafloor by diffusion and fluid flow. Methane oxidation can follow either of two pathways. Reaction with sulfate, followed by sulfide precipitation, releases methane carbon in the form of HCO_3^-, ultimately provoking the precipitation of $CaCO_3$, which remains stably sequestered in the subsurface sediment column. Oxic methane oxidation in contrast releases carbon in the form of dissolved CO_2. On a thousand-year time scale, it makes little difference whether the CO_2 is produced in the atmosphere or the ocean; in either case, it will partition itself according to the proportions reflected in Table 1.

Isotopic data from the PETM indicate a degassing time scale between these two extremes. The isotopic signature of the methane reached shallow-water $CaCO_3$ and terrestrial carbon deposits, which means that the methane did at least reach the ocean rather than precipitate as $CaCO_3$ at depth in the sediment [53]. The time scale for the $\delta^{13}C$ lightening (the release event) is estimated to be ~ 10 ky [58].

4.3.3. Significance of the time scale asymmetry

The asymmetry in the time scales of buildup and decomposition of the methane clathrate reservoir has several interesting implications. One stems from the fact that buildup takes place on a longer time scale than the silicate weathering thermostat. The sequestration of carbon in the form of methane during the charging stage of the capacitor is therefore unable to affect the pCO_2 of the atmosphere very much, because that job belongs to the silicate weathering thermostat. Discharging the methane on the other hand is faster than silicate weathering, and therefore has the capacity to affect pCO_2 on time scales shorter than that for the silicate weathering thermostat.

Another interesting potential implication of the asymmetry of buildup and breakdown of the clathrate reservoir is that the inventory of methane over the glacial/interglacial cycles ought to reflect the ocean temperature maxima, not the minimum or time mean temperature, if the time scale for degassing is fast compared to a glacial cycle but the time scale for accumulation is slower. The implication for this would be that a future degassing adjustment, as earth's climate climbs into a greenhouse it has not seen in millions of years, might be larger than any methane release seen in response to the warming associated with the glacial termination.

5. Future directions

A remarkable observation from the geologic record is the relative stability of the temperature and atmospheric pO_2 through time, crucial to nurturing complex life and civilization. As mankind increasingly takes control of the biosphere, we need to understand the mechanisms responsible for this stability. An explanation of the glacial pCO_2 cycles, of which ocean temperature plays some part, will lend confidence to the forecast of the carbon cycle in the future. Our understanding of the physics responsible for the warm deep ocean in the early Cenozoic is incomplete, as is our understanding of the physics of warm climates generally. Research into the dynamics and stability of the methane hydrate reservoir is also in its infancy. The temperature of the ocean couples together the cycles and balances of CO_2, O_2, and methane, creating new feedbacks and interactions. Ultimately, many of the outstanding research problems and questions described in this paper will be relevant to forecasting the future trajectories of climate and geochemistry of the biosphere, although perhaps on time scales longer than are typically considered in global change deliberations.

Acknowledgements

We wish to thank Katharina Billups who provided Mg paleotemperature data, and Ken Caldeira, Andy Ridgwell, and Tom Guilderson for thorough, thoughtful and intelligent reviews. *[AH]*

References

[1] R.J. Stouffer, S. Manabe, Equilibrium response of thermohaline circulation to large changes in atmospheric CO_2 concentration, Clim. Dyn. 20 (2003) 759–773.

[2] H.C. Urey, The thermodynamic properties of isotopic substances, J. Chem. Soc. (London) 297 (1947) 562–581.

[3] B.E. Bemis, H.J. Spero, J. Bijma, D.W. Lea, Reevaluation of the oxygen isotopic composition of planktonic foraminifera: experimental results and revised paleotemperature equations, Paleoceanography 13 (2) (1998) 150–160.

[4] N.J. Shackleton, Attainment of isotopic equilibrium between ocean water and the benthoic foraminifera genus *Unigerina*: isotopic changes in the ocean during the last glacial, Cent. Natl. Sci. Colloq. Int. 219 (1974) 203–209.

[5] D.P. Schrag, D.J. Depaolo, F.M. Richter, Reconstructing past sea-surface temperatures-correcting for diagenesis of bulk marine carbonate, Geochim. Cosmochim. Acta 59 (11) (1995) 2265–2278.

[6] P.N. Pearson, P.W. Ditchfield, J. Singano, K.G. Harcourt-Brown, C.J. Nicholas, R.K. Olsson, N.J. Shackleton, M.A. Hall, Warm tropical sea surface temperatures in the Late Cretaceous and Eocene epochs, Nature 413 (6855) (2001) 481–487.

[7] Y. Rosenthal, E.A. Boyle, N. Slowey, Temperature control on the incorporation of magnesium, strontium, fluorine, and cad-

mium into benthic foraminiferal shells from Little Bahama Bank: prospects for thermocline paleoceanography, Geochim. Cosmochim. Acta 61 (17) (1997) 3633–3643.

[8] T.A. Mashiotta, D.W. Lea, H.J. Spero, Glacial–interglacial changes in Subantarctic sea surface temperature and delta O-18-water using foraminiferal Mg, Earth Planet. Sci. Lett. 170 (4) (1999) 417–432.

[9] D.W. Lea, T.A. Mashiotta, H.J. Spero, Controls on magnesium and strontium uptake in planktonic foraminifera determined by live culturing, Geochim. Cosmochim. Acta 63 (16) (1999) 2369–2379.

[10] C.H. Lear, H. Elderfield, P.A. Wilson, Cenozoic deep-sea temperatures and global ice volumes from Mg/Ca in benthic foraminiferal calcite, Science 287 (5451) (2000) 269–272.

[11] K. Billups, D.P. Schrag, Paleotemperatures and ice volume of the past 27 Myr revisited with paired Mg/Ca and O-18/O-16 measurements on benthic foraminifera, Paleoceanography 17 (1).

[12] K. Billups, D.P. Schrag, Application of benthic foraminiferal Mg/Ca ratios to questions of Cenozoic climate change, Earth Planet. Sci. Lett. 209 (1–2) (2003) 181–195.

[13] P.A. Martin, D.W. Lea, Y. Rosenthal, N.J. Shackleton, M. Sarnthein, T. Papenfuss, Quaternary deep sea temperature histories derived from benthic foraminiferal Mg/Ca, Earth Planet. Sci. Lett. 198 (1–2) (2002) 193–209.

[14] J.C. Zachos, M. Pagani, L. Sloan, E. Thomas, K. Billups, Trends, rhythms, and aberrations in global climate 65 Ma to Present, Science 292 (2001) 686–693.

[15] J.C. Zachos, J.R. Breza, S.W. Wise, Early Oligocene ice-sheet expansion on Antarctica—stable isotope and sedimentological evidence from Kerguelen Plateau, Southern Indian-Ocean, Geology 20 (6) (1992) 569–573.

[16] P. Martin, Evidence for the role of deep sea temperatures in glacial climate and carbon cycle, Paleoceanography, in press.

[17] D.P. Schrag, G. Hampt, D.W. Murray, Pore fluid constraints on the temperature and oxygen isotopic composition of the glacial ocean, Science 272 (1996) 3385–3388.

[18] J.F. Adkins, K. McIntyre, D.P. Schrag, The salinity, temperature, and d18O of the glacial deep ocean, Science 298 (2002) 1769–1773.

[19] N.J. Shackleton, The 100,000-year ice-age cycle identified and found to lag temperature, carbon dioxide, and orbital eccentricity, Science 289 (5486) (2000) 1897–1902.

[20] J. Chappell, Sea level changes forced ice breakouts in the Last Glacial cycle: new results from coral terraces, Quat. Sci. Rev. 21 (10) (2002) 1229–1240.

[21] M. Huber, L.C. Sloan, Heat transport, deep waters, and thermal gradients: coupled simulation of an Eocene greenhouse climate, Geophys. Res. Lett. 28 (18) (2001) 3418–3484.

[22] R. Zhang, M. Follows, J. Marshall, Mechanisms of thermohaline mode switching with application to warm equable climates, J. Climate 15 (2001) 2056–2072.

[23] R. Zhang, M.J. Follows, J.P. Grotzinger, J. Marshall, Could the late Permian deep ocean have been anoxic? Paleoceanography 16 (3) (2001) 317–329.

[24] A. Ganopolski, S. Rahmstorf, V. Petoukhov, M. Claussen, Simulation of modern and glacial climates with a coupled global model of intermediate complexity, Nature 371 (1998) 323–326.

[25] C.D. Hewitt, A.J. Broccoli, J.F.B. Mitchell, R.J. Stouffer, A coupled model study of the last glacial maximum: was part of the North Atlantic relatively warm? Geophys. Res. Lett. 28 (2001) 1571–1574.

[26] J.C. Duplessy, N.J. Shackleton, R.G. Fairbanks, L. Labeyrie, D. Oppo, N. Kallel, Deepwater source variations during the last climatic cycle and their impact on the global deepwater circulation, Paleoceanography 3 (1988) 343–360.

[27] S.-I. Shin, Z. Liu, B. Otto-Bliesner, E.C. Brady, J.E. Kutzbach, S.P. Harrison, A simulation of the Last Glacial Maximum using the NCAR-CCSM, Clim. Dyn. 20 (2003) 127–151.

[28] R.J. Stouffer, S. Manabe, Response of a coupled ocean-atmosphere model to increasing atmospheric carbon dioxide: sensitivity to the rate of increase, J. Climate 12 (8) (1999) 2224–2237.

[29] D.E. Archer, E. Maier-Reimer, Effect of deep-sea sedimentary calcite preservation on atmospheric CO_2 concentration, Nature 367 (1994) 260–264.

[30] D. Archer, H. Kheshgi, E. Maier-Riemer, Multiple timescales for neutralization of fossil fuel CO_2, Geophys. Res. Lett. 24 (1997) 405–408.

[31] D. Archer, H. Kheshgi, E. Maier-Reimer, Dynamics of fossil fuel CO2 neutralization by marine $CaCO_3$, Glob. Biogeochem. Cycles 12 (1998) 259–276.

[32] W.H. Berger, Increase of carbon dioxide in the atmosphere during deglaciation: the coral reef hypothesis, Naturwissenschaften 69 (1982) 87–88.

[33] J.R. Petit, J. Jouzel, D. Raynaud, N.I. Barkov, J.-M. Barnola, I. Basile, M. Bender, J. Chappellaz, M. Davis, G. Delaygue, M. Delmotte, V.M. Kotlyakov, M. Legrand, V.Y. Lipenkov, C. Lorius, L. Pepin, C. Ritz, E. Saltzman, M. Stievenard, Climate and atmospheric history of the past 420,000 years from the Vostok ice core, Antarctica, Nature 399 (1999) 429–436.

[34] D.E. Archer, A. Winguth, D. Lea, N. Mahowald, What caused the glacial/interglacial atmospheric pCO$_2$ cycles? Rev. Geophys. 38 (2000) 159–189.

[35] J.C.G. Walker, P.B. Hays, J.F. Kasting, A negative feedback mechanism for the long-term stabilization of Earth's surface temperature, J. Geophys. Res. 86 (1981) 9776–9782.

[36] R.A. Berner, A.C. Lasaga, R.M. Garrels, The carbonate-silicate geochemical cycle and its effect on atmospheric carbon dioxide over the past 100 million years, Am. J. Sci. 283 (1983) 641–683.

[37] R.A. Berner, GEOCARB II: a revised model of atmospheric CO_2 over Phanerozoic time, Am. J. Sci. 294 (1994) 56–91.

[38] R.A. Berner, Z. Kothavala, GEOCARB III: a revised model of atmospheric CO_2 over Phanerozoic time, Am. J. Sci. 301 (2) (2001) 182–204.

[39] J. Kennett, B. Ingram, A 20,000-year record of ocean circulation and climate change from the Santa Barbara basin, Nature 377 (1995) 510–514.

[40] R.S. Ganeshram, T.F. Pedersen, S.E. Calvert, J.W. Murray, Large changes in oceanic nutrient inventories from glacial to interglacial periods, Nature 376 (1995) 755–758.

[41] M.A. Altabet, R. Francois, D.W. Murray, W.L. Prell, Climate-

[42] J.C.G. Walker, Evolution of the Atmosphere, Macmillan, New York, 1977, 318 pp.

[43] P.V. Cappellen, E. Ingall, Redox stabilization of the atmosphere and oceans by phosphorus-limited marine productivity, Science 271 (1996) 493–495.

[44] T.M. Lenton, A.J. Watson, Redfield revisited: 1. Regulation of nitrate, phosphate, and oxygen in the ocean, Glob. Biogeochem. Cycles 14 (2000) 225–248.

[45] I.C. Handoh, T.M. Lenton, Periodic mid-Cretaceous ocean anoxic events linked by oscillations of the phosphorus and oxygen biogeochemical cycles, Glob. Biogeochem. Cycles 17 (2003) doi:10.1029/2003GB0022039.

[46] J.B. Graham, R. Dudley, N.M. Aguilar, C. Gans, Implications of the late Palaeozoic oxygen pulse for physiology and evolution, Nature 375 (1995) 117–120.

[47] R.A. Berner, Atmospheric oxygen over Phanerozoic time, Proc. Natl. Acad. Sci. U. S. A. 96 (1999) 10955–10957.

[48] K.A. Kvenvolden, Methane hydrate—a major reservoir of carbon in the shallow geosphere, Chem. Geol. 71 (1–3) (1988) 41–51.

[49] G. MacDonald, Role of methane clathrates in past and future climates, Clim. Change 16 (1990) 247–281.

[50] V. Gornitz, I. Fung, Potential distribution of methane hydrate in the world's oceans, Glob. Biogeochem. Cycles 8 (1994) 335–347.

[51] L.D.D. Harvey, Z. Huang, Evaluation of the potential impact of methane clathrate destabilization on future global warming, J. Geophys. Res. 100 (1995) 2905–2926.

[52] G.R. Dickens, The potential volume of oceanic methane hydrates with variable external conditions, Org. Geochem. 32 (2001) 1179–1193.

[53] G.R. Dickens, Modeling the global carbon cycle with a gas hydrate capaciter: significance for the latest Paleocene thermal maximum, in: Natural Gas Hydrates: Occurrence, Distribution and Detection, AGU Geophys. Monogr. 124 (2001) 19–38.

[54] J.C.G. Walker, J.F. Kasting, Effects of fuel and forest conservation on future levels of atmospheric carbon dioxide, Palaeogeogr. Palaeoclimatol. Palaeoecol. (Glob. Planet. Change Sect.) 97 (1992) 151–189.

[55] R.D. McIver, Role of naturally occurring gas hydrates in sediment transport, AAPG Bull. 66 (1982) 789–792.

[56] R.E. Kayen, H.J. Lee, Pleistocene slope instability of gas hydrate-laden sediment of Beaufort Sea margin, Marine Geotech. 10 (1991) 125–141.

[57] M. Hovland, A.G. Judd, Seabed Pockmarks and Seepages, Graham and Trotman, London, 1988.

[58] G.R. Dickens, Rethinking the global carbon cycle with a large, dynamic and microbially mediated gas hydrate capacitor, Earth Planet. Sci. Lett. 213 (2003) 169–183.

[59] L.D. Laberyie, J.C. Duplessy, P.L. Blanc, Variations in mode of formation and temperature of oceanic deep waters over the past 125,000 years, Nature 327 (1987) 477–482.

[60] M.K. Davie, B.A. Buffett, Sources of methane for marine gas hydrate: inferences from a comparison of observations and numerical models, Earth Planet. Sci. Lett. 206 (1–2) (2003) 51–63.

[41] related variations in denitrification in the Arabian Sea from sediment 15N/14N ratios, Nature 373 (1995) 506–509.

[61] E.D. Sloan, Clathrate Hydrates of Natural Gas, Marcel Dekker, New York, 1988.

David Archer is a professor in the Geophysical Sciences Department at the University of Chicago, where he studies the carbon cycle in the ocean and in deep-sea sediments, and its relation to global climate.

Pamela Martin is an assistance professor in the Geophysical Sciences Department at the University of Chicago. Her research broadly focuses on reconstructing changes in deep-ocean temperature, chemistry, and circulation to understand oceanic controls on climate change. She is interested in the links between ocean biogeochemical cycles, atmospheric carbon dioxide and climate change on time scales ranging from orbital variations (hundreds of thousands of years) to anthropogenic variations (decades to millennium). She makes measurements of the chemical composition of fossils to document climate changes in the geologic record and use numerical models to investigate the ocean's role in the climate system.

Bruce Buffett is a professor in the Geophysical Sciences Department at the University of Chicago. His research deals with dynamical processes in the Earth's interior. This can include mantle convection, plate tectonics, and the generation of the Earth's magnetic field. A central theme that connects these processes is the thermal and compositional evolution of the Earth because it affects the energy which is available to drive the internal dynamics of the planet. I attempt to understand the structure, dynamics and evolution of the Earth's interior using theoretical models in combination with geophysical observations.

Victor Brovkin, is a Post-Doc at Potsdam Institute for Climate Impact Research, Potsdam, Germany. He has 10 years experience in biospheric modelling on a global scale with focus on developing of vegetation dynamics, terrestrial and oceanic carbon cycle models for Earth System Models of Intermediate Complexity (EMICs). His main scientific interests include stability analysis of climate–vegetation system, effect of deforestation on climate system, and integration of terrestrial and marine components of the global carbon cycle.

Stefan Rahmstorf is a research scientist at the Potsdam Institute for Climate Impact Research and a professor at the University of Potsdam. His research team studies the role of the oceans in climate change, in the past (e.g, during the last Ice Age), in the present and for a further global warming.

Andrey Ganopolski obtained his MS and PhD degrees in geophysics at the Moscow State University. Since 1994, he is research scientist at the Potsdam Institute for Climate Impact Research. His work focuses on climate modelling, climate predictions and studying of the past climate changes.

Reprinted from
Earth and Planetary Science Letters 222 (2004) 1–15

www.elsevier.com/locate/epsl

The hazard of near-Earth asteroid impacts on Earth

Clark R. Chapman *

Southwest Research Institute, Suite 400, 1050 Walnut St., Boulder CO 80302, USA

Received 15 December 2003; received in revised form 14 February 2004; accepted 4 March 2004

Abstract

Near-Earth asteroids (NEAs) have struck the Earth throughout its existence. During epochs when life was gaining a foothold ~ 4 Ga, the impact rate was thousands of times what it is today. Even during the Phanerozoic, the numbers of NEAs guarantee that there were other impacts, possibly larger than the Chicxulub event, which was responsible for the Cretaceous–Tertiary extinctions. Astronomers have found over 2500 NEAs of all sizes, including well over half of the estimated 1100 NEAs >1 km diameter. NEAs are mostly collisional fragments from the inner half of the asteroid belt and range in composition from porous, carbonaceous-chondrite-like to metallic. Nearly one-fifth of them have satellites or are double bodies. When the international telescopic Spaceguard Survey, which has a goal of discovering 90% of NEAs >1 km diameter, is completed, perhaps as early as 2008, nearly half of the remaining impact hazard will be from land or ocean impacts by bodies 70–600 m diameter. (Comets are expected to contribute only about 1% of the total risk.) The consequences of impacts for civilization are potentially enormous, but impacts are so rare that worldwide mortality from impacts will have dropped to only about 150 per year (averaged over very long durations) after the Spaceguard goal has, presumably, ruled out near-term impacts by 90% of the most dangerous ones; that is, in the mid-range between very serious causes of death (disease, auto accidents) and minor but frightening ones (like shark attacks). Differences in perception concerning this rather newly recognized hazard dominate evaluation of its significance. The most likely type of impact events we face are hyped or misinterpreted predicted impacts or near-misses involving small NEAs.
© 2004 Elsevier B.V. All rights reserved.

Keywords: asteroids; near-Earth; craters; impact; impact hazard; mass extinctions

1. Why are near-Earth asteroids important?

Although interplanetary space is very empty by human standards, Earth is in a "cosmic shooting gallery", as anyone looking up into clear, dark skies can witness: several meteors can be seen flashing across the heavens per hour—cometary and asteroidal dust grains disintegrating in the upper atmosphere. They are accompanied by a size spectrum [1] of ever larger, increasingly less common, bodies up to at least several tens of kilometers in diameter. Although occasional recovered meteorites are from the Moon and Mars (presumably other bodies are also represented by exotic celestial debris), the vast majority of near-Earth objects (NEOs) are asteroids or comets, or their smaller fragments or disintegration products

---

* Tel.: +1-303-546-9670; fax: +1-303-546-9687.
E-mail address: cchapman@boulder.swri.edu (C.R. Chapman).

0012-821X/$ - see front matter © 2004 Elsevier B.V. All rights reserved.
doi:10.1016/j.epsl.2004.03.004

called "meteoroids" (which are called "meteors" while in the atmosphere and "meteorites" when on the ground).

There is scientific consensus that the planets grew, in part, from the accumulation of much smaller objects called "planetesimals" [2]. When the epoch of planetary accretion was largely over, numerous planetesimals remained in orbit around the Sun. By convention, those in and inside of Jupiter's orbit are called "asteroids" and those farther out "comets", although each group is subdivided into specific orbital classes; comets generally have more volatiles than the more rocky or metallic asteroids, although primitive asteroids could be volatile-rich at depth. The dominant asteroid reservoirs are in a large torus called the main asteroid belt and in two groups of "Trojans" averaging 60° ahead of and behind Jupiter in its orbit. The chief known comet reservoirs are the Kuiper Belt and associated scattered disk (beyond Neptune's orbit) and the much more distant spherical halo of comets, called the "Oort Cloud".

Comets, asteroids and meteoroids slowly leak from these reservoirs, generally due to chaotic dynamics near planetary resonances (distances from the Sun where a small body has an orbital period that is a simple fraction of the orbital period of a planet), facilitated by collisions and other minor orbital perturbations (e.g., the Yarkovsky Effect, which is a force on a small body due to asymmetric re-radiation of absorbed sunlight on the warmer "afternoon" side of a spinning asteroid [3]). Some dislodged bodies soon arrive in the terrestrial planet zone, becoming NEOs. Comets rapidly disintegrate as their volatiles are exposed to the Sun. Near-Earth asteroids (NEAs), especially those with orbital aphelia (farthest point from the Sun of an elliptical orbit) remaining out in the asteroid belt, continue to suffer occasional collisional fragmentation. NEAs are in comparatively transient orbits, typically encountering the Sun, or more unusually a terrestrial body, or being ejected from the solar system on hyperbolic orbits, on time-scales of a few million years; they are continually being replenished from their reservoirs. There may be roughly equal numbers of comets and NEAs among NEOs larger than several kilometers in diameter, and comets and asteroids may be roughly equal sources of meteor-producing interplanetary dust particles (see [4] for a critical evaluation). But throughout the

enormous size range that yields recovered meteorites up to NEAs that threaten civilization, the asteroid belt's inner half is the overwhelmingly dominant reservoir [5]. Therefore, comets, which are estimated to contribute only 1% of the total risk [6], are not emphasized in this review.

This impact environment has existed for the past 3.5 Gyr, according to the terrestrial crater record as well as the lunar chronology derived from associating datable lunar samples with cratered units on the Moon. Generally, the average Earth/Moon impact rate has varied little more than a factor of two during that time, although brief spikes in cratering rate (e.g., by a "comet shower") must have happened (e.g., [7]). Prior to 3.8 Ga, the impact environment was very different. About a dozen huge lunar basins formed from the time of Nectaris (dated at 3.90–3.92 Ga, although possibly as old as 4.1 Ga) until the last one (Orientale) at ~ 3.82 Ga, implying an abrupt decay and cessation of whatever source of objects produced that Late Heavy Bombardment (LHB). (About twice as many observable basins, and presumably others now erased, formed before Nectaris, but it is controversial whether there was a lull in impact rate before a "cataclysm" [8] or instead a generally high bombardment rate persisted since lunar crustal solidification, followed by a rapid decline from ~ 3.9 to ~ 3.82 Ga [9].) In any case, there can be no remnant today of any such short-lived population and, presumably, other now-decayed populations of NEOs may have existed during the first aeon of Earth's existence, due to rearrangements or late formations of planets, which could have stirred up small-body populations [10] or due to tidal or collision break-ups of an Earth-approaching body. Whatever else might have happened, the observed lunar LHB alone would have subjected the Earth, for ~ 50 Myr, to a bombardment rate *thousands* of times that of today, with pivotal implications for the origin and early evolution of life.

Even at the low modern impact rate, impacts happen often enough to affect profoundly the evolution of life (e.g., the Chicxulub impact 65 Myr ago, dominantly responsible for the K–T mass extinction). Because of the comparatively short timespan of human lives and even of civilization, the importance of impacts as a modern hazard is debatable. Below, I argue that the impact hazard is significant in the context of other man-made and natural hazards that society takes seriously, although impacts are obviously far less important than the chief issues affecting our lives. Of course, NEAs have other vital virtues. They and their accompanying meteoroids bring samples from far-flung locations in the solar system to terrestrial laboratories for analysis and they leave traces in ancient impact craters and basins on the Earth and the Moon, permitting broad insights into primordial and recent processes operating in the inner solar system. In the future, NEAs may provide way-stations for astronauts en route to Mars or elsewhere; they also may provide raw materials for utilization in space.

2. Historical recognition of the impact hazard

Ideas that comets might be dangerous date back at least to the 17th century, when Edmond Halley is said to have addressed the Royal Society and speculated that the Caspian Sea might be an impact scar [11]. The physical nature of comets remained poorly understood, however, until the mid-20th century. The first NEA (Eros) was not discovered until 1898 and the first NEA that actually crosses Earth's orbit (Apollo) was not found until 1932. By the 1940s, three Earth-crossing NEAs had been found, their basic rocky nature and relationship to meteorites was appreciated, and it was possible to estimate, albeit crudely, their impact rate [12]. The actual damage that a NEA impact might cause on Earth was concretely described by Baldwin [13], a leading advocate for the impact origin of lunar craters. Later, Öpik [14] (who understood both orbital dynamics and impact physics) proposed that NEA impacts might account for mass extinctions in the Earth's paleontological record. Around the same time, Shoemaker [15] firmly established the impact origin of Meteor Crater in Arizona.

Despite the prescience of these early planetary science pioneers, it was not only a cultural but a scientific shock when Mariner 4's first photographs of the Martian surface revealed it to be covered by craters [16]; a decade later, Mariner 10 found the same on Mercury. Although some fictional accounts of impact catastrophes were published in the 1970s, it was not until 1980/1981 that two events crystallized in the minds of many scientists both the dramatic effects on Earth history and the modern threat posed by impacts. First was publication [17] of

the Alvarez et al.'s hypothesis for the K–T boundary and second was a Snowmass, CO, NASA-sponsored workshop entitled "Collision of Asteroids and Comets with the Earth: Physical and Human Consequences", chaired by Eugene Shoemaker. In 1979 and 1980, the Voyagers first encountered Jupiter and Saturn, demonstrating that cratered surfaces extended from Mercury at least out through the giant planets' satellite systems. After a quarter century of space exploration, the particulars of a few NEAs, a few craters on the Earth, and the familiar cratered lunar surface had been linked and generalized to the solar system as a whole. Like any other planet, Earth's surface certainly has been bombarded over the aeons by the same cosmic projectiles.

Even after discovery of the Chicxulub impact structure in Mexico and its temporal simultaneity with the Cretaceous–Tertiary (K–T) boundary and mass extinctions [18], it has taken some earth scientists a while to recognize and accept the statistical inevitability that Earth is struck by asteroids and comets. Each impact, occurring on timescales of tens to hundreds of Myr, liberates tens of millions to billions of megatons (Mt, TNT-equivalent) of energy into the fragile ecosphere, which *must* have had dramatic consequences every time. A few researchers still consider the Chicxulub impact to be only one of several contributing factors to the K–T extinctions (e.g., [19]) and direct evidence firmly linking other mass extinctions to impacts is so far either more equivocal than for the K–T, or altogether lacking. Some geoscientists still think of asteroid impacts as ad hoc explanations for paleontological changes and they resist the logic that earlier, even greater impact catastrophes surely occurred. If the great mass extinctions are not attributed to impacts (e.g., explained instead by episodes of volcanism or sea regressions), one must ask how the huge impacts that must have occurred failed to leave dramatic evidence in the fossil record.

A new thread in public awareness of the modern impact hazard developed in the late 1980s when advanced telescopic search techniques identified NEAs passing by the Earth at distances comparable to that of the Moon. Such "near misses" made headlines and also inspired an aerospace organization and the U.S. Congress to mount a political mandate that NASA examine the impact threat and methods for mitigating it. This led to the definition [20] (and redefinition [21] after the dramatic 1994 impacts of Comet Shoemaker-Levy 9 fragments into Jupiter) of the Spaceguard Survey, which NASA formally endorsed in 1998 and committed to discovering 90% of NEAs >1 km diameter within one decade. (Spaceguard is a network of professional observatories, dominated by two 1-m aperture telescopes near Socorro, NM, operated by MIT Lincoln Laboratory (LINEAR), plus amateur and professional observers who follow up the discoveries in order to refine knowledge of NEA orbits.) As larger NEAs are discovered and their orbital paths extrapolated ahead one century are found to pose zero danger of impact, then we are safer: only the remaining, undiscovered asteroids pose a threat. In 2000, the British government established a Task Force on Potentially Hazardous NEOs, which led to a report [11] and the establishment of the first governmental organization solely devoted to the impact hazard, the NEO Information Centre. Most recently, as the Spaceguard Survey approaches its goal, NASA tasked a new group (NEO Science Definition Team, SDT) to advise on possibilities of extending NEA searches down to smaller sizes; it reported [6] in August 2003.

Funding for research on the modern impact hazard has been minimal, so much of the thinking has taken place in the context of conferences and committee studies rather than comprehensive research programs; reports from these activities, some published as "grey literature", others in professional series, constitute the chief sources of information on the topic [6,11,20–26]. An extensive literature exists on the role of impacts in Earth's geological and paleontological history; the most recent compendium [27] is the fourth in a series of "Snowbird Conferences", which commenced in 1981 soon after the publication [17] of the Alvarez et al.'s hypothesis. The dynamical and physical properties of NEAs were recently reviewed in several chapters of "Asteroids III" [28].

3. Physical and dynamical properties of NEAs

NEAs are defined, somewhat arbitrarily, as asteroids whose perihelia (closest orbital distance to the Sun) are < 1.3 AU (1 AU = the mean distance of Earth from the Sun). About 20% of NEAs are currently in orbits that can approach the Earth's orbit to within < 0.05 AU; these are termed potentially hazardous objects (PHOs). In terms of their origin and physical nature, PHOs are no different from other NEAs; they just happen to come close enough to Earth at the present time so that close planetary encounters could conceivably perturb their orbits so as to permit an actual near-term collision, hence they warrant careful tracking. The Spaceguard search programs (chiefly LINEAR; Lowell Observatory's LONEOS in Flagstaff, AZ; Jet Propulsion Laboratory's Near-Earth Asteroid Tracking [NEAT] in Maui and on Mt. Palomar, CA; and Spacewatch on Kitt Peak, AZ [29]) continue to discover a new NEA every few days. As of February 2004, nearly 2670 NEAs were known (of which nearly 600 were PHOs), which compares with only 18 when the 1981 Snowmass conference met. The census is believed to be complete for NEAs >3 km diameter. The estimated number of NEAs >1 km in diameter (the size for which NASA established Spaceguard's 90% completeness goal by 2008) is ~ 1100 ± 200 [6], of which about 55% had been found by early 2004. As shown in Fig. 1, there is a roughly power-law increase in numbers of NEAs with decreasing size (differential power law exponent = − 3.35) down to the billion-or-so NEAs ≥ 4 m diameter; 4 m constitutes the annual impact event on Earth with an energy ~ 5 kt [30]. Frequencies are least secure, with the uncertainties approaching an order of magnitude, for NEAs too rare to be witnessed as bolides (brilliant meteors) but too small to be readily discovered telescopically, e.g., ~ 10–200 m diameter. This includes objects of the size (~ 50 m) that produced the dramatic 15 Mt Tunguska lower atmospheric explosion in Siberia as recently as 1908. The expected frequency of Tunguskas is less than once per thousand years; it is odd that the last one was so recent. An alternative possibility is that the destruction of thousands of square kilometers of forest was accomplished by a blast much less energetic than 15 Mt, due to a more common, smaller object (see appendix 4 in [6]).

The observed distribution of NEO orbits (characterized by semi-major axis a, eccentricity e and inclination i), after correction for observational biases in discovery, has been modeled in terms of source regions for these bodies within and beyond the aster-

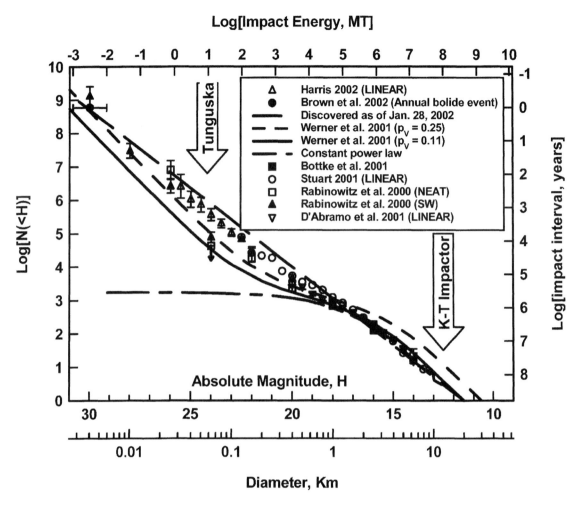

Fig. 1. Size distribution for cumulative number N of NEAs larger than a particular size, estimated in several ways, chiefly from telescopic search programs [6]. H is the stellar magnitude of an asteroid at 1 AU distance from both the Sun and the Earth (courtesy of A. Harris).

oid belt [5]. The result is that $37 \pm 8\%$ are derived from the ν_6 secular resonance (involving the precession rate of Saturn's orbit), which shapes the inner edge of the asteroid torus; $23 \pm 9\%$ come from the 3:1 resonance with Jupiter (where an asteroid orbits the Sun three times during 1 Jovian year), near 2.5 AU. And $33 \pm 3\%$ are derived from hundreds of weaker resonances throughout the asteroid belt (weighted toward the inner belt), in which asteroids gradually drift (by the Yarkovsky Effect) into numerous, weak orbital commensurabilities (ratios of orbital periods) involving Mars, Jupiter, and Saturn; they then chaotically diffuse, often becoming Mars-crossing and finally, after a few tens of Myr, NEAs. Lastly,

$6 \pm 4\%$ come from the short-period, Jupiter-family comet population (most of these are dormant, inactive comets, but a few have shown cometary activity). The contribution of long-period or new comets (e.g., from the Oort Cloud) to the threatening NEO population has recently [6] been assessed to be very low ($\sim 1\%$). Smaller, meteorite-sized bodies are also derived from the inner half of the asteroid belt, by similar dynamical mechanisms, although the Yarkovsky Effect is more important for smaller bodies, which accumulate much of their measured exposures to cosmic rays while moving within the asteroid belt rather than after they are in short-lived orbits that cross terrestrial planet orbits [31].

Mineralogical compositions of NEAs are assessed from absorption bands and other spectral signatures in reflected visible and near-infrared sunlight, after accounting for modification of the optical properties of surface minerals by the solar wind and micrometeoroid bombardment ("space weathering", [32]). These spectra are summarized by a colorimetric taxonomy [33]; the majority are divided between low-albedo types inferred to resemble carbonaceous chondritic meteorites and moderate-albedo types inferred to resemble ordinary chondrites and other stony meteorites. There are some more exotic types, like nickel–iron (metallic) meteorites and basaltic achondrites. Such inferences have been augmented by radar reflection [34] (which is especially sensitive to metal content) and confirmed by more detailed close-up examination of the large NEA, Eros, by the NEAR Shoemaker spacecraft [35,36]. Briefly, NEA colors and spectra, and inferred compositions, appear in proportions similar to those for asteroids in the inner half of the asteroid belt [37], consistent with the calculations of main-belt source regions for NEAs summarized above. Thus, accounting for physical processes (like weeding out of weak materials by the Earth's atmosphere), there appears to be compatibility between the compositions of inner-main-belt asteroids, NEAs and the meteorites that fall to Earth. No doubt, some NEAs are made of materials too weak to survive atmospheric passage; these might be dormant or dead comets or unevolved, primitive asteroids, and would be important targets for future investigations by spacecraft.

Our appreciation of the physical configurations of NEAs has undergone a revolution in the past decade. Although it was surmised several decades ago that some NEAs might be double (double craters are common on the Earth, and to varying degrees on the Moon, Venus and Mars [38,39]), only lately has it become clear that nearly 20% of NEAs have satellites or are double bodies. Definitive proof comes from radar delay-Doppler mapping [34] (see Fig. 2). But as adaptive optics and other modern techniques (including analysis of "eclipsing binary" lightcurves) discover common asteroid duplicity also among main-belt asteroids, Trojans, and Kuiper Belt objects [40], it is clear that small bodies must no longer be assumed to be simple objects. Several independent modes of formation are required to explain all of the double or satellite-containing small-body systems. But break-up by tidal disruption during close passage to a planet (as exemplified by Comet Shoemaker-Levy 9's break-up in 1992 [41]) seems to be the chief process accounting for the high fraction of NEO satellites and double bodies [40,42]. Clearly, the potential for a threatening NEA to have one or more satellites may complicate a deflection operation.

Tidal break-up is facilitated by another geophysical attribute of NEAs. Long ago, it was proposed [43] that larger main-belt asteroids might be "rubble piles" because inter-asteroidal collisions sufficiently energetic to fragment them would be insufficient to launch the fragments onto separate heliocentric orbits; instead, the pieces would reaccumulate into a rubble pile. It is now clear, from advanced modelling, that

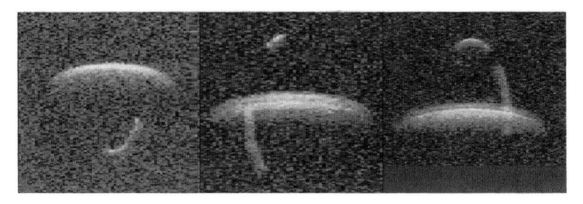

Fig. 2. Three radar delay-Doppler "images", taken over several hours, of the binary asteroid 1999 KW4. The motion of the satellite is evident. Such radar presentations should not be interpreted literally as an image. Nevertheless, it is evident that a smaller object is orbiting the main body (courtesy of Steve Ostro).

most asteroids, including those of sub-kilometer sizes, should be rubble piles or at least battered and badly fractured [42]. Lightcurves confirm that most NEAs >200 m diameter are weak or cohesionless, while smaller ones are monolithic "rocks". No NEA larger than 200 m (nor any main-belt asteroid) rotates faster than ~ 2.2 h, at which a cohesionless, fragmental body would fly apart by centrifugal force. However, all of the nearly dozen NEAs <200 m diameter with measured lightcurves have spin periods <2 h, ranging downward to just a couple minutes [44]. Clearly, the latter are strong, monolithic rocks, while larger NEAs are rubble piles, susceptible to disaggregation by tidal forces during a close passage to Earth or another large planet. Numerical simulations [42] show that some such tidal encounters result in double bodies or a dominant body with one or more satellites.

Small-scale surface properties of small, nearly gravitationless NEAs, below the resolution of radar delay-Doppler mapping, remain conjectural, except for Eros, which was imaged down to cm-scales near NEAR Shoemaker's landing site. Pre-NEAR Shoemaker predictions about the small-scale structure of Eros' surface were dramatically incorrect [45,46]: unlike the lunar regolith, small (<10 m diameter) craters are very rare on Eros, whereas boulders and rocks are extremely common. The character of surficial soils and regoliths on smaller NEAs is difficult to predict, but it is important, as all proposed deflection technologies (and mining operations) would have to interact with an NEA's surface, whether to attach a device, burrow into the object, or affect the surface remotely (e.g., by neutron bomb detonation or laser ablation). Probably, the surfaces of rapidly spinning small bodies <200 m diameter are composed of hard rock (or metal), with only an extremely thin layer of surficial particulates (e.g., bound by electrostatic forces).

4. Past history of impacts on Earth

I now discuss briefly, from a planetary science perspective, the role of impacts on the geological and biological history of our planet. By considering the past, I set the stage for the modern impact hazard. Clearly, impacts dominated the early geological evolution of the surface of our planet until at least 3.8 Ga.

It is almost equally incontrovertible that impacts have continued to interrupt more quiescent evolution of our planet's ecosphere well into the Cenozoic; such impacts will continue, even though other processes (e.g., plate tectonics, volcanism, weathering and erosion) are now more important than localized impacts in shaping geomorphology.

About 170 impact craters have been recognized on Earth [47], and perhaps double that number according to private, commercial records. They range from recent, small (tens to hundreds of meters in diameter) impact craters to multi-hundred km structures expressed in the geologic record although lacking crater-like morphology, which has been eroded away. Published ages for some of these impact scars are precise, but others are poorly characterized and often turn out to be erroneous (raising doubts about alleged periodicities in impact rates). The Earth's stratigraphic history is increasingly incomplete for older epochs, but the virtual total loss of datable rocks back toward 4 Ga is consistent with the inferences from the lunar LHB that Earth was pummeled by a couple lunar-basin-forming projectiles every Myr for 50–100 Myr, which would have boiled away any oceans and completely transformed the atmospheric, oceanic and crustal environment of the planet. Additionally, *thousands* of K–T boundary level events, one every 10,000 years, must have had profound repercussions.

The LHB would certainly have "frustrated" the origin of life on Earth [48,49]. Yet, some impacting projectiles might have contributed life-enhancing, volatile-rich substances to our planet. Moreover, impacts necessarily eject small fractions of excavated materials at greater than escape velocity. Any simple, extant lifeforms might survive in such ejecta (orbiting in geocentric or heliocentric orbits), and "re-seed" life upon re-impacting Earth after terrestrial environments had relaxed from the violent aftereffects of such impacts [50,51]. As I noted above, the Earth's impact environment became similar to today's by ~ 3.5 Ga. Dozens of K–T level impacts would have happened since that time, several of which were at least an order-of-magnitude even more devastating. Momentous events, like "Snowball Earth" [52,53], have been hypothesized to have occurred in pre-Phanerozoic times; the inevitable cosmic impacts must be considered as plausible triggers for such dramatic climatic changes, or their cessation, during those aeons.

During the Phanerozoic, there must have been several K–T (or greater) impact events, roughly equaling the number of major mass extinctions recorded in the fossil record. Only the K–T boundary extinction is now accepted as being largely, or exclusively, due to impact (the formation of Chicxulub). Evidence accumulates that the greatest mass extinction of all, the Permian–Triassic event, was exceptionally sudden [54] and is associated with evidence for impact [55], but generally the search for evidence as robust as what proved the impact origin of the K–T has been unproductive. Perhaps the K–T impact was exceptionally efficient in effecting extinction (e.g., because of the composition of the rocks where it hit, or if it were an oblique impact or augmented by accompanying impacts). However, straightforward evaluations of the expected physical [56] and biological [57] repercussions of massive impacts suggest that any such impact should result in such extreme environmental havoc that a mass extinction would be plausible, although conditions may cause consequences to vary from impacts of similar magnitude [58].

I have argued [59] that impacts must be exceptionally more lethal globally than any other proposed terrestrial causes for mass extinctions because of two unique features: (a) their environmental effects happen essentially instantaneously (on timescales of hours to months, during which species have little time to evolve or migrate to protective locations) and (b) there are compound environmental consequences (e.g., broiler-like skies as ejecta re-enter the atmosphere, global firestorm, ozone layer destroyed, earthquakes and tsunami, months of ensuing "impact winter", centuries of global warming, poisoning of the oceans). Not only the rapidity of changes, but also the cumulative and synergistic consequences of the compound effects, make asteroid impact overwhelmingly more difficult for species to survive than alternative crises. Volcanism, sea regressions, and even sudden effects of hypothesized collapses of continental shelves or polar ice caps are far less abrupt than the immediate (within a couple of hours) worldwide consequences of impact; lifeforms have much better opportunities in longer-duration scenarios to hide, migrate, or evolve. The alternatives also lack the diverse, compounding negative global effects. Only the artificial horror of global nuclear war or the consequences of a very remote possibility of a stellar explosion near the Sun could compete with impacts for immediate, species-threatening changes to Earth's ecosystem. Therefore, since the NEA impacts inevitably happened, it is plausible that they—and chiefly they alone—caused the mass extinctions in Earth's history (as hypothesized by Raup [60]), even though proof is lacking for specific extinctions. What other process could possibly be so effective? And even if one or more extinctions *do* have other causes, the largest asteroid/comet impacts during the Phanerozoic cannot avoid having left traces in the fossil record.

5. The impact hazard: consequences for society in the 21st century

Cosmic projectiles rain down on us, ranging from the frequent flashes of meteoroids, to less frequent meteorite-producing bolides, to the even less common A-bomb level upper-atmospheric explosions recorded by Earth-orbiting surveillance satellites, to the historically rare Tunguska-level events, and finally to the still rarer but potentially extremely destructive impacts of bodies >100 m diameter, which must be considered not in terms of their frequency but instead in terms of their low but finite *probabilities* of impacting during the timeframe that is important to us, our children, and our grandchildren—the 21st century.

The statistical frequency of impacts by bodies of various sizes is fairly well known (Fig. 1). Less well understood are the physical and environmental consequences of impacts of various sizes. The most thorough evaluation of the environmental physics and chemistry of impacts is by Toon et al. [56]; later research has elucidated the previously poorly understood phenomena of impact-generated tsunami [6,61,62]. I have evaluated numerous impact scenarios (see [63]), emphasizing their potential consequences on human society, which are even less well understood than environmental effects. The most comprehensive analysis of the risks of NEA impacts is that of the NASA NEO Science Definition Team (SDT) [6].

Although giant impacts are very rare, when the threshold for globally destructive effects is exceeded

(NEAs >1.5–3 km diameter) then the potential mortality is unprecedentedly large, so such impacts dominate mortality [64], perhaps 3000 deaths per year worldwide, comparable with mortality from other significant natural and accidental causes (e.g., fatalities in airliner crashes). This motivated the Spaceguard Survey. Now the estimated mortality is somewhat lower, ~ 1000 annual deaths [6] due to somewhat lower estimates of the number of NEAs >1 km diameter and somewhat higher estimates of the threshold size for destructive global effects. Since most of that mortality has been eliminated by discovery of 55% of NEAs >1 km diameter and demonstration that none of them will encounter Earth in the next century, the remaining global threat is from the 45% of yet-undiscovered large NEAs plus the minor threat from comets. Once the Spaceguard Survey is complete, the residual global threat will be < 100 annual fatalities worldwide, see Table 1 [6].

The SDT [6] also evaluated two other sources of mortality due to NEO impactors smaller than those that would cause global effects: (a) impacts onto land, with local and regional consequences analogous to the explosion of a bomb and (b) impacts into an ocean, resulting in inundation of shores by the resulting tsunamis. The SDT evaluated fatalities for land impacts using (a) a model for the radius of destruction by impactors >150 m diameter [65] that survive atmospheric penetration with most of their cosmic velocity (although 220 m may be more nearly correct [66]) and (b) a map of population distribution across the Earth. A thorough analysis of the tsunami hazard [67], based on reanalysis of wave and run-up physics combined with analysis of coastal populations, provided an estimated number of "people affected per year" by impact-generated tsunami. As the SDT notes, historically only ~ 10% of people in an inundation zone die, thanks to advance warning and evacuation. Hence, in Table 1, which summarizes

mortality from land impacts, ocean impacts, and globally destructive impacts, I divide the SDT's estimated tsunami hazard by a factor of 10.

In Table 1, the "overall hazard" is that posed by nature, before the Spaceguard Survey started to certify that a fraction of NEAs (more larger ones than smaller ones) will not hit. The "residual hazard" (see Fig. 3) is what is expected after about 2008. Whereas non-global impacts constitute < 10% of the natural impact hazard, they are nearly half of the residual hazard. The land-impact hazard is chiefly due to bodies 70–200 m diameter (indeed, the chances are better than 1% that such an impact will kill ~ 100,000 people during the 21st century; larger bodies, 150–600 m are mainly responsible for the somewhat smaller tsunami hazard.

The SDT's main goal was to derive the cost–benefit ratio for building an augmented Spaceguard Survey, so they emphasized property damage rather than mortality, which gives greater weight to destruction by tsunamis compared with land impacts. On that basis, they calculated the costs of various ground- and space-based telescope systems that might retire 90% of the residual non-global impact hazard in the next decade or two. The SDT's final recommendation was to proceed, beginning in 2008, with what they calculated would be a cost-effective 7–20-year program, costing between US$236 and US$397 million, designed to discover 90% of PHOs >140 m diameter.

It is subjective to compare the impact hazard, given its inherent low-probability high-consequence character, with other societal hazards. I consider mortality rather than property damage as being more central to fears of impacts. But neither mortality nor economic loss estimates provide a good forecast of how societies respond to different kinds of hazards. The ~ 3000 deaths from the terrorist attacks of September 11, 2001 had dramatic national and international consequences (involving economics, politics, war, etc.), while a similar number of U.S. highway fatalities during the same month were hardly noticed, except by family members and associates of the deceased. Risk perception expert Paul Slovic believes that asteroid impacts have many elements of a "dreadful" hazard (being perceived as being involuntary, fatal, uncontrollable, catastrophic and increasing [increasing in news reports,

Table 1
Estimated annual worldwide deaths from impacts

	Overall hazard	Residual hazard			
	Total	Total	Land	Tsunami	Global
Minimum	363	36	28	5	3
Nominal	1090	155	51	16	88
Maximum	3209	813	86	32	695

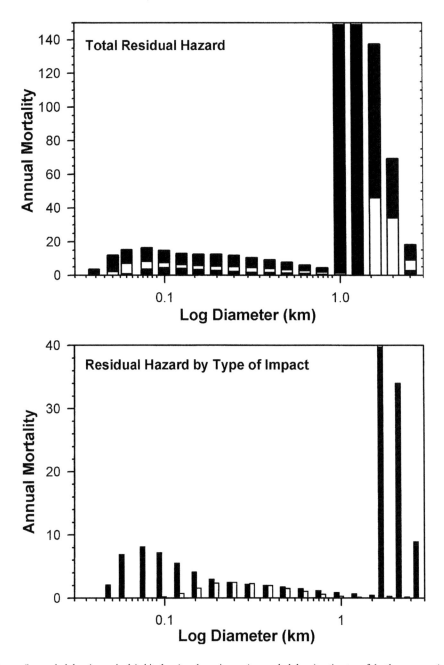

Fig. 3. Top: Minimum (lower dark bars), nominal (white bars) and maximum (upper dark bars) estimates of deaths per year (averaged over long durations) for the total residual impact hazard as a function of impactor diameter. Bottom: The nominal (white) bars in the top graph are broken down by type of impact. The three tall black bars on the right represent the globally destructive impacts. The remaining black bars show land impacts while the white bars show tsunami mortality. (Based on the data from the SDT [6]).

anyway]), like terrorism or nuclear threats, in contrast with more mundane hazards that may be more serious measured by objective criteria [67]. Society

often spends much—even orders of magnitude—more per life saved to reduce "dreadful" hazards than mundane ones. For this reason, efforts to reduce the impact hazard and to plan for mitigation (e.g., evacuation of ground zero, storing food supplies in order to survive a global agricultural disaster or developing capabilities to deflect a threatening NEO) may be perceived by many citizens as money well spent. On the other hand, Slovic's public opinion polls show that many others regard the impact hazard as being trivial.

6. Three representative scenarios of NEA consequences

I briefly summarize three scenarios (drawn from many more in [63]), which illustrate the breadth of issues that must be confronted in managing potential consequences of NEA impacts. For each impact disaster scenario, I consider the nature of the devastation, the probability that the event will happen, the likely warning time, the possibilities for post-warning mitigation, the nature of issues to be faced in after-event disaster management, and—of most practical interest—what can be done *now* to prepare in advance.

6.1. 2–3 km diameter civilization destroyer

A million-megaton impact, even though ~ 100 times less energetic than the K–T impact, would probably destroy civilization as we know it. The dominant immediate global effect would be sudden cooling, lasting many months, due to massive injection of dust into the stratosphere following impact. Agriculture would be largely lost, worldwide, for an entire growing season. Combined with other effects (a firestorm the size of India, destruction of the ozone layer, etc.), it is plausible that billions might die from collapse of social and economic institutions and infrastructure. No nation could avoid direct, as well as indirect, consequences of unprecedented magnitude. Of course, because civilization has never witnessed such an apocalypse, predictions of consequences are fraught with uncertainty.

As discussed earlier, few bodies of these sizes remain undiscovered, so the chances of such an event are probably < 1-in-100,000 during the next century. The warning time would almost certainly be long, in the case of a NEA, but might be only

months in the case of a comet. With years or decades of advance warning, a technological mission might be mounted to deflect the NEA so that it would miss the Earth; however, moving such a massive object would be very challenging. In any case, given sufficient warning, many immediate fatalities could be avoided by evacuating ground zero and longer-term casualties could be minimized by storing food supplies to survive the climate catastrophe. Susceptible infrastructure (transportation, communications, medical services) could be strengthened in the years before impact. However, no preparation for mitigation is warranted for such a rare possibility until a specific impact prediction is made and certified. The only advance preparations that might make sense would be *at the margins* of disaster planning developed for other purposes: considering such an apocalypse might foster "out-of-the-box" thinking about how to define the outer envelope of disaster contingencies, and thus prove serendipitously useful as humankind faces an uncertain future.

6.2. Once-in-a-century mini-Tunguska atmospheric explosion

Consider a 30–40-m office-building-sized object striking at 100 times the speed of a jetliner. It would explode ~ 15 km above ground, releasing the energy of ~ 100 Hiroshima-scale bombs. Weak structures would be damaged or destroyed by the blast wave out to 20 km. The death toll might be hundreds; although casualties would be far higher in a densely populated place, they would much more likely be zero (i.e., if the impact were in the ocean or a desolate location). Such an event is likely to occur in our grandchildren's lifetime, although most likely over the ocean rather than land. Even with the proposed augmented Spaceguard Survey, it is unlikely that such a small object would be discovered in advance; impact would occur without warning. Since it could occur literally anywhere, there are no location-specific kinds of advance measures that could or should be taken, other than educating people (perhaps especially military forces that might otherwise mistake the event as an intentional attack) about the possibilities for such atmospheric explosions. In the lucky circumstance that the object is discovered years in advance, a relatively modest space mission could deflect such a small body, preventing impact [26].

6.3. Prediction (or media report) of a near-term impact

This NEA scenario is the one most likely to become an urgent issue for public officials. Indeed, such events have already happened. The problem, which can develop within hours in the 24-h global news media, is that something possibly real about a NEA is twisted by human fallibility and/or hyperbole. Hypothetical examples include: (a) a prediction, a few days in advance, of an actual near-miss ("just" 60,000 km from Earth) by a >100-m asteroid, which might be viewed with alarm by a distrustful public who would still fear an actual impact; (b) the reported (or mis-reported) prediction by a reputable (but mistaken or misquoted) astronomer that a huge impact will occur on a specific day in the future in a particular country, resulting in panic for several days until the report is withdrawn; or (c) a prediction, officially endorsed by an entity like the International Astronomical Union, of a one-in-a-few-hundred impact *possibility* on a specific date decades in the future (Torino Scale = 2; see below), which because of circumstances cannot be refined for months. On January 13/14, 2004, some NEA experts believed for a few hours that there was a 10–25% chance that a just-detected NEA, 30 m in size, would strike the Earth's northern hemisphere just a few days later [68]; a public announcement of this possible "mini-Tunguska" was being considered, but then an amateur astronomer made observations that discounted any imminent impact and the real object was later verified as being much farther away.

Ways to eliminate instances of hype and misunderstanding involve public education about science, critical thinking and risk; familiarizing science teachers, journalists and other communicators with the impact hazard might be especially effective. One approach that has evolved since a 1999 conference in Torino (Turin), Italy, is promulgation of the Torino Scale [63,69,70], which attempts to place impact predictions into a sober, rational context (on a 10-point Richter-like scale, predicted impact possibilities usually rate a 0 or 1, and are unlikely to exceed 4 during our lifetimes).

7. Evaluation of the modern impact hazard

Unlike other topics in astronomy (except solar flares and coronal mass ejections), only the impact hazard presents serious practical issues for society. Contrasting with most practical issues involving meteorology, geology and geophysics, the impact hazard is both more extreme in potential consequences and yet so rare that it has not even been experienced in more than minor ways in historical times. It has similarities to natural hazards in that its practical manifestations mainly involve familiar destructive processes, such as fire, high winds, earthquakes, falling debris and floods. The impact hazard also ranks with other natural disasters in the mid-range of risks of death [67]: much less important than war, disease, famine, automobile accidents or murder but much more important than shark attacks, botulism, fireworks accidents or terrorism. Yet, impacts differ from natural disasters because the hazard is mainly not location-dependent (impacts happen anywhere, not just along faults, although ocean impact effects are amplified along coastlines) and there are no precursor or after-shock events.

There are also similarities and differences compared with terrorism and other human-caused calamities. Like terrorism, the impact hazard is "dreadful" (in Slovic's nomenclature), it seems to strike randomly (at least unexpectedly) in time and location, and few have been (or, in my estimation, are likely to be) killed, although in each case many *could* be killed. Dissimilar attributes include the essential "act of God" nature of impacts, whereas terrorism involves willful acts of evil, inspiring retribution. Also, we can probably *do something* about most impact threats, whereas terrorism and threats of nuclear war are dealt with by such imperfect human endeavors as diplomacy. Another disproportionate comparison involves past public expenditures: hundreds of billions of dollars are being allocated to the war on terrorism compared with a few million spent annually on the impact hazard (mostly funding the Spaceguard Survey).

The practical, public implications and requirements of the impact hazard are characterized by its *uncertainty* and "iffy" nature. Yet, the chief scientific evaluations of the hazard, and thus (because of the subject's popularity) its public promulgation in the news, is skewed with respect to reality. In the last few years, many peer-reviewed papers have been published (often with popular commentaries and even CNN crawlers) about how many >1-km NEAs there are, ranging from lows of ~ 700 [71] to highs approaching 1300. Yet far less attention is paid (although not quite none at all (e.g., [72]) to the much greater uncertainties in environmental effects of impacts. And there is essentially no serious, funded research concerning the largest sources of uncertainty—those concerning the psychology, sociology and economics of such extreme disasters—which truly determine whether this hazard is of academic interest only or, instead, might shape the course of history. For example, many astronomers and geophysicists, who are amateurs in risk perception and disaster management, assume that "panic" is a probable consequence of predicted or actual major asteroid impacts. Yet some social scientists (e.g., [73]) have concluded that people rarely panic in disasters. Such issues, especially in a post-September 11th terrorism context, could be more central to prioritizing the impact hazard than anything earth and space scientists can do. If an actual Earth-targeted body is found, it will be the engineers and disaster managers whose expertise will suddenly be in demand.

I have noted the primacy of psychological perceptions in characterizing the impact hazard. Since impact effects (other than the spectacle of meteors and occasional meteorite falls) have never been experienced by human beings now alive, we can relate to this hazard only theoretically. Since it involves very remote possibilities, the same irrationality applies that governs purchases of lottery tickets or re-building in 100-year floodplains just after a recent 100-year flood. Because society fails to apply objective standards to prioritizing hazard mitigation funding, it is plausible that the residual risks of this hazard might be altogether ignored (the Spaceguard Survey has been cheap, but it becomes increasingly costly to search for the remaining, small NEAs); or society may instead over-react and give "planetary defense" more priority than battling such clear-and-present dangers as influenza. Yet, contrasting with the irrational perceptions of the impact hazard, it potentially can be mitigated in much more concrete ways than is true of most hazards. An impact can be predicted in advance in ways that remain imperfect [70] but are much more reliable than predictions of earthquakes or even storms, and the components of technology exist—at affordable costs given the consequences of an actual impact—to move any threatening object away and avoid the disaster altogether. In contrast with the dinosaurs, human beings have the insight and capability to avoid extinction by impacts.

Acknowledgements

I thank Chris Koeberl, Rick Binzel and Larry Nittler for helpful reviews. Preparation of this review was assisted by the Alan Harris, David Morrison and members of the B612 Foundation. *[AH]*

References

[1] Z. Ceplecha, Meteoroids: an item in the inventory, in: T.W. Rettig, J.M. Hahn (Eds.), Completing the Inventory of the Solar System, Astron. Soc. Pac. Conf. Ser., vol. 107, 1996, pp. 75–84.

[2] S.J. Weidenschilling, Formation of planetesimals and accretion of the terrestrial planets, Space Sci. Rev. 92 (2000) 295–310.

[3] W.F. Bottke Jr., D. Vokrouhlický, D.P. Rubincam, M. Brož, The effect of Yarkovsky thermal forces on the dynamical evolution of asteroids and meteoroids, in: W.F. Bottke Jr., A. Cellino, P. Paolicchi, R.P. Binzel (Eds.), Asteroids, vol. III, Univ. Arizona Press, Tucson, 2002, pp. 395–408.

[4] S.F. Dermott, D.D. Durda, K. Grogan, T.J.J. Kehoe, Asteroidal dust, in: W.F. Bottke Jr., A. Cellino, P. Paolicchi, R.P. Binzel (Eds.), Asteroids, vol. III, Univ. Arizona Press, Tucson, 2002, pp. 423–442.

[5] W.F. Bottke, A. Morbidelli, R. Jedicke, J.-M. Petit, H.F. Levison, P. Michel, T.S. Metcalfe, Debiased orbital and absolute magnitude distribution of the near-Earth objects, Icarus 156 (2002) 399–433.

[6] Near-Earth Object Science Definition Team, Study to Determine the Feasibility of Extending the Search for Near-Earth Objects to Smaller Limiting Diameters, NASA Office of Space Science, Solar System Exploration Div., Washington, DC, 2003, 154 pp., http://neo.jpl.nasa.gov/neo/neoreport030825.pdf.

[7] T.S. Culler, T.A. Becker, R.A. Muller, P.R. Renne, Lunar

impact history from ^{40}Ar/^{39}Ar dating of glass spherules, Science 287 (2000) 1785–1788.

[8] G. Ryder, Mass flux in the ancient Earth–Moon system and benign implications for the origin of life on Earth, J. Geophys. Res. 107 (E4) (2002) 5022 (doi: 10.1029/2001JE001583).

[9] W.K. Hartmann, Megaregolith evolution and cratering cataclysm models—lunar cataclysm as a misconception (28 years later), Meteorit. Planet. Sci. 38 (2003) 579–593.

[10] J.E. Chambers, J.J. Lissauer, A. Morbidelli, Planet V and the origin of the lunar Late Heavy Bombardment, Bull.-Am. Astron. Soc. 33 (2001) 1082.

[11] H. Atkinson, C. Tickell, D. Williams (Eds.), Report of the Task Force on Potentially Hazardous Near Earth Objects, British National Space Centre, London, 2000, 54 pp.

[12] F. Watson, Between the Planets, The Blakiston Co., Philadelphia, 1941, 222 pp.

[13] R.B. Baldwin, The Face of the Moon, Univ. Chicago Press, Chicago, 1949, 239 pp.

[14] E.J. Öpik, On the catastrophic effect of collisions with celestial bodies, Ir. Astron. J. 5 (1958) 36.

[15] E.M. Shoemaker, Impact mechanics at Meteor Crater, Arizona, in: B.M. Middlehurst, G.P. Kuiper (Eds.), The Moon Meteorites and Comets, Univ. Chicago Press, Chicago, 1963, pp. 301–336.

[16] R.B. Leighton, B.C. Murray, R.P. Sharp, J.D. Allen, R.K. Sloan, Mariner IV photography of Mars: initial results, Science 149 (1965) 627–630.

[17] L.W. Alvarez, W. Alvarez, F. Asaro, H.V. Michel, Extraterrestrial cause for the Cretaceous–Tertiary extinction, Science 208 (1980) 1095–1108.

[18] A.R. Hildebrand, W.V. Boynton, Proximal Cretaceous–Tertiary boundary impact deposits in the Caribbean, Science 248 (1990) 843–847 (see also);
A.R. Hildebrand, W.V. Boynton, Cretaceous ground zero, Nat. Hist. (6) (1991) 47–53;
G.T. Penfield, Pre-Alvarez impact, Nat. Hist., (6) (1991) 4.

[19] J.D. Archibald, Dinosaur extinction: changing views, in: J.G. Scotchmoor, D.A. Springer, B.H. Breithaupt, A.R. Fiorillo (Eds.), Dinosaurs: The Science Behind the Stories, American Geological Institute, 2002, pp. 99–106.

[20] D. Morrison (Ed.), The Spaceguard Survey: Report of the NASA International Near-Earth Object Detection Workshop, NASA, Washington, DC, 1992, http://impact.arc.nasa.gov/reports/spaceguard/index.html.

[21] E.M. Shoemaker (Ed.), Report of the Near-Earth Objects Survey Working Group, NASA, Washington, DC, 1995, http://impact.arc.nasa.gov/reports/neoreport/index.html.

[22] G.H. Canavan, J.C. Solem, J.D.G. Rather (Eds.), Proceedings of the Near-Earth-Object Interception Workshop, Los Alamos Natl. Lab. LA-12476-C, Los Alamos NM, 1993, 296 pp.

[23] Proceedings of the Planetary Defense Workshop, Lawrence Livermore Natl. Lab. CONF-9505266, Livermore CA, 1995, 513 pp.

[24] T. Gehrels (Ed.), Hazards due to Comets and Asteroids, Univ. Arizona Press, Tucson, 1994, 1300 pp.

[25] J.L. Remo (Ed.), Near-Earth Objects: The United Nations

International Conference, Annals of the New York Academy of Sciences, vol. 822, 632 pp.

[26] M.J.S. Belton (Ed.), Mitigation of Hazardous Comets and Asteroids, Cambridge Univ. Press, Cambridge, 2004, in press.

[27] C. Koeberl, K.G. MacLeod (Eds.), Catastrophic Events and Mass Extinctions: Impacts and Beyond, Geological Soc. America, Boulder CO, Special Paper, vol. 356, 746 pp.

[28] W.F. Bottke Jr., A. Cellino, P. Paolicchi, R.P. Binzel (Eds.), Asteroids, vol. III, Univ. Arizona Press, Tucson, 2002, 785 pp.

[29] G.H. Stokes, J.B. Evans, S.M. Larson, Near-Earth asteroid search programs, in: W.F. Bottke Jr., A. Cellino, P. Binzel, R.P. Binzel (Eds.), Asteroids, vol. III, Univ. Arizona Press, Tucson, 2002, pp. 45–54.

[30] P. Brown, R.E. Spalding, D.O. ReVelle, E. Tagliaferri, S.P. Worden, The flux of small near-Earth objects colliding with the Earth, Nature 420 (2002) 294–296.

[31] D. Vokrouhlický, P. Farinella, Efficient delivery of meteorites to the Earth from a wide range of asteroid parent bodies, Nature 407 (2000) 606–608.

[32] C.R. Chapman, Space weathering of asteroid surfaces, Annu. Rev. Earth Planet. Sci. (2004) (in press) doi: 10.1146/annurev.earth.32.101802.120453.

[33] S.J. Bus, F. Vilas, M.A. Barucci, Visible-wavelength spectroscopy of asteroids, in: W.F. Bottke Jr., A. Cellino, P. Paolicchi, R.P. Binzel (Eds.), Asteroids, vol. III, Univ. Arizona Press, Tucson, 2002, pp. 169–182.

[34] S.J. Ostro, R.S. Hudson, L.A.M. Benner, J.D. Giorgini, C. Magri, J.-L. Margot, M.C. Nolan, Asteroid radar astronomy, in: W.F. Bottke, A. Cellino, P. Paolicchi, R.P. Binzel (Eds.), Asteroids, vol. III, Univ. Arizona Press, Tucson, 2002, pp. 151–168.

[35] J.F. Bell III, et al., Near-IR reflectance spectroscopy of 433 Eros from the NIS instrument on the NEAR mission, Icarus 155 (2002) 119–144.

[36] T.J. McCoy, et al., The composition of 433 Eros: a mineralogical–chemical synthesis, Meteorit. Planet. Sci. 36 (2001) 1661–1672.

[37] R.P. Binzel, D.F. Lupishko, M. DiMartino, R.J. Whiteley, G.J. Hahn, Physical properties of near-Earth objects, in: W.F. Bottke Jr., A. Cellino, P. Paolicchi, R.P. Binzel (Eds.), Asteroids, vol. III, Univ. Arizona Press, Tucson, 2002, pp. 255–271.

[38] V.R. Oberbeck, M. Aoyagi, Martian doublet craters, J. Geophys. Res. 77 (1972) 2419–2432.

[39] C.M. Cook, H.J. Melosh, W.F. Bottke Jr., Doublet craters on Venus, Icarus 165 (2003) 90–100.

[40] W.J. Merline, S.J. Weidenschilling, D.D. Durda, J.L. Margot, P. Pravec, A.D. Storrs, Asteroids do have satellites, in: W.F. Bottke Jr., A. Cellino, P. Paolicchi, R.P. Binzel (Eds.), Asteroids, vol. III, Univ. Arizona Press, Tucson, 2002, pp. 289–312.

[41] E. Asphaug, W. Benz, Size, density, and structure of Comet Shoemaker-Levy 9 inferred from the physics of tidal breakup, Icarus 121 (1996) 225–248.

[42] D.C. Richardson, Z.M. Leinhardt, H.J. Melosh, W.F. Bottke Jr., E. Asphaug, Gravitational aggregates: evidence and evolution, in: W.F. Bottke Jr., A. Cellino, P. Paolicchi, R.P. Binzel

(Eds.), Asteroids, vol. III, Univ. Arizona Press, Tucson, 2002, pp. 501–515.

[43] C.R. Chapman, The evolution of asteroids as meteorite parent-bodies, in: A.H. Delsemme (Ed.), Comets Asteroids Meteorites: Interrelations, Evolution and Origins, Univ. of Toledo Press, Toledo OH, 1977, p. 265.

[44] R.J. Whiteley, D.J. Tholen, C.W. Hergenrother, Lightcurve analysis of four new monolithic rapidly-rotating asteroids, Icarus 157 (2002) 139–154.

[45] C.R. Chapman, W.J. Merline, P.C. Thomas, J. Joseph, A.F. Cheng, N. Izenberg, Impact history of Eros: craters and boulders, Icarus 155 (2002) 104–118.

[46] C.R. Chapman, What we know and don't know about surfaces of potentially hazardous small bodies, in: M.J.S. Belton (Ed.), Mitigation of Hazardous Comets and Asteroids, Cambridge Univ. Press, Cambridge, 2004, in press.

[47] Earth Impact Database, http://www.unb.ca/passc/ImpactDatabase/, accessed 2003 Dec. 12.

[48] K.A. Maher, D.J. Stevenson, Impact frustration of the origin of life, Nature 331 (1988) 612–614.

[49] N.H. Sleep, K.J. Zahnle, J.F. Kasting, H.J. Morowitz, Annihilation of ecosystems by large asteroid impacts on the early Earth, Nature 342 (1989) 139–142.

[50] N.H. Sleep, K. Zahnle, Refugia from asteroid impacts on early Mars and the early Earth, J. Geophys. Res. 103 (1998) 28529–28544.

[51] J.C. Armstrong, L.E. Wells, G. Gonzalez, Rummaging through Earth's attic for remains of ancient life, Icarus 160 (2002) 183–196.

[52] P.F. Hoffman, A.J. Kaufman, G.P. Halverson, D.P. Schrag, A Neoproterozoic snowball Earth, Science 281 (1998) 1342–1346.

[53] P.F. Hoffman, D.P. Schrag, The snowball Earth hypothesis: testing the limits of global change, Terra Nova 14 (2002) 129–155.

[54] D.H. Erwin, S.A. Bowring, J. Yugan, End-Permian mass extinctions: a review, in: C. Koeberl, K.G. MacLeod (Eds.), Catastrophic Events and Mass Extinctions: Impacts and Beyond, Geological Soc. America, Boulder CO, Special Paper, vol. 356, 2002, pp. 363–383.

[55] A.R. Basu, M.I. Petaev, R.J. Poreda, S.B. Jacobsen, L. Becker, Chondritic meteorite fragments associated with the Permian–Triassic boundary in Antarctica, Science 302 (2003) 1388–1392.

[56] O.B. Toon, K. Zahnle, D. Morrison, R.P. Turco, C. Covey, Environmental perturbations caused by the impacts of asteroids and comets, Rev. Geophys. 35 (1997) 41–78.

[57] D.S. Robertson, M.C. McKenna, O.B. Toon, S. Hope, J.A. Lillegraven, Survival in the first hours of the Cenozoic, Geol. Soc. Amer. Bull. (2004) (in press).

[58] D.A. Kring, Environmental consequences of impact cratering events as a function of ambient conditions on Earth, Astrobiology 3 (2003) 133–152.

[59] C.R. Chapman, Impact lethality and risks in today's world: lessons for interpreting Earth history, in: C. Koeberl, K.G. MacLeod (Eds.), Catastrophic Events and Mass Extinctions: Impacts and Beyond, Geological Society of America, Boulder CO, Special Paper, vol. 356, 2002, pp. 7–19.

[60] D.M. Raup, Bad Genes or Bad Luck? Norton, New York, 1991, 210 pp.

[61] J.G. Hills, C.L. Mader, Tsunami produced by the impacts of small asteroids, in: J.L. Remo (Ed.), Near-Earth Objects: The United Nations International Conference, Annals of the New York Academy of Sciences, vol. 822, 1997, pp. 381–394.

[62] S.R. Chesley, S.N. Ward, A quantitative assessment of the human and economic hazard from impact-generated tsunami, Environ. Hazards (2003) (submitted for publication).

[63] D.M. Morrison, C.R. Chapman, D. Steel, R. Binzel, Impacts and the public: communicating the nature of the impact hazard, in: M.J.S. Belton (Ed.), Mitigation of Hazardous Comets and Asteroids, Cambridge Univ. Press, Cambridge, 2004, in press.

[64] C.R. Chapman, D. Morrison, Impacts on the Earth by asteroids and comets: assessing the hazard, Nature 367 (1994) 33–40.

[65] J.G. Hills, M.P. Goda, The fragmentation of small asteroids in the atmosphere, Astron. J. 105 (1993) 1114–1144.

[66] P.A. Bland, N.A. Artemieva, Efficient disruption of small asteroids by Earth's atmosphere, Nature 424 (2003) 288–291.

[67] D. Morrison, C.R. Chapman, P. Slovic, in: T. Gehrels (Ed.), Hazards due to Comets and Asteroids, Univ. Arizona Press, Tucson, 1994, pp. 59–91.

[68] C.R. Chapman, NEO impact scenarios, American Inst. of Aeronautics and Astronautics 2004 Planetary Defense Conference: Protecting Earth from Asteroids, Garden Grove, Calif. (2004 23 Feb.).

[69] R.P. Binzel, The Torino impact hazard scale, Planet. Space Sci. 48 (2000) 297–303.

[70] C.R. Chapman, The asteroid/comet impact hazard: *Homo sapiens* as dinosaur? in: D. Sarewitz, et al. (Ed.), Prediction: Science, Decision Making, and the Future of Nature, Island Press, Washington, DC, 2000, pp. 107–134.

[71] D. Rabinowitz, E. Helin, K. Lawrence, S. Pravado, A reduced estimate of the number of kilometre-sized near-Earth asteroids, Nature 403 (2000) 165–166.

[72] K.O. Pope, Impact dust not the cause of the Cretaceous–Tertiary mass extinction, Geology 30 (2002) 99–102.

[73] L. Clarke, Panic: Myth or Reality, Contexts (American Sociological Association, Univ. Calif. Press) v. 1 #3 (2002) (http://www.contextsmagazine.org/content_sample_v1-3.php).

Dr. Clark R. Chapman is a planetary scientist at the Boulder, CO, office of Southwest Research Institute. He is past chair of the Division for Planetary Sciences of the American Astronomical Society, first editor of Journal of Geophysical Research-Planets, and has been on the science teams of the Galileo, NEAR-Shoemaker and MESSENGER deep space missions. His PhD (1972) is from the Earth and Planetary Sciences Dept. of MIT.

Reprinted from
Earth and Planetary Science Letters 220 (2004) 231–245

www.elsevier.com/locate/epsl

The terrestrial Li isotope cycle: light-weight constraints on mantle convection

Tim Elliott [a,*], Alistair Jeffcoate [a], Claudia Bouman [b,1]

[a] *Department of Earth Sciences, Wills Memorial Building, Queens Road, University of Bristol, Bristol BS8 1RJ, UK*
[b] *Faculteit der Aard- en Levenswetenschappen, Vrije Universiteit, De Boelelaan 1085, 1081 HV Amsterdam, The Netherlands*

Received 15 September 2003; received in revised form 29 January 2004; accepted 1 February 2004

Abstract

The two stable isotopes of Li are significantly fractionated by exchange reactions with clays near the Earth's surface. The isotopic legacy of this process provides a robust tracer of surface material that is returned (recycled) to the mantle. Altered oceanic crust has a heavy Li isotopic composition (high $^7Li/^6Li$). Heterogeneous distribution of subducted, altered oceanic crust in the mantle will result in variations in Li isotope ratios. A rapidly accumulating dataset of Li isotope analyses on mantle-derived materials indeed indicates a significant range in Li isotope ratios. This observation provides powerful evidence for the widespread distribution of recycled material in the convecting mantle. There is substantial overlap in Li isotopic compositions of ocean island basalts (OIB) and mid-ocean ridge basalts (MORB). Some OIB, however, have slightly heavier Li compositions than typical, depleted MORB. At face value this suggests a larger contribution of recycled oceanic crust in the sources of some OIB than in the upper mantle. Yet recent evidence implies that heavy Li is lost from the slab at subduction zones and the recycled residual crust is left with an isotopically light signature. Extremely light Li isotope ratios are observed in some continental mantle xenoliths but not in OIB proposed to contain large proportions of recycled crust. Studies of subduction zone lavas imply the heavy Li isotope signature of altered oceanic crust is transferred to the mantle above the subducting plate. Thus cycling of sub-arc mantle may be an important process in forming chemical heterogeneities sampled by OIB. Inferences from initial Li isotope results are thus notably different from those from the longer established but more equivocal radiogenic isotope tracers. Li isotopes promise to provide significant new constraints on the distribution of recycled material in the mantle and its implications for mantle convection.
© 2004 Elsevier B.V. All rights reserved.

Keywords: Li isotopes; mantle evolution; recycling

* Corresponding author. Tel.: +44-117-954-5426; Fax: +44 117 925 3385.
E-mail addresses: tim.elliott@bris.ac.uk (T. Elliott), a.jeffcoate@bris.ac.uk (A. Jeffcoate), claudia.bouman@thermo.com (C. Bouman).

[1] Present address: Finnigan Advanced MassSpectrometry, Thermo Electron Corperation, Barkhausenstr. 2, 28197 Bremen, Germany.

0012-821X / 04 / $ – see front matter © 2004 Elsevier B.V. All rights reserved.
doi:10.1016/S0012-821X(04)00096-2

1. Introduction

Mantle convection stirs the Earth's interior. Understanding this first-order planetary process has been improved by increasingly sophisticated fluid dynamic models [1] and better resolved seismic images of the mantle's velocity structure [2]. It is also necessary to develop in tandem geochemical tracers of convection. The importance of geochemistry is that it provides a time-integrated signal of the effects of convection and can trace the fate of material with distinctive compositional signatures. In partnership with the complementary present-day picture from seismology, geochemistry provides the observables for convection model testing [3,4].

The formation and subduction of oceanic plates is an inherent part of mantle convection. The return to the mantle of oceanic crust (and its depleted residue), variably influenced by its near-surface residence, is frequently termed 'recycling'. Determining how recycled material is distributed in the present-day mantle informs on the style of mantle convection. In seeking this elusive goal, a number of isotope systems have been employed as tracers of recycled plates. Radiogenic isotope systems are most commonly used (e.g. [3,4]), but can be strongly influenced by processes other than oceanic crustal formation, alteration and subduction. Although many popular interpretations of radiogenic isotope variations implicitly invoke recycled material, the evidence is equivocal [5].

Stable isotope tracers are highly desirable to ground-truth inferences from radiogenic isotopes. Stable isotope fractionation is dominantly the result of low-temperature processes near the Earth's surface. Thus significant variations in the stable isotope ratios of mantle-derived samples unambiguously reflect variable contributions of recycled material. This logic has prompted a careful re-examination of the oxygen isotope signatures of mantle-derived basalts using high-precision laser fluorination techniques (see [6], for recent summary). Such oxygen isotope studies have identified some lavas that clearly contain recycled material in their source [7]. Yet most mantle-derived melts show a rather small range in oxygen isotope compositions. Whilst this may reflect a lack of

recycled input, it may also reflect a lack of sensitivity of oxygen isotopes to some recycling processes.

Here we highlight the potential of the Li isotope system in tracking the fate of subducted material and its distribution by mantle convection. We initially summarise the behaviour of Li isotopes in the near-surface environment to understand the controls on Li isotope fractionation. We then consider the effects of subduction zone processes on the Li isotope systematics of the altered oceanic crust. Finally we assess the burgeoning, but still modest, Li isotope data on mantle-derived samples. Li isotope geochemistry is a juvenile research area, dominated by work in the last decade. Systematic study of the Li isotope ratios of mantle-derived rocks is even more immature. Thus we can only present a tantalising aperitif of work that may follow and note that our interpretations are still at the frontier stage.

2. Preliminaries

Lithium has two stable isotopes, 6 and 7, which have the biggest relative mass difference of any isotope pair aside from hydrogen–deuterium. The potential for mass-dependent fractionation is thus obvious. Unlike the better established light stable isotope systems (e.g. C, O, S), Li is a trace cation and does not form an integral part of hydrological, atmospheric or biological cycles. Hence measurements of Li isotopes are likely to provide rather different information than is available from the more commonly used isotope systems.

As with other stable isotope measurements, Li isotope ratios are typically expressed in a delta notation:

$$\delta^7\text{Li} = [(^7\text{Li}/^6\text{Li}_{\text{sample}})/(^7\text{Li}/^6\text{Li}_{\text{standard}})-1]\times 1000$$

where the standard is conventionally the National Institute of Standards highly purified Li_2CO_3 reference material NIST L-SVEC [8] which has $^7\text{Li}/^6\text{Li} \approx 12.15$ [9]. As has occurred in the early stages of development of other isotope systems, some confusion has arisen over the form in which data are reported. Both $\delta^7\text{Li}$ and $\delta^6\text{Li}$ have been used in the literature, but increasingly the δ^7Li notation has become standard (e.g. [10]). In the δ^7Li formulation, more positive values are also isotopically *heavier*, in keeping with other stable isotope systems. Measurements reported as δ^6Li can be readily converted to δ^7Li. Simply changing the sign of δ^6Li measurements yields a value that generally closely approximates the corresponding δ^7Li.

Measurements of Li isotope ratios have traditionally been made using thermal ionisation mass spectrometry [11–13]. Although Li is efficiently ionised by solid source mass spectrometers, highly reproducible isotope ratios are more difficult to obtain by this technique than for the gas-source machines used to measure oxygen isotopes. The reproducibility of Li isotope measurements has thus lagged behind oxygen isotope measurements, typically ~ 1‰ compared to ~ 0.05‰ (all errors in this contribution are cited as 2σ standard deviations). This has hampered application of Li isotopes to studying mantle processes, where variations in Li isotope ratios are only a few per mil (see below), but it has been sufficiently precise for documenting the larger isotopic variations in the near-surface environment. Thermal ionisation measurements are also time-consuming and so the Li isotope database has remained relatively limited. The advent of plasma ionisation multicollector mass spectrometry has allowed a considerably faster throughput of smaller samples [14,15]. New-generation plasma ionisation multicollector mass spectrometers have also enabled a major improvement in reproducibility, typically ± 0.2‰ [16], much closer to oxygen isotope measurements and better than the best thermal ionisation data (± 0.5‰) [10]. This advance is highly significant for documenting Li isotopic variations in the mantle.

3. Behaviour of Li in the hydrological cycle

The fractionation of Li isotopes is dominated by partitioning of Li between clays and water [17–21]. It is important to review briefly the fundamentals of these low-temperature processes in order to understand the potential of Li isotopes as a tracer of subducted material. This pioneering work of Chan and co-workers in the last decade has laid the essential foundations for exploiting Li isotopes as high-temperature tracers [17,22,23]. Readers interested in more detail on this and other aspects of Li geochemistry are referred to a recent and comprehensive review by Tomascak [24].

Weathering of the continents both releases Li and isotopically fractionates it [23]. Minerals are characteristically enriched in ^{6}Li relative to co-existing aqueous fluid [21,25]. Thus river waters have heavy Li isotope compositions compared to the original bed-rock and associated suspended load [20]. The Li isotopic composition of river water is not highly sensitive to the Li isotope composition of the bed-rock of the catchment area, in contrast to some radiogenic isotope ratios used to monitor chemical weathering [23]. There is thus promise in using the evolution of the Li isotope ratio of seawater to assess past changes in the intensity of continental weathering, with its attendant implications for atmospheric CO_2 and long-term climate change (e.g. [26]). In a simplistic scenario it might be anticipated that a change to lighter Li isotope compositions in ancient oceans would reflect lower weathering rates in the past. Whilst the general process of Li isotopic fractionation at the surface is clear, a more detailed understanding of the behaviour of Li isotopes during weathering, transport and incorporation into the geological record is required before Li isotope ratios can be used as weathering proxy (for recent review on chemical proxies of past conditions see [27]). However, an increasing number of groups are investigating this potential and this is an exciting prospect for the future [21,28–32].

In the marine system too, partitioning of Li between water and clays results in major isotopic fractionation [17,18,25]. This results in an initially puzzling observation. Unlike radiogenic isotope systems, such as ^{87}Sr/^{86}Sr, where seawater represents a intermediate mix of a hydrothermal input (unradiogenic, mantle-dominated signature) and river water (radiogenic, continental signature), seawater has a Li isotope ratio heavier than its inputs (Fig. 1). This paradox is explained by the

A

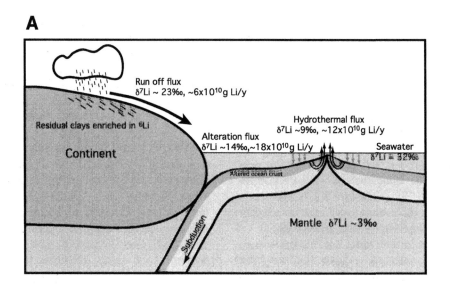

B

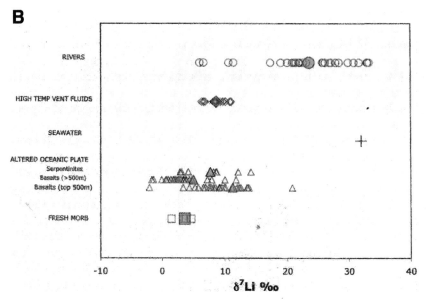

Fig. 1. (A) Cartoon of the main fluxes that control Li isotope mass balance in the oceans. Flux estimates taken from [23]. (B) Analyses that constrain the ocean Li isotope mass balance. Individual analyses are shown as open symbols with weighted means indicated by larger symbols. Typical uncertainties are ±1‰. Data sources: rivers [23], high-temperature vents [22], seawater compilation from [13], fresh MORB [17,34]. The oceanic plate is divided into three stratigraphic components to illustrate the upper (top 500 m), high-δ^7Li, low-temperature altered basalts [17,33]; lower (> 500 m) basalts, some of which represent the light residue of high-temperature hydrothermal fluids (note these are also depleted in Li and only weakly affect the weighted mean of this section) [33]; serpentinised upper mantle [74]. The low-temperature altered basalts have strongly elevated Li concentrations and dominate the Li isotope budget of the altered crust.

low-temperature alteration of the oceanic crust to hydrous minerals (such as smectite and phillipsite). These alteration products are important sinks for oceanic Li and also preferentially remove ^{6}Li from seawater during this process [17,33]. Enhanced removal of ^{6}Li from the oceans results in a complementary, heavy seawater composition of δ^7Li ≈ 32 (e.g. compilation in [13]), heavier than its average input. Although the Li gained by the alteration material is lighter than mean river water, δ^7Li < 23 ‰ [23], it is nevertheless heavier than fresh oceanic basalts, δ^7Li ≈ 3 [17,34].

In summary, weathering of silicates on the continents produces river water with isotopically heavy Li which feeds the oceans. Low-temperature alteration of oceanic crust serves to make seawater even heavier than river water. Even though altered oceanic crust is isotopically light compared to mean river water, it is notably heavier than mantle values (Fig. 1B).

The process of increasing the δ^7Li of altered oceanic crust as part of the low-temperature Li cycle thus imparts a distinctive signature to the oceanic plate that can be used to trace it. We first, however, assess if the magnitude of this heavy, recycled flux is sufficient to affect the Li isotopic composition of the mantle as a whole. The flux might seem most straightforwardly calculated by using an estimate for average altered oceanic crust and present-day rates of plate subduction. Although it is clear that altered oceanic crust is heavy [17,33], Li isotope analyses of altered oceanic crust are only available from a few localities worldwide. Only three samples have been analysed that are older than 10 Myr. Constructing a representative value for average altered oceanic crust that has experienced a full range of alteration processes is therefore difficult. Thus we prefer to estimate the net Li isotope flux to the oceanic crust, and ultimately the subduction zone, indirectly by examining the Li isotope mass balance in the oceans. Using the commonly made assumption of a steady-state seawater composition, the Li isotope flux to the altered oceanic crust must balance the input flux from rivers and hydrothermal vents. Since the hydrothermal flux is itself dominantly derived from the oceanic crust itself,

the *net* flux to the oceans equals the riverine flux of $\sim 6 \times 10^{10}$ g/yr with δ^7Li = 23 ‰ [23].

The ramifications of returning this flux of heavy Li to the mantle via subduction of the altered oceanic crust are explored in Fig. 2. It is evident that the magnitude of the recycled flux is just sufficient to influence the composition of the mantle as a whole. Moreover, if recycled material is not efficiently mixed throughout the whole mantle it can significantly change (~ 1 ‰) the Li isotope composition of a more restricted portion of the mantle, e.g. the upper mantle (Fig. 2). The relative Li isotopic compositions of different mantle reservoirs thus have clear implications for assessing the style of mantle convection.

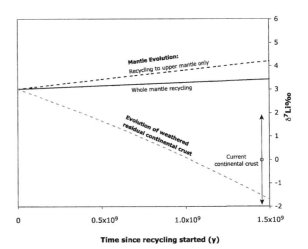

Fig. 2. Li isotopic evolution of crustal and mantle reservoirs in response to weathering and subduction of Li. The simplistic model illustrates the cumulate effect of removing the riverine Li flux (5.6×10^{10} g/yr, δ^7Li = 23 ‰ [23]) from the continental crust and adding it to the mantle via subduction. The graph shows the effects of two scenarios: adding the subducted flux to the whole mantle (4×10^{27} g) and just upper mantle (1.33×10^{27} g). Both reservoirs are initially assumed to have [Li] = 1 µg/g and δ^7Li = 3 ‰. The continental crust (2×10^{25} g) is assumed to have an initial δ^7Li = 3 ‰ (since it is ultimately derived from the mantle) and [Li] = 22 µg/g [35]. Each reservoir is assumed to be homogeneous and therefore changes are averaged over the whole reservoir. The model is run in 1×10^8 yr time steps for 1.5×10^9 yr. After this time, the composition of the model continental crust composition is at the lower end of continental composites analysed by [35] and so the model is run no further. The composition for current continental crust is from [35], illustrating a preferred average for whole crust and 1 standard deviation of all upper crustal samples measured.

The increase in δ^7Li of the mantle sketched in Fig. 2 ultimately results from weathering of the continents, followed by cycling heavy Li through the oceans and altered oceanic crust into the mantle. Continental crust is originally derived from the mantle and these two reservoirs should initially have had the same Li isotope ratio. Continued addition of heavy Li into the mantle by subduction must have a complementary effect on the Li isotope ratio of the continents. The Li isotope composition of the continental crust thus provides an important constraint on how heavy the mantle can have become through time. Valuable recent studies report a light Li isotopic composition (δ^7Li ≈ 0) for both the upper [35,36] and lower [37] continental crust with a mean [Li] = 22 ppm [35]. The light Li isotope ratio of the present-day continental crust can be produced in only 1 Gyr of recycling at current rates. This suggests that the recycled Li flux calculated from present-day oceanic mass balance may be anomalously high, possibly as a result of enhanced current weathering rates from Himalayan uplift (e.g. [26]) or simply due to inaccuracies in the approach. Regardless of the rate at which the continents have become light, however, the Li isotope composition of the continental crust limits the change in the δ^7Li of the whole mantle to be $\sim 0.3‰$.

Before considering how subduction modifies the Li alteration flux into the mantle it is worth briefly contrasting the Li isotope signature of altered oceanic crust with longer established measurements of oxygen isotopes. Oxygen is nearly equally abundant in most silicate rocks, from clays to mantle minerals, whereas altered upper oceanic crust has a Li concentration some 10 times greater than the mantle. Recycling of altered oceanic crust to the mantle will thus have a greater leverage on the Li isotope system. Moreover, heavy oxygen isotope ratios in the upper oceanic crust are compensated by light oxygen isotope ratios in the lower oceanic crust [38]. Thus the oxygen isotopic composition of the complete sequence of altered oceanic crust is the same as pristine mantle. Nevertheless, oxygen isotopes have provided striking evidence for the presence of specific portions of recycled crust in some ocean island basalt (OIB) sources, notably sedi-

mentary material [39] and lower portions of the altered oceanic crust [40]. Yet oxygen isotopes are insensitive to the wholesale addition of altered oceanic crust to a mantle source. Interpretation of the similarity of most peridotite, mid-ocean ridge basalt (MORB) and OIB oxygen isotope ratios [7] is therefore ambiguous.

4. The subduction zone filter

Plate tectonics inevitably mixes altered oceanic crust, with its budget of heavy Li, back into the mantle. However, the passage of Li from the surface to depth is not simple and is strongly affected by the processes occurring during subduction (Fig. 3A). Water is successively lost during subduction, first through porosity reduction during compaction and then through a series of prograde dehydration reactions as the descending slab becomes hotter (e.g. [41,42]). As in the near-surface environment, the partitioning of Li between residual solid and liquid will result in isotopic fractionation. Although isotopic fractionation diminishes with increasing temperature, there is still significant inferred [22,33] and measured [25] Li isotopic fractionation between mineral phases and water during high-temperature hydrothermal circulation at $\sim 350°$C. Thus the potential for isotopic fractionation extends well into the subduction zone.

Fluid is expelled out of the subduction zone as a result of compaction of the sedimentary pile. These fluids that return to the ocean along the so-called décollement are isotopically heavier than the associated sediments [19]. Further insight into the shallow subduction zone is provided by serpentine diapirs which sample the highly hydrated fore-arc. Analyses of Mariana serpentinites show a wide range of δ^7Li from light ($-0.5‰$) to heavy values ($+10‰$) [43], although finer-grained fractions show dominantly heavier values (δ^7Li 5–9‰) [44]. Thus some of the heavy subducted Li is clearly lost to the oceans and shallow wedge, but the amount of Li lost from the slab in these early dehydration processes is estimated to be a minor fraction of the total Li subducted [19,45]. In contrast, the budget of B is significantly de-

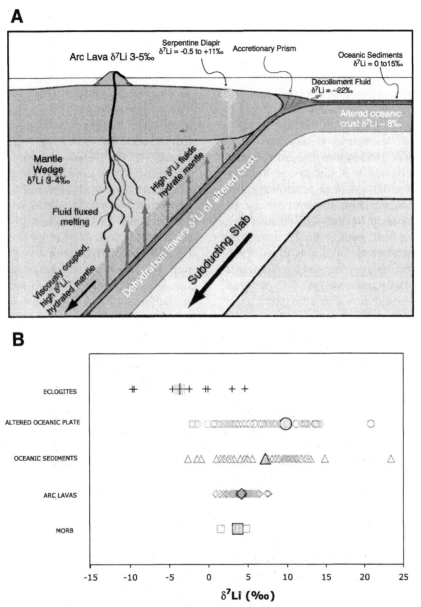

Fig. 3. (A) Cartoon of the Li isotope behaviour in the subduction zone, illustrating processes discussed in the text. Sketch indicates transfer of fluids with heavy Li to hydrate the adjacent mantle (hatched). The hydrated mantle is viscously coupled and down-dragged with the complementary, isotopically light, dehydrated oceanic crust. Fluids that traverse into the hotter interior mantle wedge and induce melting have already isotopically equilibrated their Li with the preceding mantle. Thus arc lavas have δ^7Li comparable to the pristine mantle wedge. (B) Li isotopic composition of inputs and outputs from subduction zone. Individual data are shown with weighted means represented as larger symbols. Note that despite the heavy mean compositions of both incoming sediments and altered ocean crust, the vast majority of arc lavas have δ^7Li within error of MORB (uncertainties for most arc measurements ± 1‰). Samples sources: eclogites [52], altered oceanic crust (see Fig. 1), oceanic sediments [18,19,34,49,75], arc lavas [34,48,49,76]. The older Panama samples reported in [76] are not plotted due to the possible effects of 10 Myr of weathering in a tropical environment, which can increase δ^7Li of samples [10].

pleted by fore-arc dehydration [45,46] which compromises its potential as a tracer of recycled material.

A major mass flux at subduction zones is associated with magmatism at the volcanic front. Subduction zone magmas are characterised by some distinctive chemical signatures. Notably arc lavas have elevated concentrations of elements empirically observed to be mobile in aqueous fluids (such as Li) relative to elements that form highly charged ions and have low solubilities in aqueous fluids (e.g. Y) but otherwise behave similarly during magmatic processes (for a recent review see [47]). Thus arc lavas have high Li/Y ratios (e.g. [48]). Analogous ratios, such as Ba/Th, have long been used to implicate the addition of aqueous-rich fluids from the down-going slab to the source of arc lavas (in the mantle wedge). The elevated Li/Y of arc lavas would hence be anticipated to be associated with heavy Li, fluxed from the subducting slab. It is surprising that very few arc lavas have Li isotope ratios significantly heavier than MORB [48] (see Fig. 3B). Two studies of arc lavas show variations of Li isotope ratios with some key trace element ratios, suggesting a trend to heavier Li ratios with larger slab-derived inputs [34,49]. Such systematic variations are not observed, however, in comprehensive studies of a number of other arcs ([48], Bouman, unpublished data).

That few arc lavas have heavy Li isotope ratios implies that little of the budget of heavy Li that enters the subduction zone is lost to the surface as a result of arc magmatism. Yet, as discussed above, several 'fluid-mobile' tracers of the slab are evident in arc lavas. The lack of an associated heavy Li isotopic signature can be related to the relative affinity of Li for mantle phases. Li has an ionic radius comparable to Mg and can occupy the abundant Mg lattice sites available in the mantle. Although apparently preferentially transported by fluids, high-temperature partitioning experiments show that Li is only moderately incompatible in either melt or fluid relative to mantle phases [50,51]. Thus it is likely that heavy Li initially carried from the slab in aqueous fluids isotopically re-equilibrates with the mantle [48]. The heavy Li isotopic signature from the slab is effec-

tively transferred to the cold portion of mantle lying above the slab (Fig. 3A).

Striking evidence for the loss of heavy Li from the subducted oceanic crust itself is found in the extremely light Li isotopic ratios (δ^7Li as low as $-11‰$) of Alpine eclogites [52] (see Fig. 3B). These high-pressure metamorphic rocks are believed to be fragments of deeply subducted oceanic crust that have been rapidly exhumed by thrusting during the Alpine orogeny. Zack et al. [52] argue that the loss of Li during pro-grade metamorphism during subduction is associated with significant isotopic fractionation. As in the surface environment, Li in the fluid phase is enriched in the heavy isotope leaving the dehydrated slab isotopically light. If accumulated effects of dehydration are modelled as a Rayleigh distillation process, the final 'anhydrous' residue at the end of subduction zone processing can have extremely light Li isotope ratios. This mechanism can be invoked to account for the very light Li isotope ratios observed in the eclogites (see Fig. 3B). Comparable isotopic fractionation during subduction has been noted in the boron isotope system [53–55]. The Alpine eclogites thus represent the light complement of the heavy isotopic signature that has left the slab to reside in the overlying, viscously coupled mantle.

The effects of subduction thus provide additional interest in the use of Li isotopes as tracers of mantle convection. The deep recycled, dehydrated altered oceanic crust becomes isotopically light. Meanwhile the cool mantle adjacent to the slab becomes heavy. Thus we have two inputs to the mantle as a result of subduction that are isotopically distinct from the mantle itself. Moreover, since these are stable isotope variations, these signatures can be unambiguously linked to low-temperature processes near the surface of the Earth. Li isotopes therefore provide a means to trace the mixing of subducting slab and adjacent mantle wedge back into the convecting mantle. It is of great interest to see how Li isotope signatures of recycling compare to radiogenic isotope measurements. Although the latter have long been used to infer the distribution of recycled material, their variations can be interpreted in other ways and are much more equivocal.

5. Mantle signatures

MORB and OIB represent our primary sample of the mantle (for a review see [56]). The melting process averages melts from a large volume of source rock and so studying melts is an efficient approach to sample the mantle. MORB are the product of shallow (dominantly less than 60 km) decompression melting in response to plate spreading. MORB thus effectively sample the ambient shallow mantle. OIB are thought to be derived from melting of mantle that is hotter, or with a low solidus, such that it commences melting deeper than MORB mantle. Thus OIB can be generated by upwelling even beneath thick oceanic lithosphere. The upwellings required to produce OIB are frequently attributed to mantle plumes. Since plumes need to be sourced at a major thermal boundary, most appealingly the core–mantle boundary, they are generally thought to sample deeper mantle than the source of MORB.

OIB are typically more 'enriched' than MORB, with higher incompatible element contents and isotope ratios that reflect long-term elevation of more incompatible relative to less incompatible elements (e.g. high $^{87}Sr/^{86}Sr$). Maintaining this contrast in geochemical signatures between MORB and OIB sources is an important constraint on mantle convection models [3,4,57,58]. The nature of the 'enrichment' in different OIB is not uniform and several end-member compositions have been identified. The role of recycled material has long been invoked to account for several of these end-member compositions. In a popular general scenario, manifest in a variety of different specific forms (see [59]), a deep mantle layer is fed by subducted plates and ultimately forms the source for OIB. From the behaviour of Li in the subduction zone discussed above, this lower mantle layer would likely evolve with a lighter Li isotope ratio than the upper mantle. Thus OIB might be expected to have a lighter Li isotope ratio than MORB. This should be particularly marked for the so-called HIMU OIB which are argued to have a large portion of recycled mafic crust in their sources (see [56]).

Before examining the Li isotope data for oceanic basalts, several important caveats need to be addressed. Whilst variations in stable isotope ratios in mantle-derived rocks represent the least ambiguous signal of the (variable) presence of material that has been affected by processes close to the Earth's surface, it is vital to assess if this is from deep recycled material or simply shallow-level contamination during magmagenesis [60,61] or even recent weathering [10]. Conclusively ruling out a role of crustal assimilation is difficult, but the potential effects can be minimised by analysing relatively 'primitive' samples. In the case of Li, it is possible to measure the isotope ratio of the early crystallising phase olivine, which should be least affected by contamination processes. The ability to analyse olivines is also an asset in studying older samples where identification of possible alteration in the whole rock is difficult [10]. Fresh olivines can be readily picked from many OIB in the same way that fresh glass can be prepared from MORB to produce reliable sample material. Although fractionation of stable isotopes is diminished at high temperatures, it is important to quantify the possible isotope fractionation induced by magmatic processes. Neither variable melting [62] nor basaltic differentiation processes [63], however, appear able to generate significant variations in Li isotope ratios.

Li isotope data remain sparse for samples from our planet's interior. In the case of MORB, there are only six measurements of fresh samples in the literature and these range in δ^7Li from 1.5 to 4.7‰ [17,34]. Five of these six measurements range from 3.2 to 4.7‰, which is within error of the techniques used in these analyses. A recent high-precision study (± 0.3‰) of hand-picked, fresh glasses from the East Pacific Rise (EPR) yields a range of δ^7Li of 3.1–5.2‰ [64]. Significantly, the Li isotope compositions in this study correlate with radiogenic isotope and highly incompatible element ratios. Covariation of Li isotopes with $^{143}Nd/^{144}Nd$, for example, makes contamination an unlikely explanation for the range in Li isotopes. The EPR dataset supports other unpublished studies that argue for a 'depleted', N-MORB δ^7Li of 3–4‰ [17,65] and significant Li isotope variations within MORB globally [66].

The spread of δ^7Li in MORB data requires recycled material to be mixed variably back into the

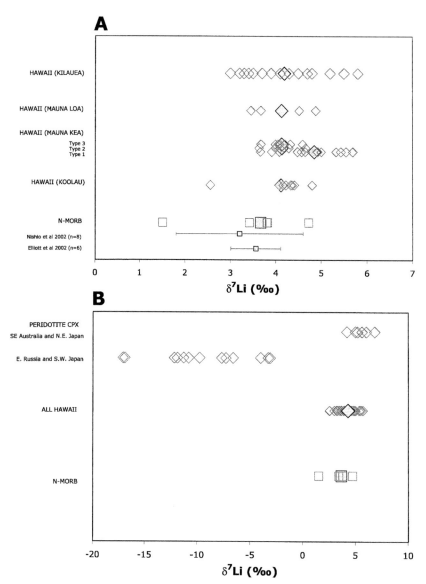

Fig. 4. (A) Li isotopic compositions of all published MORB [17,34] and Hawaiian lavas [10,13,21,63]. The published MORB samples are all N-MORB (depleted compositions) and can be compared with other unpublished N-MORB averages [64,65]. The Hawaiian samples have uncertainties of 0.5‰, except for Kilauea where measurements have errors of ∼1‰. MORB samples likewise have errors of ∼1‰ apart from the analyses of [64] which have a reproducibility of <0.3‰. Some of the Hawaiian samples reported in [10] are altered and we have only plotted samples that have K/Rb ratios that fall within the range of 380–630, cited as representative of pristine samples [10]. The samples from Mauna Kea have been sub-divided into types 1–3 that display distinct chemical signatures and variable contributions from different source components (see [10]). (B) δ7Li of clinopyroxenes from several mantle nodule localities [71] compared to MORB and Hawaiian data reported in A.

upper mantle. This has long been proposed as a logical conceptual corollary of plate tectonics [67]. Yet only very recently has the first unequivocal evidence of this process been presented, in the form of subtle variations in oxygen isotope ratios [68]. The Li isotope data provide strong additional support to this idea.

Li isotope analyses of OIB are also scarce. An

important OIB datum is that of the international standard BHVO, a recent eruption from Hawaii. It is well established that this basalt has $\delta^7Li >$ 4.7‰ [10,13,21] and so is *heavier* than N-MORB. This lonely data point has been considerably augmented by further analyses from Hawaii [10,63] (see Fig. 4A). The latter study is at higher precision (± 0.5‰ compared to typical ± 1‰ of previous studies) and also includes analyses of the older Koolau lavas. The Koolau lavas are remarkable for an enriched, heavy oxygen isotope component in their source, clearly implicated as sedimentary in origin [39]. Additionally samples from Mauna Kea can be sub-divided into groups that reflect different proportions of distinct components in their sources. For example, the type 1 lavas have high SiO_2 and have been argued to contain recycled, mafic oceanic crust whereas type 3 lavas, with the highest $^3He/^4He$, contain the largest proportion of 'undegassed' mantle.

There is considerable overlap between the δ^7Li of MORB and Hawaiian lavas (Fig. 4A). Nevertheless, the Hawaiian lavas are on average heavier than N-MORB. Despite the notable geochemical contrasts between the different Hawaiian compositions (Koolau and various Mauna Kea types) their differences in δ^7Li are subtle. It is worth noting, however, that the Mauna Kea lavas invoked to contain the largest component of recycled mafic, oceanic crust (type 1) have heavier δ^7Li than the other lava types. This is in keeping with unpublished data [69,70] that document heavy Li isotope ratios (5–7‰) in end-member 'HIMU' lavas, which are traditionally thought to contain large amounts of recycled mafic oceanic crust [56].

Recent analyses of clinopyroxenes from mantle nodules [71] provide an exciting counter-poise to the data from mantle-derived lavas (Fig. 4B). Whilst some samples have δ^7Li that overlap OIB values, the clinopyroxenes from Far East Russia and southwestern Japan display δ^7Li as low as -17‰. These extremely light Li isotope ratios implicate the involvement of material from subduction-processed oceanic crust in the sub-continental mantle lithosphere represented by these nodules [71]. The samples also clearly document the possibility for large mantle δ^7Li variations. In

a similar vein, Kobayashi et al. [72] report ion-probe data from Hawaiian olivine-hosted melt inclusions with δ^7Li as low as -10‰. Nevertheless, this isotopically light component does not appear to represent a significant proportion of the source since it is not evident in the bulk lava compositions (Fig. 4A).

6. Implications

The observations above lead to some valuable inferences. For OIB, like MORB, a range of δ^7Li points to variable amounts of recycled material in their sources. Very light Li isotope ratios from subduction-processed slab is dramatically evident in some mantle nodules and melt inclusions, but not in OIB lava compositions. Indeed, samples proposed to contain a large proportion of ancient basaltic slab are generally heavier than N-MORB. Likewise there is no evidence of lighter Li in samples frequently linked to a lower mantle reservoir. High $^3He/^4He$ samples from Hawaii (type 3 Mauna Kea) have marginally higher δ^7Li than N-MORB (Fig. 4A). Thus there is no indication of the drastic division in Li isotope signature between OIB and MORB that should result from subduction zone processing and preferential accumulation of subducted slabs in the lower mantle (e.g. [73]).

Clearly much more work is required to build on these initial observations. From the tantalising glimpse into the mantle provided by the current, we suggest that the trend to heavy δ^7Li in OIB (and enriched MORB) sources has a common origin in the down-dragged mantle adjacent to the subducting slab (Fig. 3A). Fluids released from the down-going plate will be heavy, but variable in composition, depending on the prior history of dehydration. This process could account for the large variability in δ^7Li not clearly related to other chemical tracers. Material from the slab has to pass into a hot portion of the mantle wedge before it triggers large-scale melting. The colder mantle layer overlying the slab will acquire a recycled signature, such as heavy δ^7Li, from the passage of slab-derived fluids through it (Fig. 3A). The mantle layer may also similarly acquire a sedimentary signature. This overlying mantle will be viscously coupled to the slab and so carried down beyond the subduction zone. As the slab descends and warms, the overlying enriched layer will cease to be viscously coupled. Depending on the thermal and dynamic history of the plate, the heavy δ^7Li layer of mantle may be carried to variable depths to contribute to both MORB and OIB sources.

This scenario of recycling is still highly speculative. What is clear, however, from our current understanding of the behaviour of Li isotopes is that although there is ample evidence for the role of recycled material in some OIB and MORB sources, this does not appear to be in the form of subduction zone-processed mafic oceanic crust. Interestingly these sentiments are in keeping with other recent discussions on the geochemistry of OIB [5].

Acknowledgements

The manuscript was considerably improved by the reviews from Peter van Keken, Paul Tomascak and an anonymous reviewer. Comments from Chris Hawkesworth and George Helffrich also helped point things in the right direction. The editorial encouragement and patience of Alex Halliday is gratefully acknowledged. Our developments in Li isotope geochemistry have been generously supported by the Leverhulme Trust, NWO and NERC.*[AH]*

References

[1] R. Trompert, U. Hansen, Mantle convection simulations with rheologies that generate plate-like behaviour, Nature 395 (1998) 686–689.

[2] H. Bijwaard, W. Spakman, E.R. Engdahl, Closing the gap between regional and global travel time tomography, J. Geophys. Res. 103 (1998) 30055–30078.

[3] P.J. Tackley, S.X. Xie, The thermochemical structure and evolution of Earth's mantle: constraints and numerical models, Phil. Trans. R. Soc. London A 360 (2002) 2593–2609.

[4] P.E. van Keken, E.H. Hauri, C.J. Ballentine, Mantle mixing: The generation, preservation, and destruction of chemical heterogeneity, Annu. Rev. Earth Planet. Sci. 30 (2002) 493–525.

[5] Y. Niu, M.J. O'Hara, Origin of ocean island basalts: a new persepctive from petrology, geochemistry and mineral physics considerations, J. Geophys. Res. 108 (2003) 2002JB002048.

[6] J.M. Eiler, Oxygen isotope variations of basaltic lavas and upper mantle rocks, in: Stable Isotope Geochemistry, Rev. Mineral. Geochem. 43 (2001) 319–364.

[7] J.M. Eiler, K.A. Farley, J.W. Valley, E. Hauri, H. Craig, S.R. Hart, E.M. Stolper, Oxygen isotope variations in ocean island basalt phenocrysts, Geochim. Cosmochim. Acta 61 (1997) 2281–2293.

[8] G.D. Flesch, A.R. Anderson, H.J. Svec, A secondary isotopic standard for ^{6}Li/^{7}Li determinations, Int. J. Mass Spectrom. Ion Process. 12 (1973) 265–272.

[9] H.P. Qi, P.D.P. Taylor, M. Berglund, P. DeBièvre, Calibrated measurements of the isotopic composition and atomic weight of the natural Li isotopic reference material IRMM-016, Int. J. Mass Spectrom. Ion Process. 171 (1997) 263–268.

[10] L.H. Chan, F.A. Frey, Lithium isotope geochemistry of the Hawaiian plume: Results from the Hawaii Scientific Drilling Project and Koolau volcano, Geochem. Geophys. Geosyst. 4 (2003) 8707.

[11] L.H. Chan, Lithium isotope analysis by thermal ionization mass-spectrometry of lithium tetraborate, Anal. Chem. 59 (1987) 2662–2665.

[12] T. Moriguti, E. Nakamura, High-yield lithium separation and the precise isotopic analysis for natural rock and aqueous samples, Chem. Geol. 145 (1998) 91–104.

[13] R.H. James, M.R. Palmer, The lithium isotope composition of international rock standards, Chem. Geol. 166 (2000) 319–326.

[14] P.B. Tomascak, R.W. Carlson, S.B. Shirey, Accurate and precise determination of Li isotopic compositions by multi-collector sector ICP-MS, Chem. Geol. 158 (1999) 145–154.

[15] Y. Nishio, S. Nakai, Accurate and precise lithium isotopic determinations of igneous rock samples using multi-collector inductively coupled plasma mass spectrometry, Anal. Chim. Acta 456 (2002) 271–281.

[16] A.B. Jeffcoate, T. Elliott, A. Thomas, C. Bouman, Precise, small sample size determinations of lithium isotopic compositions of geological reference materials and modern seawater by MC–ICPMS, Geostand. Newsl. (in press).

[17] L.H. Chan, J.M. Edmond, G. Thompson, K. Gillis, Lithium isotopic composition of submarine basalts – implications for the lithium cycle in the oceans, Earth Planet. Sci. Lett. 108 (1992) 151–160.

[18] L.B. Zhang, L.H. Chan, J.M. Gieskes, Lithium isotope geochemistry of pore waters from Ocean Drilling Program Sites 918 and 919, Irminger Basin, Geochim. Cosmochim. Acta 62 (1998) 2437–2450.

[19] L.H. Chan, M. Kastner, Lithium isotopic compositions of pore fluids and sediments in the Costa Rica subduction

zone: Implications for fluid processes and sediment contribution to the arc volcanoes, Earth Planet. Sci. Lett. 183 (2000) 275–290.

[20] Y. Huh, L.H. Chan, J.M. Edmond, Lithium isotopes as a probe of weathering processes: Orinoco River, Earth Planet. Sci. Lett. 194 (2001) 189–199.

[21] J.S. Pistiner, G.M. Henderson, Lithium-isotope fractionation during continental weathering processes, Earth Planet. Sci. Lett. 214 (2003) 1–13.

[22] L.H. Chan, J.M. Edmond, G. Thompson, A lithium isotope study of hot-springs and metabasalts from midocean ridge hydrothermal systems, J. Geophys. Res. 98 (1993) 9653–9659.

[23] Y. Huh, L.H. Chan, L. Zhang, J.M. Edmond, Lithium and its isotopes in major world rivers: Implications for weathering and the oceanic budget, Geochim. Cosmochim. Acta 62 (1998) 2039–2051.

[24] P.B. Tomascak, Developments in the understanding and application of lithium isotopes in the Earth and planetary sciences, in: Geochemistry of Non-Traditional Stable Isotopes, Mineralogical Society of America, Washington, DC, in press.

[25] W.E. Seyfried, X. Chen, L.H. Chan, Trace element mobility and lithium isotope exchange during hydrothermal alteration of seafloor weathered basalts: an experimental study at 350°C, 500 bars, Geochim. Cosmochim. Acta 62 (1998) 959–960.

[26] M.E. Raymo, W.F. Ruddiman, P.N. Froelich, Influence of late Cenozoic mountain building on ocean geochemical cycles, Geology 16 (1988) 649–653.

[27] G.M. Henderson, New oceanic proxies for paleoclimate, Earth Planet. Sci. Lett. 203 (2002) 1–13.

[28] J. Hoefs, M. Sywall, Lithium isotope composition of Quaternary and Tertiary biogene carbonates and a global lithium isotope balance, Geochim. Cosmochim. Acta 13 (1997) 2679–2690.

[29] J. Kosler, M. Kucera, P. Sylvester, Precise measurement of Li isotopes in planktonic formainiferal tests by quadrapole ICPMS, Chem. Geol. 181 (2001) 169–179.

[30] N. Vigier, K.W. Burton, S.R. Gislason, N.W. Rogers, B.F. Schaefer, R.H. James, Constraints on basalt erosion from Li isotopes and U-series nuclides measured in Iceland rivers, Geochim. Cosmochim. Acta 66 (2002) A806–A806.

[31] C.S. Marriott, N.S. Belshaw, G.M. Henderson, Lithium and calcium isotope fractionation in inorganically precipitated calcite: Assessing their potential as a paleothermometer, Geochim. Cosmochim. Acta 66 (2002) A485–A485.

[32] E.C. Hathorne, R.H. James, N.B. Harris, The lithium isotope composition of planktonic foraminifera, EOS Trans. AGU 84 (2003) PP11A-0214.

[33] L.H. Chan, J.C. Alt, D.A.H. Teagle, Lithium and lithium isotope profiles through the upper oceanic crust: a study of seawater-basalt exchange at ODP Sites 504B and 896A, Earth Planet. Sci. Lett. 201 (2002) 187–201.

[34] T. Moriguti, E. Nakamura, Across-arc variation of Li

isotopes in lavas and implications for crust/mantle recycling at subduction zones, Earth Planet. Sci. Lett. 163 (1998) 167–174.

[35] F.-Z. Teng, W.F. McDonough, R.L. Rudnick, C. Dalpé, P.B. Tomascak, B.W. Chappell, S. Gao, Lithium isotopic composition and concentration of the upper continental crust, Geochim. Cosmochim. Acta (submitted).

[36] F.-Z. Teng, W.F. McDonough, R. Rudnick, P.B. Tomascak, C. Dalpé, Lithium content and isotopic composition of the upper continental crust, EOS Trans. AGU 83 (2002) V61D-12.

[37] F.-Z. Teng, W.F. McDonough, R. Rudnick, P.B. Tomascak, A.E. Saal, Lithium isotopic composition of the lower continental crust: a xenolith perspective, EOS Trans. AGU 84 (2003) V51A-02.

[38] R.T. Gregory, H.P. Taylor, An oxygen isotope profile in a section of Cretaceous oceanic crust, Samail Ophiolite, Oman: evidence for $\delta^{18}O$ buffering of the oceans by deep (>5 km) seawater-hydrothermal circulation at mid-ocean ridges, J. Geophys. Res. 86 (1981) 2737–2755.

[39] J.M. Eiler, K.A. Farley, J.W. Valley, A.W. Hofmann, E.M. Stolper, Oxygen isotope constraints on the sources of Hawaiian volcanism, Earth Planet. Sci. Lett. 144 (1996) 453–467.

[40] J.C. Lassiter, E.H. Hauri, Osmium-isotope variations in Hawaiian lavas: evidence for recycled oceanic lithosphere in the Hawaiian plume, Earth Planet. Sci. Lett. 164 (1998) 483–496.

[41] D.M. Kerrick, J.A.D. Connolly, Metamorphic devolatilization of subducted marine sediment and the transport of volatiles into the Earth's mantle, Nature 411 (2001) 293–296.

[42] D.M. Kerrick, J.A.D. Connolly, Metamorphic devolatilization of subducted oceanic metabasalts: implications for seismicity, arc magmatism and volatile recycling, Earth Planet. Sci. Lett. 189 (2001) 19–29.

[43] J.G. Ryan, L.D. Benton, I. Savov, Isotopic and elemental signatures of the forearc and impacts on subduction recycling: evidence from the Marianas, EOS Trans. AGU 82 (2001) V11B-11.

[44] L.D. Benton, I. Savov, J.G. Ryan, Recycling of subducted lithium in forearcs: insights from a serpentine seamount, EOS Trans. AGU 80 (1999) V21B-07.

[45] J.G. Ryan, J. Morris, G.E. Bebout, W.P. Leeman, Describing chemical fluxes in subduction zones: insights from 'depth profiling' studies of arcs and forearc rocks, in: G.E. Bebout, D.W. Scholl, S.H. Kirby, J.P. Platt (Eds.), Subduction Top to Bottom, American Geophysical Union, Washington DC, 1996, pp. 263–267.

[46] A. Kopf, A. Deyhle, Back to the roots: boron geochemistry of mud volcanoes and its implications for mobilization depth and global B cycling, Chem. Geol. 192 (2002) 195–210.

[47] T. Elliott, Tracers of the slab, in: J.M. Eiler (Ed.), Inside the Subduction Factory, AGU Geophys. Monogr. 138 (2003) 23–45.

[48] P.B. Tomascak, E. Widom, L.D. Benton, S.L. Goldstein,

J.G. Ryan, The control of lithium budgets in island arcs, Earth Planet. Sci. Lett. 196 (2002) 227–238.

[49] L.H. Chan, W.P. Leeman, C.F. You, Lithium isotopic composition of Central American volcanic arc lavas: implications for modification of subarc mantle by slab-derived fluids: correction, Chem. Geol. 182 (2002) 293–300.

[50] J.M. Brennan, E. Neroda, C.C. Lundstrom, H.F. Shaw, F.J. Ryerson, D.L. Phinney, Behaviour of boron, beryllium and lithium during melting and crystallisation: constraints from mineral-melt partitioning experiments, Geochim. Cosmochim. Acta 62 (1998) 2129–2141.

[51] J.M. Brennan, F.J. Ryerson, H.F. Shaw, The role of aqueous fluids in the transfer of boron, beryllium and lithium during subduction: experiments and models, Geochim. Cosmochim. Acta 62 (1998) 3337–3347.

[52] T. Zack, P.B. Tomascak, R.L. Rudnick, C. Dalpe, W.F. McDonough, Extremely light Li in orogenic eclogites: The role of isotope fractionation during dehydration in subducted oceanic crust, Earth Planet. Sci. Lett. 208 (2003) 279–290.

[53] S.M. Peacock, R.L. Hervig, Boron isotopic composition of subduction-zone metamorphic rocks, Chem. Geol. 160 (1999) 281–290.

[54] L.D. Benton, J.G. Ryan, F. Tera, Boron isotope systematics of slab fluids as inferred from a serpentine seamount, Mariana forearc, Earth Planet. Sci. Lett. 187 (2001) 273–282.

[55] E.F. Rose, N. Shimizu, G.D. Layne, T.L. Grove, Melt production beneath Mt. Shasta from boron data in primitive melt inclusions, Science 293 (2001) 281–283.

[56] A.W. Hofmann, Mantle geochemistry: the message from oceanic volcanism, Nature 385 (1997) 219–229.

[57] F. Albarède, R.D. van der Hilst, Zoned mantle convection, Phil. Trans. R. Soc. London A 360 (2002) 2569–2592.

[58] D. Bercovici, S.-I. Karato, Whole mantle convection and the transition-zone water filter, Nature 425 (2003) 39–44.

[59] P.J. Tackley, Mantle convection and plate tectonics toward an integrated physical and chemical theory, Science 288 (2000) 2002–2007.

[60] M.A.M. Gee, M.F. Thirlwall, R.N. Taylor, D. Lowry, B.J. Murton, Crustal processes: major controls on Reykjanes Peninsula lava chemistry, SW Iceland, J. Petrol. 39 (1998) 819–839.

[61] J.M. Eiler, K. Gronvold, N. Kitchen, Oxygen isotope evidence for the origin of chemical variations in lavas from Theistareykir volcano in Iceland's northern volcanic zone, Earth Planet. Sci. Lett. 184 (2000) 269–286.

[62] A.B. Jeffcoate, T. Elliott, D.A. Ionov, Li isotope fractionation in the mantle, Geochim. Cosmochim. Acta 66 (2002) A364–A364.

[63] P.B. Tomascak, F. Tera, R.T. Helz, R.J. Walker, The absence of lithium isotope fractionation during basalt differentiation: New measurements by multicollector sector ICP-MS, Geochim. Cosmochim. Acta 63 (1999) 907–910.

[64] T. Elliott, A.L. Thomas, A.B. Jeffcoate, Y. Niu, Li isotope variations in the upper mantle, Geochim. Cosmochim. Acta 66 (2002) A214–A214.

[65] Y. Nishio, S. Nakai, K. Hirose, T. Ishii, Y. Sano, Li isotopic systematics of volcanic rocks in marginal basins, Geochim. Cosmochim. Acta 66 (2002) A556–A556.

[66] P.B. Tomascak, C.H. Langmuir, Lithium isotope variability in MORB, EOS Trans. AGU 80 (1999) V11E-10.

[67] C.J. Allègre, D.L. Turcotte, Implications of a 2-component marble-cake mantle, Nature 323 (1986) 123–127.

[68] J.M. Eiler, P. Schiano, N. Kitchen, E.M. Stolper, Oxygen-isotope evidence for recycled crust in the sources of mid-ocean-ridge basalts, Nature 403 (2000) 530–534.

[69] Y. Nishio, S. Nakai, T. Kogiso, H.G. Barsczus, Lithium isotopic composition of HIMU oceanic basalts: implications for the origin of HIMU component, XXIII General Assembly of the International Union of Geodesy and Geophysics, 2003, p. 178.

[70] A.B. Jeffcoate, T. Elliott, Tracing recycled Li in the mantle: insights into mantle heterogeneities, EOS Trans. AGU 84 (2003) V52A-0416.

[71] Y. Nishio, S. Nakai, J. Yamamoto, H. Sumino, T. Matsumoto, V.S. Prikhod'ko, S. Arai, Lithium isotopic systematics of the mantle-derived ultramafic xenoliths: implications for EM1 origin, Earth Planet. Sci. Lett. 217 (2004) 245–261.

[72] K. Kobayashi, R. Tanaka, T. Moriguti, K. Shimizu, E. Nakamura, Lithium, boron and lead isotope systematics on glass inclusions in olivine phenocrysts from Hawaiian lavas, (2003) A223.

[73] L.H. Kellogg, B.H. Hager, R.D. van der Hilst, Compositional stratification in the deep mantle, Science 283 (1999) 1881–1884.

[74] S. Decitre, E. Deloule, L. Reisberg, R. James, P. Agrinier, C. Mevel, Behavior of Li and its isotopes during serpentinization of oceanic peridotites, Geochem. Geophys. Geosyst. 3 (2002) 1007.

[75] C.F. You, L.H. Chan, A.J. Spivack, J.M. Gieskes, Lithium, boron, and their isotopes in sediments and pore waters of ocean drilling program site-808, Nankai Trough – implications for fluid expulsion in accretionary prisms, Geology 23 (1995) 37–40.

[76] P.B. Tomascak, J.G. Ryan, M.J. Defant, Lithium isotope evidence for light element decoupling in the Panama subarc mantle, Geology 28 (2000) 507–510.

Tim Elliott graduated from Cambridge in 1987 to undertake a PhD at the Open University in Milton Keynes, finishing in 1991. Having had his fill of the delights of new towns he left the country to spend 2 years at Lamont-Doherty Geological Observatory, New York, USA and 6 years at the Faculteit der Aardwetenschappen, Vrije Universiteit, Amsterdam, The Netherlands. He became a lecturer at the Department of Earth Sciences, University of Bristol in 1999.

Alistair Jeffcoate graduated from Royal Holloway College, University of London in 2001. He joined the nascent isotope lab at the Department of Earth Sciences in Bristol where he is currently finishing his PhD.

Claudia Bouman graduated from the Universiteit van Utrecht, The Netherlands in 1998. She has just submitted her doctoral thesis from the Vrije Universiteit, Amsterdam and currently works as an application specialist for Thermo-Finnigan, Bremen.

Reprinted from
Earth and Planetary Science Letters 220 (2004) 3–24

www.elsevier.com/locate/epsl

African climate change and faunal evolution during the Pliocene–Pleistocene

Peter B. deMenocal

Lamont-Doherty Earth Observatory, Palisades, NY 10964, USA

Received 12 August 2003; received in revised form 16 December 2003; accepted 21 December 2003

Abstract

Environmental theories of African faunal evolution state that important evolutionary changes during the Pliocene–Pleistocene interval (the last ca. 5.3 million years) were mediated by changes in African climate or shifts in climate variability. Marine sediment sequences demonstrate that subtropical African climate periodically oscillated between markedly wetter and drier conditions, paced by earth orbital variations, with evidence for step-like (± 0.2 Ma) increases in African climate variability and aridity near 2.8 Ma, 1.7 Ma, and 1.0 Ma, coincident with the onset and intensification of high-latitude glacial cycles. Analysis of the best dated and most complete African mammal fossil databases indicates African faunal assemblage and, perhaps, speciation changes during the Pliocene–Pleistocene, suggesting more varied and open habitats at 2.9–2.4 Ma and after 1.8 Ma. These intervals correspond to key junctures in early hominid evolution, including the emergence of our genus *Homo*. Pliocene–Pleistocene shifts in African climate, vegetation, and faunal assemblages thus appear to be roughly contemporary, although detailed comparisons are hampered by sampling gaps, dating uncertainties, and preservational biases in the fossil record. Further study of possible relations between African faunal and climatic change will benefit from the accelerating pace of important new fossil discoveries, emerging molecular biomarker methods for reconstructing African paleovegetation changes, tephra correlations between terrestrial and marine sequences, as well as continuing collaborations between the paleoclimatic and paleoanthropological communities.
© 2004 Elsevier B.V. All rights reserved.

Keywords: Africa; paleoclimate; orbital forcing; monsoon; eolian dust; savannah hypothesis; human evolution; speciation

1. Introduction

Recent extraordinary fossil discoveries, advances in the analysis of extant fossil collections, and the emergence of detailed paleoclimatic records have focused new attention on the possible role that changes in African climate may have had in the evolutionary history of African mammalian fauna spanning the last 5–6 million years. The basic premise is that large-scale shifts in climate alter the ecological composition of a landscape which, in turn, present specific faunal adaptation or speciation pressures leading to genetic selection and innovation.

This review explores the African faunal and paleoclimatic evidence which constrains current environmental hypotheses of African faunal evolution. Still hotly debated, these hypotheses now draw upon a wealth of new fossil and paleocli-

E-mail address: peter@ldeo.columbia.edu (P.B. deMenocal).

matic evidence which have been used to explore possible temporal and causal relationships between known changes in climate and observed changes in faunal diversity and adaptation. Beginning with an overview of some leading climate–evolution hypotheses, this review discusses recent developments in what is known about the Pliocene–Pleistocene development of African paleoclimates based on terrestrial and marine paleoclimatic data. The hominid fossil record of the human family tree is now viewed to be considerably more complex than previously thought [1], although key transitional intervals of evolutionary and behavioral change are apparent. Advances in the analysis of large fossil mammal databases from collections in Kenya, Tanzania, and Ethiopia have led to new perspectives on the timing, signatures, and possible causes of key steps in African faunal evolution and assemblage changes. Taken together, these new data narrow the range of possible scenarios whereby changes in climate may have led to changes in fauna.

2. African climate – faunal evolution hypotheses

Natural selection, the process by which new adaptive structures evolve and persist, is viewed as the primary mechanism by which organisms change in relation to their environment. Long-term shifts in climate and secular shifts in climate variability are two separate ecological signals which can establish natural selection opportunities. Most environmental hypotheses of African faunal evolution are 'habitat-specific' [2] in that they consider faunal adaptations to a specific environment, most commonly the emergence of grassland savannah which occurred after the late Miocene [3–5]. The 'variability selection' hypothesis [2,6,7] incorporates observations about patterns of African paleoclimatic variability as recorded by deep-sea sediments, and it emphasizes the importance of climatic instability as a mechanism for natural selection. A review article by Richard Potts [2] presents a superb overview of the historical contexts, and faunal and ecological evidence for many environmental hypotheses of African faunal evolution.

2.1. 'Habitat-specific' hypotheses

The savannah hypothesis is perhaps the best known and most widely studied of the habitat-specific hypotheses of African faunal evolution [8–11]. In an early and unmistakably habitat-specific account of human evolution, Raymond Dart [8] posited that key human traits such as bipedality and larger brains were consequences of life on the open savannah. It was very provocative and influential (it is still found in some textbooks), but it was largely unconstrained by data as the fossil record at that time consisted of very few specimens with little paleoclimatic context. Current evidence indicates that bipedality was established millions of years before the widespread expansion of savannah grasslands [3,7,12–14].

Current interpretations of the savannah hypothesis state that the evolution of African mammalian fauna, including early hominids, was primarily attributable to the step-like development of cooler and drier and more open conditions which occurred since the late Miocene. The mid-Pliocene aridification shift near 3.2–2.6 Ma [15–17], in particular, is viewed to have favored the evolution of arid-adapted fauna and to have influenced early hominid evolution and behavior [15,18–26].

The turnover pulse hypothesis is one variant of the savannah hypothesis initially proposed by Elisabeth Vrba [21,27,28]. The essence of this view is that focused bursts of biotic change (or 'turnover', quantified in terms of clustered first-and/or last-appearance datums) were initiated by fundamental shifts in African climate which occurred roughly near 2.8 Ma, 1.8 Ma, and 1.0 Ma [15,28–31]. An analysis of fossil bovid (antelope) evolution from the largest African collections revealed clusters of first- and last-appearance datums which occurred near 2.7–2.5 Ma, with secondary clusters near ca. 1.8 Ma and ca. 0.7 Ma [21]. Many of the first appearances between 3.0 and 2.5 Ma were grazing species, supporting the view that the faunal changes were linked to aridity and expanding grasslands [21,26,32]. Although elegant and influential, the turnover pulse hypothesis has also been challenged recently (see Section 4.1) [33–35].

Other habitat-specific hypotheses argue for the

adaptive importance of environments other than open grasslands. Based on fossil pollen records from a 3.3–3.0 Ma fossil hominid site in South Africa, Rayner and others have alternatively proposed the forest hypothesis [36] which argues that closed vegetation, not savannah, was the adaptive environment for early bipeds. Blumenschine [37] has proposed that early tool-making hominids exploited the resources of grassland–woodland mosaic zones.

2.2. The variability selection hypothesis

This view of African faunal change accommodates one of the more obvious yet also curious features of African faunal and paleoclimatic records [2,6]. Individual fossil hominid and other mammal lineages typically persisted over long durations (10^5–10^6 years) yet they are preserved within sediment sequences which indicate much shorter-term (10^3–10^4 year) alternations between wetter and drier paleoenvironments [38,39]. Similarly, marine records of African paleoclimatic variability demonstrate that faunal lineages must have survived repeated alternations of wetter and drier periods paced by earth orbital variations of the African monsoon [15,17].

The 'variability selection' hypothesis emphasizes the importance of secular changes in climatic variability (amplitudes and durations of departures from the mean climatic state) on faunal adaptation, selection, and evolution [2,6]. In contrast to habitat-specific hypotheses, variability selection calls upon environmental instability (such as changes in the amplitudes and durations of orbital-scale wet–dry cycle amplitudes) as agents for introducing genetic variance, natural selection, and faunal innovation. Variability selection would argue that the largest faunal speciation and innovation pressures should occur during those periods when the amplitudes of climate variability changed markedly. This hypothesis proposes that African hominids and other fauna would have occupied increasingly diverse habitats following increases in paleoclimatic variability [2]. Many of the largest African faunal evolution events occurred when there were increases in the amplitudes of paleoclimatic variability [2].

3. African paleoclimate

3.1. Subtropical North African climate

North African vegetation zones range impressively from tropical rainforests to hyperarid subtropical deserts (Fig. 1a). The tremendous seasonal range in North African rainfall is a consequence of the seasonal migration of the intertropical convergence zone associated with the African monsoon. During boreal summer, sensible heating over the North African land surface centered near 20°N draws moist maritime air from the equatorial Atlantic into western and central subtropical Africa [40] which nourishes the grassland and woodland savannahs (Fig. 1b). East African summer rainfall is also related to the westerly airstream of the African monsoon, but is highly variable due in part to topographic rainshadow effects [41]. The Kenyan and Ethiopian Highlands orographically capture some of this moisture and this summer monsoonal runoff feeds the many subtropical rivers draining East Africa such as the Nile and Omo rivers. During boreal winter, the African and Asian landmasses cool relative to adjacent oceans and the regional atmospheric circulation reverses; dry and variable northeast trade winds predominate and ocean temperatures warm off Somalia and Arabia Fig. 1c).

The large seasonality in rainfall promotes the production and transport of atmospheric mineral dust which is exported to the adjacent Atlantic and NW Indian oceans. Interannual to multidecadal changes in West African dust export to the Atlantic are very closely related to changes in subtropical African summer rainfall [42], which in turn have been linked to tropical Atlantic sea-surface temperature (SST) anomalies [43]. A prominent summer dust plume off West Africa is carried at mid-tropospheric levels (700 mbar) by the African Easterly Jet and its load is mainly derived from western Saharan sources (Fig. 1b). The winter African dust plume originates from seasonally dry sub-Saharan and Sahelian soils which are carried by northeast trade winds (Fig. 1c). Summer dust plumes off Arabia and northeast Africa (Fig. 1b) are associated with Indian monsoon surface winds and they transport abun-

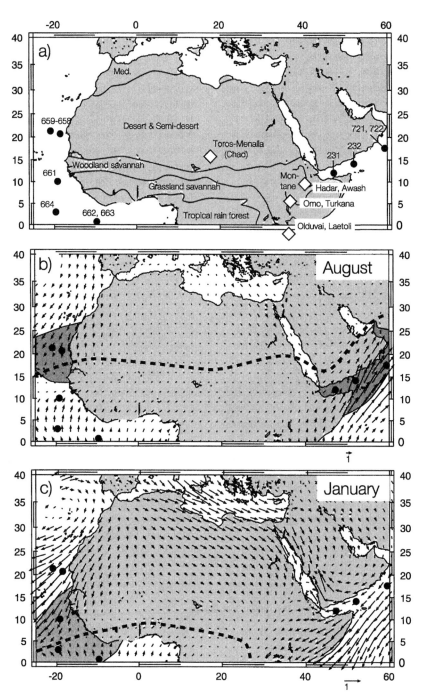

Fig. 1. (a) Regional map of North African vegetation zones, locations of DSDP and ODP drill sites (filled circles), and locations of selected African mammal fossil localities (open diamonds). (b) Boreal summer (August) surface wind stress (unit vector = 1 dyne/cm^2), intertropical convergence zone (ITCZ, heavy dashed line) location, and boundaries of seasonal tropospheric dust plumes off NW Africa and NE Africa/Arabia. Dust plume contours were derived from haze frequency data [46]. (c) Boreal winter (January) surface wind stress (unit vector = 1 dyne/cm^2), ITCZ location, and boundary of the seasonal tropospheric dust plume off NW Africa.

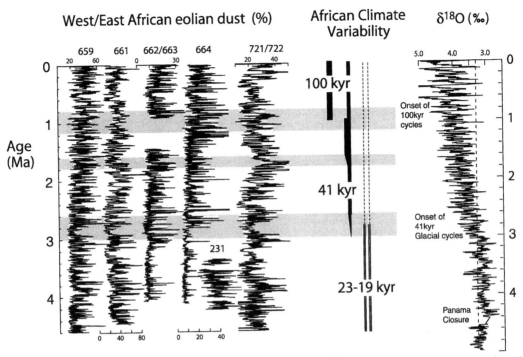

Fig. 2. Pliocene–Pleistocene records of eolian dust deposition at seven DSDP/ODP sites off western and eastern subtropical Africa [15] (Fig. 1). Terrigenous sedimentation at these sites is dominated by wind-borne eolian dust contributions from source areas in the Sahara/Sahel regions off West Africa, or source areas in the Arabian peninsula and northeast Africa. The terrigenous percentage records were determined by chemical removal of carbonate, opal, and organic fractions and most records were constrained by detailed oxygen isotopic stratigraphies [15,30,31,56]. The records collectively document progressive shifts in African climate variability and increasing aridity after 3.0–2.6 Ma, 1.8–1.6 Ma, and 1.2–0.8 Ma. These African aridity shifts were coincident with the onset and subsequent amplification of high-latitude glacial cycles [15,30,31,56]. The composite benthic foraminifer oxygen isotope record illustrates the evolution of high-latitude climate over the study interval [15,146,147].

dant mineral dust to the Arabian Sea and the Gulf of Aden [44–46]. Sediment trap and mineralogic studies demonstrate that wind-borne detritus from these source areas comprises the dominant source of terrigenous sediment to the eastern equatorial Atlantic and the Arabian Sea [45,46].

3.2. Marine paleoclimatic records

Marine sediments accumulating off the western and eastern margins of subtropical North Africa have provided some of the most compelling evidence for recurrent arid–humid climate cycles and progressive step-like increases in African aridity during the late Neogene (last ca. 5 Myr). These records document how African climate varied during a period of profound global climate shifts associated with the gradual onset of high-latitude

glacial cycles at 3.2–2.6 Ma which followed the isolation of the Atlantic basin resulting from the closure of the isthmus of Panama after 4.4–4.6 Ma [47] (Fig. 2). Pliocene–Pleistocene cooling at high latitudes occurred as a series of steps commencing with the onset of glacial ice rafting and modest 41 kyr glacial cycles after ca. 2.8 Ma, a shift toward cooler conditions and higher-amplitude 41 kyr cycles after ca. 1.8–1.6 Ma, and a pronounced shift toward still larger 100 kyr glacial cycles after ca. 1.2–0.8 Ma [48–51] (Fig. 2).

3.2.1. Patterns of subtropical African paleoclimatic variability

Marine sediment records of Pliocene–Pleistocene eolian export from West and East subtropical Africa (Fig. 1) reveal several consistent patterns of variability (Fig. 2) [15].

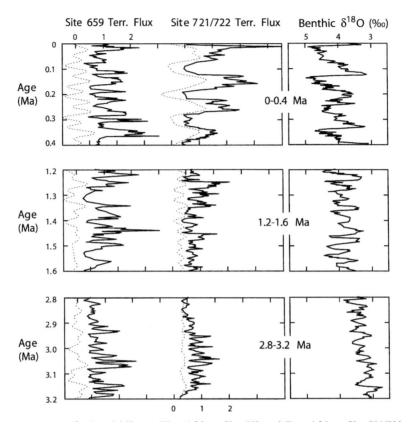

Fig. 3. Detail of terrigenous input (flux) variability at West African Site 659 and East African Site 721/722 spanning three intervals with differing patterns of eolian variability. The terrigenous (eolian) flux records (in units of $g/cm^2/kyr$) were calculated at each site as the product of terrigenous percentage data (Fig. 2), and interval sedimentation rate and dry bulk density data [15,30,31,54,56]. Dashed curves in each panel show the combined 100 kyr and 41 kyr period bandpass filters of the terrigenous flux series to illustrate changes in the amplitude and dominant period of eolian supply during the Pliocene–Pleistocene. Note the predominance of precessional (23–19 kyr) variability for the 3.2–2.8 Ma interval, whereas larger 41 kyr cycles dominate the 1.6–1.2 Ma interval, and still larger 100 kyr and 41 kyr cycles dominate the 0.4–0 Ma interval (particularly at Site 721/722). Long-term trends towards increased eolian flux variability are most evident at Site 721/722, although they are also apparent at West African Sites 659, 662, and 664 [15,30,31,54,56].

1. Orbital-scale African climate variability persisted throughout the entire interval, extending in some cases into the Miocene and Oligocene [52,53].
2. The onset of large-amplitude African aridity cycles was closely linked to the onset and amplification of high-latitude glacial cycles [15,30, 31,54–57].
3. Eolian concentration and supply (flux) increased gradually after 2.8 Ma [30,31,54,55].
4. Step-like shifts in the amplitude and period of eolian variability occurred at 2.8 (± 0.2) Ma, 1.7 (± 0.1) Ma, and 1.0 (± 0.2) Ma [15,30,31].
5. Evidence for 10^4–10^5 year 'packets' of high-

and low-amplitude paleoclimatic variability which were paced by orbital eccentricity [15,30].

The marine record of African climate variability is perhaps best described as a succession of wet–dry cycles with a long-term shifts toward drier conditions, punctuated by step-like shifts in characteristic periodicity and amplitude. Prior to 2.8 Ma subtropical African climate varied at the 23–19 kyr period (Figs. 2 and 3) which has been interpreted to reflect African monsoonal variability resulting from low-latitude (precessional) insolation forcing of monsoonal climate [15]. After 2.8 (± 0.2) Ma, African climate varied primarily at

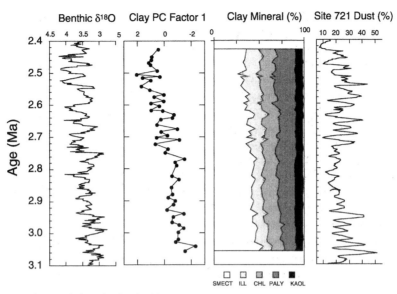

Fig. 4. Clay mineral assemblage variations in the Arabian Sea are used to reconstruct past variations in regional climate by taking advantage of the very different source vectors of specific clay minerals [148–150]. Presently, smectite derives from weathering and runoff of the Deccan traps of western India, and chlorite contributions similarly reflect northern source clay minerals associated with Indus River drainage. In contrast, variations in the fibrous Mg-rich clay mineral palygorskite reflect climate changes in southern source regions in Somalia and Arabia, where it forms in arid, alkaline environments [149]. The clay mineral percentage data [150] reveal a trend toward increasing palygorskite and illite, and decreasing smectite abundances after 2.8 Ma which are interpreted to reflect enhanced aridity (or transport) from source areas in Somalia and southern Arabia [149,150]. This shift is apparent in first principal component extracted from the clay mineral abundance matrix (PCA factor 1, 53% of variance).

the longer 41 kyr period and the amplitude of these cycles increased again after 1.7 (±0.1) Ma. Eolian variability shifted towards longer and larger-amplitude 100 kyr cycles after 1.0 (±0.2) Ma. These shifts in African eolian variability were synchronous with shifts in the onset and amplification of high-latitude ice sheets and cooling of the subpolar oceans [49,58], suggesting a coupling between high- and low-latitude climates after the onset of glaciation near 2.8 Ma [15,30,31,56] (Figs. 2 and 3). Three separate 0.4 Myr intervals from West African Site 659 [30] and East African Site 721/722 [15] are shown in Fig. 3 to illustrate these shifts in eolian variability.

African climate sensitivity to high-latitude glacial boundary conditions has been examined using general circulation model experiments [59–63] which indicate shifts to cooler, drier African conditions during glacial extrema. A detailed Pliocene–Pleistocene pollen record from Site 658 documents a progressive expansion of xeric vege-

tation and greater variability after ca. 3 Ma [17]. Changes in clay mineral assemblages at Site 721/722 similarly indicate increases in NE African aridity after 2.8 Ma (Fig. 4). Based on analysis of the terrigenous (eolian) grain size record over the last 3.5 Myr at Site 722, Clemens and others [31] additionally documented discrete shifts in the intensity and phase of the Indian monsoon at ca. 2.6 Ma, 1.7 Ma, and 1.2 Ma, and 0.6 Ma.

Most eolian concentration records display long-term trends toward greater dust concentrations after 2.8 Ma (Fig. 2), and this is particularly evident in calculated dust flux records [15,30,31,56]. A post-2.8 Ma shift toward enhanced eolian variability is most dramatic at Sites 662, 663, and 664 which are furthest from dust source areas (Fig. 1a). In addition to enhanced source area aridity, the abrupt dust increases at these most distal sites are likely reflecting glacial trade wind invigoration [17,64–72] and/or a southward shift of the intertropical convergence zone [73,74]. The ultimate cause for the onset of glaciations and African aridity after ca. 2.8 Ma remains a mystery, although several plausible mechanisms have been proposed [50,75–78].

At Arabian Sea Site 721/722, eolian grain size measurements document gradual weakening of the monsoonal dust-transporting winds after ca. 2.8 Ma, whereas the fluxes of eolian sediment to this same site increased markedly over the same time [31]. Thus, for East Africa and Arabia the post-2.8 Ma increase in eolian supply most likely reflects real increases in source area aridity. These sites also indicate an interval of pronounced aridity between 1.6 and 1.8 Ma which is not expressed at other locations (Fig. 2), perhaps suggesting a regionally specific signal. A pronounced shift toward more open conditions in East Africa at this time is indicated from stable isotopic records of soil carbonates recovered from several hominid fossil localities [3,4].

The eolian records exhibit 'variability packets' – long-term (10^4–10^5 year) modulations in the amplitudes of African paleoclimatic variability. These modulations have been attributed to orbital eccentricity modulation of Earth precession which regulates seasonal insolation and, consequently, the strength of the African and Indian monsoons [15,30,31]. These eccentricity-modulated 'variability packets' represent 10^4–10^5 year intervals of exceptionally high- or low-amplitude paleoclimatic variability, and are evident in both West and East African eolian records (Fig. 5). The impact of these paleoclimatic variability packets on African faunal evolution remains to be investigated, but at least one view, the variability selection hypothesis, cites the potential importance of such extended periods of paleoclimatic stability and instability to faunal adaptation and evolution [6,7].

3.2.2. On the fundamental pacing of African paleoclimatic variability

The fundamental pacing of African climate remains curiously unresolved [66]. From a marine sediment perspective there is abundant evidence that once high-latitude ice sheets became sufficiently large to sustain glacial–interglacial oscillations after 2.8 Ma, African climate covaried with

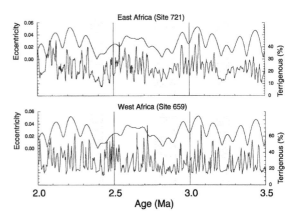

Fig. 5. Evidence for high- and low-amplitude African climate 'variability packets' from West and East African Sites 659 and 721/722 [15,30]. African paleoclimate variability over this interval was strongly regulated by orbital precession which is modulated by orbital eccentricity (shown). Prolonged (10^4–10^5 year) intervals of exceptionally high- or low-amplitude paleoclimate variability are apparent off both margins of subtropical Africa. Highest variability occurs during periods of maximum orbital eccentricity (e) when modulation of the precession index ($\Pi = e\sin(\omega)$) is greatest [63,151].

these high-latitude climate cycles at the characteristic 41 kyr and 100 kyr periodicities. However, the Pliocene–Pleistocene succession of sapropel layers in the Mediterranean Sea suggests that orbital precession was the fundamental tempo of African paleoclimatic changes [79–81]. These dark, organic-rich sapropel layers were deposited during African humid periods when enhanced monsoonal rainfall and Nile River runoff led to increased Mediterranean stratification and reduced ventilation of its deep eastern basins [79,82]. Dramatic cliffside exposures of uplifted Pliocene–Pleistocene marine sediments in southern Sicily and Calabria (Fig. 6) and sediment cores from eastern Mediterranean basins confirm the predominant influence of orbital precession on African climate since at least the late Miocene to the present [79–81,83].

It is possible to reconcile these two views by acknowledging that precession was the fundamental driver of African monsoonal climate throughout the late Neogene, but that high-latitude glacial cooling and drying effects were superimposed on this signal only after 2.8 Ma [15,56]. The 'glacial tempo' apparent in the marine dust and pol-

Fig. 6. Photograph of late Miocene (9.3–8.4 Ma) sapropel bedding cycles from the Gibliscemi A section in south central Sicily [53,83]. The darker strata are organic-rich sapropel layers which were deposited during African humid periods when enhanced monsoonal rainfall and Nile River runoff led to increased Mediterranean stratification and reduced ventilation of its deep eastern basins [79,82]. The regular bundling of sapropel layers into groups of three to four cycles reflects the 100 kyr eccentricity modulation of the precessional African humid periods. The predominant influence of orbital precession and eccentricity on African monsoonal rainfall persists throughout the Pliocene and Pleistocene [81,152], although there is evidence for some obliquity forcing as well [153]. Photograph courtesy of Frits Hilgen and Utrecht research group on astronomical climate forcing and timescales.

len records thus reflects glacial-stage increases in source area aridity and/or increases in the strength of the transporting winds [17,56,66,84], both of which are known subtropical climate responses to imposed glacial boundary conditions in climate models [15,59,61]. Prior to 2.8 Ma, the marine sediment eolian records and the sapropel records tell a consistent story, namely that changes in African monsoonal climate were paced mainly by precession [15] (e.g. Fig. 5), as would be expected in the absence of large ice sheets [85].

3.2.3. Paleoclimatic variability of southern Africa

Comparatively little was known about the paleoclimatic history of southern Africa during this same period, but a recently developed SST record off the coast of Namibia (25°S) documents profound changes over the past 5 Myr. An alkenone-derived SST record and other paleoproductivity indices from ODP Site 1084 [86] demonstrate that the wind-driven Benguela Current upwelling system steadily intensified and SSTs decreased

markedly after 3.2 Ma (Fig. 7), with subsequent sudden periods of intensification (and cooler SSTs) near 2.0 Ma and 0.6 Ma. The greatly di-

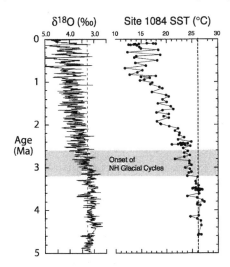

Fig. 7. Marine sediment record of SSTs at South Atlantic ODP Site 1084 (26°S). Alkenone measurements document large (10–15°C) and progressive cooling since 3.2 Ma of the wind-driven Benguela Current upwelling system [86].

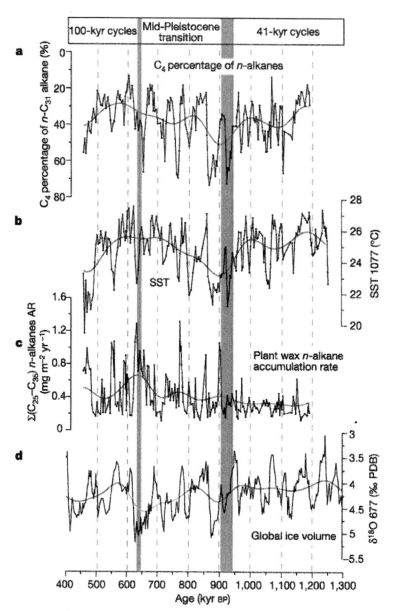

Fig. 8. Middle Pleistocene changes in (a) subtropical South African vegetation, (b) tropical Atlantic SSTs, (c) subtropical South African *n*-alkane fluxes of plant leaf wax compounds, and (d) a benthic oxygen isotopic record of high-latitude ice volume variability. Subtropical South African vegetation changes closely covaried with tropical Atlantic SSTs, and both signals exhibit the same 100 kyr and 41 kyr pacing apparent in high-latitude paleoclimatic records. Increases in arid-adapted (C₄) vegetation are associated with cooler SSTs and glacial maxima [87]. Note the increase in large-amplitude 100 kyr vegetation and SST cycles after ca. 900 ka. Figure reproduced from [87] with permission from *Nature* (Volume 422, pages 418–421) © 2003 Macmillan Publishers Ltd.

minished upwelling and warmer SSTs prior to 3.2 Ma were interpreted to reflect warmer and more stable conditions, with wetter and more mesic environments in southern Africa [86], a conclusion

that is broadly supported by South African paleoecological data [25].

Changes in tropical and south Atlantic SSTs have been shown to be a key determinant regulating African vegetation changes during the Pleistocene. Schefuss and others [87] used molecular biomarker analyses of plant leaf waxes (*n*-alkanes) preserved in deep-sea sedimentary organic matter at ODP Site 1077 (11°S) to quantify changes in the proportions of arid- (C_4) and humid- (C_3) adapted vegetation in subtropical southwest Africa (Fig. 8). They found that the reconstructed percentage of C_4 vegetation varied between 20 and 70% and that the mid-Pleistocene floral variations were strongly linked to glacial–interglacial variations in tropical SSTs which were also measured at this site using the alkenone method. The records reveal a close correspondence between cooler tropical SSTs and higher percentages of arid-adapted C_4 vegetation during glacial periods (Fig. 8).

3.3. Terrestrial paleoclimatic records

Continuous records of African paleoclimatic change are rare from terrestrial sequences in East Africa because active faulting, erosion, and non-deposition punctuate the geologic record. However, these data provide critical ground truth information on the general Pliocene–Pleistocene development of African climate, as well as specific paleoenvironmental contexts associated with fossil localities. These records generally support the view that East African climate changed from warmer, wetter conditions in the late Miocene and early Pliocene [5,88–90] to a more seasonally contrasted, cooler, and drier and perhaps more variable climate during the late Pliocene (after ca. 3 Ma). Pollen spectra from various fossil sites in NE Africa indicate shifts to cooler and drier vegetation types (increase in shrubs, heath, and grasses) after ca. 2.5 Ma [16]. Stable isotopic analyses of pedogenic carbonates from the Turkana and Olduvai basins indicate gradual replacement of woodland by open savannah grasslands between 3 Ma and 1 Ma, with step-like increases in savannah vegetation near 1.8 Ma, 1.2 Ma and 0.6 Ma [3,4].

The Turkana Basin (Fig. 1a) is one of the best studied depositional basins in East Africa due, in part, to the richness of its fossil record which spans nearly 4.5 Myr. Although it is tempting to imagine that this basin should have recorded the successions of wet–dry cycles so evident in the marine records, there is little expression of such simple depositional cycles [91,92]. Lake Turkana is presently a closed basin fed by monsoonal runoff from the Ethiopian highlands, but only about 10% of the geologic record there is lacustrine. Still, the Turkana Basin may have recorded some of the climatic oscillations which are evident in deep-sea records. Temporal correlations exist between the Site 721/722 dust record and stratigraphic occurrences of specific moist-indicator lithofacies such as molluscan, ostracod, and other bioclastic sediments in Turkana Basin sequences [93]. There were also several prolonged intervals of continuous lacustrine deposition in the Turkana Basin which persisted for 0.1–0.2 Myr and is represented by interbedded diatomite and claystone deposits. One such lacustrine interval, paleolake Lokochot [93], occurred between 3.4 and 3.5 Ma and is characterized by several decimeter-thick interbeds of diatomite (*Aulacosira* spp.) and clayey siltstone, reflecting oscillations in lake level or sediment supply. Radiometric dates of bounding tephra layers [94] suggest that these bedding cycles were about 20–25 kyr in duration which compares favorably to the precessional cycles evident in the marine paleoclimate records during this time (Fig. 3). Similar bedding cycles for this interval have been noted also in the Middle Awash sequences in Ethiopia [95,96].

4. Pliocene–Pleistocene African faunal evolution

The fossil record of African mammalian evolution suggests important steps in faunal speciation, migration, and adaptation throughout the Pliocene–Pleistocene. In the specific case of early human evolution, behavioral developments are also evident in the archeological record of stone tool use and development. Taphonomic (fossil preservation) and taxonomic (fossil identification) biases can complicate interpretation of faunal trends within and between localities [2], and some localities are additionally impacted by regional tectonic uplift histories which complicate paleoenvironmental interpretations [97,98].

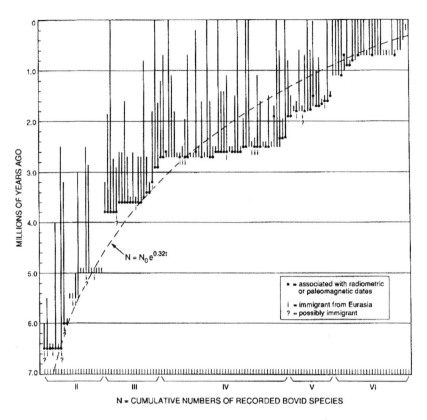

Fig. 9. Range chart of Africa-wide occurrences of fossil bovids (antelope) spanning the last 7 Myr. The dashed line represents a theoretical 'null hypothesis' assuming a uniform rate of faunal turnover (speciation) set at 32% per million years. The pronounced faunal 'turnover pulses', clusters of origination and extinction events, which occurred near 2.8 Ma and 1.8 Ma were also associated with appearances of arid-adapted fauna. A recent reanalysis of a subset of these data, the Omo–Turkana fossil collection, indicates that faunal turnover there was much more gradual and muted [33]. Figure reproduced from [21] with permission from Yale University Press.

4.1. The fossil record of (non-hominid) African mammalian evolution

The fossil record of African bovids (antelope family) is particularly useful for investigating changes in African environments. Bovid fossils typically comprise a large proportion of the total fossil record at any given location, they are highly diverse and represent nearly every environment, and some species are indicative of specific paleoenvironments. Three main approaches have been used to explore the paleoclimatic significance of fossil bovid evolutionary changes: shifts in speciation and extinction rates (turnover) [18,21,28, 33,99], shifts in relative abundances of habitat-specific taxa [25,26,32], and changes in the mor-

phology of locomotor limb bones (ecomorphological analyses) [100–102].

An analysis of the entire African bovid fossil record spanning the late Neogene revealed that many first and last appearances of the 127 separate bovid lineages were clustered at 2.7–2.5 Ma, with secondary groupings near ca. 1.8 Ma and ca. 0.7 Ma [21] (Fig. 9). Many of the first appearances near 2.7 Ma were arid-adapted species. In a separate analysis of fossil African micromammal (e.g. rodents) assemblages from the Shungura Formation (Ethiopia), Wesselmann [103] reported a moist-adapted rodent assemblage for Member B (ca. 3 Ma) whereas the fauna in Member F (ca. 2.3 Ma) were dry-adapted.

The turnover pulse hypothesis has been challenged based on a recent analysis of the Omo–Turkana fossil mammal collection (Fig. 1), one of the richest and best-dated collections in Africa [33]. Calculated faunal turnover in this basin was much less pronounced than in the all-Africa compilation [21], and turnover was found to have occurred more gradually across the 3–2 Ma interval [33]. The absence of a large faunal turnover signal in the Turkana Basin has been attributed to differing analytical protocols, treatment of rare species, and the regional buffering effect of the ancestral Omo River [91]. The evolution of fossil pigs (suids), hominids [34], and carnivores [35] shows similarly little evidence for such focused episodes of faunal turnover. Furthermore, some terrestrial paleoclimatic data indicate that true savannah grasslands were only established after the early and middle Pleistocene [3,4].

Rather than using faunal turnover as the hallmark of environmental change, several recent studies have used the fossil record to examine changes in the abundances of habitat-specific taxa to infer paleoenvironmental shifts. A comprehensive examination of bovid (antelope), cercopithicid (monkey), and suid (pig) fauna in the Omo sequence in southern Ethiopia [26] (Fig. 1) documented a remarkable decrease in closed woodland and forest species and an increase in grassland species between 3.6 and 2.4 Ma, with a marked rise between 2.6 and 2.4 Ma (Fig. 10). They concluded that "climate change caused significant shifts in vegetation in the Omo paleo-ecosystem" and that "climate forcing in the late Pliocene is more clearly indicated by population shifts within the Omo mammal community than by marked turnover at species level" [26]. This analysis also noted that faunal assemblages became much more variable between adjacent depositional members after 2.6–2.5 Ma. Analysis of East and South African fossil mammal collections also indicates aridification and increased percentages of grazing animals (expanded grasslands), but only after 1.8 Ma [25].

4.2. The fossil record of African hominid evolution

Important Pliocene–Pleistocene events in human evolution post-date the divergence between

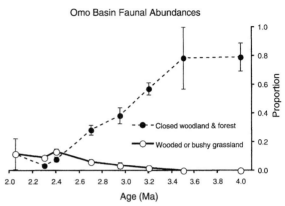

Fig. 10. Relative abundances of two endmember habitat groupings of fossil mammal taxa from the Omo collection during the late Pliocene [26]. Between 3.6 and 2.4 Ma there was a gradual reduction in moist-indicator species corresponding to closed woodland/forest habitats (dashed line), with a significant increase in the proportion of dry-indicator species corresponding to wooded or busy grasslands (solid line) between 2.8 and 2.4 Ma. This study also noted that mammalian faunal assemblages became much more variable between depositional members after 2.6–2.5 Ma, indicating greater long-term ecological instability after this time [2].

ape and human lineages which molecular clock studies estimate to have occurred at about 4.5–6.5 Ma [104–106]. This split is now thought to have occurred significantly earlier based on the recent discovery of *Sahelanthropus tchadensis*, a fossil hominid faunally dated at 6–7 Ma which exhibits both primitive and derived features and is thought to be close to the purported 'last common ancestor' [107]. Bipedality is a fundamental characteristic of the human family (Hominidae) and early bipedality is evident in the latest Miocene (ca. 5.8 Ma) hominid *Orrorin tugensis* [12] and in earliest Pliocene hominid specimens dating to 4.4–4.2 Ma [108–112]. The hominid footprint trackways of three individuals preserved in a volcanic ash bed at Laetoli, northern Tanzania (Fig. 1), provide striking confirmation that obligate bipedality was established by at least 3.6 Ma [13], well before the expansion of savannah grasslands [3].

Fossil hominids are, arguably, the least diagnostic faunal group for investigating past relationships between African climate and faunal change. They are extremely rare (typically less than 1% of a typical assemblage), not particularly useful paleoecologically, and they are not very diverse relative to other taxa (only a few different species coexisted at any one time). Still, the Pliocene–Pleistocene fossil record documents fundamental changes in hominid morphology and behavior which can be evaluated within the broader contexts of other faunal and paleoenvironmental changes [1,2,20,23,26,99,113,114].

4.2.1. Junctures in early human evolution

Hominid evolution over the Pliocene–Pleistocene interval was punctuated by significant changes in species morphology, innovation, and diversity near 2.8 Ma, 1.7 Ma, and ca. 1 Ma [21,28,114,115]. Radiometric dates constraining most fossils are typically very precise [94,116, 117], but the ages of these hominid faunal transitions carry significant uncertainties (± 0.2 Myr) due to gaps in the fossil record, taxonomic uncertainties [118,119], and taphonomic (preservational) biases [114].

The earliest Pliocene hominid species presently documented is the 4.4 Ma *Ardipithecus ramidus* (Fig. 11) based on fragmentary but well-dated fossil evidence from Aramis, Middle Awash in northern Ethiopia [120] (Fig. 1). Dental morphology suggests that *A. ramidus* is the most 'apelike' hominid of the ancestral australopithecine lineage. Sedimentological evidence at the site indicates a well-watered wooded habitat in this now hyperarid terrain [116] which broadly agrees with soil carbonate isotopic evidence for woodland vegetation at many East African fossil localities during the early Pliocene [3,121] (Fig. 11). *Australopithecus anamensis* has been described at several sites near Lake Turkana, Kenya, spanning the interval 4.2–3.9 Ma [122].

Many specimens of *Australopithecus afarensis* ('Lucy') have now been described spanning 3.9–2.9 Ma [112,114]. They collectively indicate a single, ecologically diverse, highly sexually dimorphic, bipedal Pliocene hominid whose known range encompassed Ethiopia to Tanzania [112]. Cranial measurements indicate that *A. afarensis* was relatively small-brained, even relative to its body size which was also small relative to subsequent hominid taxa. Analysis of other faunal remains associated with *A. afarensis* specimens from

many regions reveals that members of this taxon were consistently associated with well-watered, wooded paleoenvironments [25].

The recent discovery of a new hominid (*Kenyanthropus platyops*) which was contemporary with *A. afarensis* near 3.5 Ma weakens the case that *A. afarensis* was the root taxon leading to all subsequent hominid taxa [123,124]. *K. platyops* was a particularly important find because it effectively makes the human family tree more complex than previously believed (Fig. 11), and because it possesses characteristics (human-like facial morphology and smaller cheek teeth) which are not seen in the fossil record until the emergence of the first representatives of the *Homo* clade nearly one million years later. A well-watered woodland and woodland mosaic paleoenvironment was indicated from faunal remains at the fossil site (western Turkana Basin, Fig. 1).

4.2.2. Hominid evolution between 2.9 and 2.4 Ma

Several fundamental faunal speciation and hominid behavioral changes occur in the fossil record between 2.9 and 2.4 Ma, although the fossil record within this interval is notably poor [114]. At least two new hominid lineages emerged from the ancestral lineage which itself became extinct, and this period marks the first appearance of stone tools (Fig. 11). Earliest members of the 'robust' australopithecine genus *Paranthropus* first occur in the fossil record near 2.8 Ma [22,114, 125,126]. Among many characteristic anatomical differences, the robustly framed *Paranthropus* were distinguished by uniquely large cheek teeth and strong jaw musculature which apparently reflects a highly specialized masticatory adaptation for processing coarse vegetable matter [23, 125,127]. A second lineage, represented by the earliest members of our genus *Homo*, first occurred near 2.3–2.5 Ma [114,115,128] (Fig. 11). Earliest fossils of the *Homo* clade are characterized by a more gracile frame, smaller cheek teeth, and much larger absolute cranial volumes than any prior hominid species [115,118].

The earliest known occurrence of stone tools (the first crude choppers and scrapers comprising the Oldowan complex) is now well dated near 2.3–2.6 Ma [129–133] (Fig. 11). Whether tool

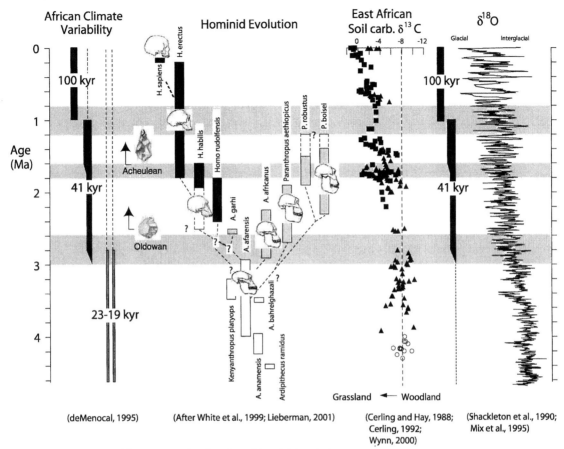

Fig. 11. Summary diagram of important paleoclimatic and hominid evolution events during the Pliocene–Pleistocene. Marine paleoclimatic records indicate that African climate became progressively more arid after step-like shifts near 2.8 (±0.2) Ma, and subsequently after 1.7 (±0.1) Ma and 1.0 (±0.2) Ma, coincident with the onset and intensification of high-latitude glacial cycles [49,50,146]. These events are associated with changes toward dry-adapted African faunal compositions, including important steps in hominid speciation, adaptation, and behavior (see text for discussion). Soil carbonate carbon isotopic data from East African hominid fossil localities document the Pliocene–Pleistocene progressive shifts from closed woodland forest C_3-pathway vegetation (-9 to -12‰) to arid-adapted C_4-pathway savannah grassland vegetation ($+2$ to 0‰) (data from Cerling and Hay [4] (solid boxes), Cerling [2] (solid triangles), and Wynn [121] (open circles)).

use was practised exclusively by *Homo* is not known, but the recent discovery of tool cut marks on mammalian bones dated near 2.5 Ma heralds the important behavioral development of meat processing and marrow extraction [130].

The synchronous existence of (at least) two distinct hominid lineages and the emergence of lithic technology has been interpreted to reflect adaptations to a more arid, varied environment [2,22,125]. The 2.9–2.5 Ma interval corresponds

to a period of either modestly [33] or greatly [21] enhanced turnover in fossil bovids (Fig. 9) with associated increases in the proportions of arid-adapted fauna (Fig. 10) [21,26,103]. Analysis of associated faunal remains found with *Paranthropus* fossils from multiple sites indicates they occurred in both wooded and more open paleoenvironments, but always habitats that included wetland fauna [25]. Similar analyses for earliest *Homo* fossils indicate that they were associated with more varied habitats that any prior hominids [25]. Earliest *Homo erectus* fossils were associated with arid, open grassland environments [25].

4.2.3. Hominid evolution between 1.8 and 1.6 Ma

By 1.8–1.6 Ma, *Homo habilis* became extinct and its immediate successor, and our direct ancestor, *H. erectus*, first occurs in the fossil record near 1.8 Ma [114]. *H. erectus* may have migrated to southeast Asia as early as 1.9–1.8 Ma [134]. Near 1.7 Ma, South African [135] and East African bovid assemblages shift toward further absolute increases in the abundance of arid-adapted species [21] (Fig. 9). Enhanced East African aridity near 1.8–1.6 Ma is supported by soil carbonate stable isotopic evidence for broadly expanded savannah vegetation in East Africa [3,4,136] (Fig. 11). Increased aridity and variability is also evident in the marine paleoclimatic records, particularly those off East Africa (Fig. 2). Earliest occurrences of the more sophisticated Acheulean tool kit (bifacial blades and hand axes) occurred near 1.7–1.6 Ma [132,137] (Fig. 11).

4.2.4. Hominid evolution between 1.2 and 0.8 Ma

The hominid fossil record between 1.4 and 0.8 Ma is notably poor. Available evidence suggests that by 1 Ma the *Paranthropus* lineage had become extinct [114] and *H. erectus* had broadly expanded its geographic range and occupied sites in North Africa, Europe, and western Asia [138,139]. Hominid brain size increased rapidly in later Pleistocene hominid specimens after 780 ka [140,141]. The fossil record of African bovidae suggests a final phase of increased arid-adapted species turnover near 1.2 Ma and 0.6 Ma [21,28,99] (Fig. 9). Soil carbonate isotopic evidence indicates that pure C_4 grasslands, such as the vast savannahs of subtropical Africa today, were only established after ca. 1.2 Ma and after 0.6 Ma [3] (Fig. 11). The marine paleoclimatic records indicate that conditions were not just generally drier but were punctuated by increasingly longer and more severe arid episodes (Figs. 2 and 3). At Olorgesailie (Kenya), successions of fluviolacustrine sediments deposited at about 1.0–0.5 Ma reveal recurrent episodes of large-scale landscape remodeling due to persistent shifts in regional hydrology [2].

5. Summary of major Pliocene–Pleistocene events in African paleoclimatic and faunal change

Taken together, the African faunal and paleoclimatic records suggest three restricted intervals – 2.9–2.4 Ma, 1.8–1.6 Ma, and 1.2–0.8 Ma – when shifts toward increasingly variable, drier African conditions were accompanied by some changes in African faunal assemblages and, perhaps, speciation. Long-term trends toward increased abundances of arid-adapted taxa during the Pliocene–Pleistocene are evident, with indications for faunal changes between 2.9 and 2.4 Ma (the age range encompassing major faunal transitions) and after 1.8 Ma. Challenging many habitat-specific hypotheses, the paleoclimatic record does not support unidirectional shifts to permanently drier conditions. The marine paleoclimate records additionally indicate prolonged (10^4–10^5 year) intervals of exceptionally high- and low-amplitude variability, or 'variability packets', which may have also presented adaptation pressures [2]. It seems prudent that African climate and faunal evolution hypotheses address the adaptive significance of the higher-frequency (orbital-scale) remodeling of African landscapes [2] as well as the longer-term trends toward more open and more variable environments [2,21].

6. Research frontiers

As evolutionary theorists continue to investigate the possible role of past climate change on African faunal populations there are several promising areas of research which may significantly constrain the problem. These developments include efforts to more thoroughly integrate faunal and paleoclimatic records, and the applications of new analytical tools for reconstructing African vegetation shifts.

Recent years have seen great strides toward more and better integration of African faunal and paleoclimatic datasets. Although well-dated fossil faunal collections have existed for decades in some cases, these datasets have been reanalyzed only very recently for the potential paleoecological content they possess [25,26,100,102]. These newer studies are distinguished by their application of more sophisticated statistical approaches and classification protocols which have resulted in new records which constrain the timing, amplitude, and ecological contexts of faunal changes during the Pliocene–Pleistocene. The faunal assemblage change evident near 2.8 Ma in the Omo sequence is one such example of the kinds of evidence which may better constrain possible relationships between climate change and evolution [26] (Fig. 10). These studies have benefited from vast improvements in radiometric dating [94,117] and stratigraphic contexts [38,90,92,116, 142,143] of fossil material. Dating and correlation of volcanic ash horizons define much of the chronological control for fossil assemblages throughout East Africa [143]. These same ash-fall events are recorded in deep-sea sediment sequences of the Gulf of Aden and Arabian Sea [142,144, 145], providing firm stratigraphic ties between the fossil and paleoclimatic records.

The emergence of organic geochemical analytical techniques for isolating and classifying vascular plant leaf wax compounds presents promising opportunities for constraining how and when African vegetation cover changed throughout the Pliocene–Pleistocene. Long-chain, odd-numbered *n*-alkanes are lipid constituents of epicuticular plant leaf waxes which are abraded and transported by winds and are preserved as trace components in the organic fraction of deep-sea sediments [87]. Carbon isotopic analysis of the *n*-alkane fraction is used to quantify the relative proportion of arid-adapted C_4 vegetation sources (e.g. savannah grasses and sedges) [87] (Fig. 8). The application of this method to late Neogene marine sediment sequences off East Africa could settle debates about the timing and nature of African paleoenvironmental shifts and the extent to which these signals impacted African biota.

Acknowledgements

I'd like to especially thank Alex Halliday who invited this contribution and for his excellent editorial guidance and patience. Thorough reviews by Frank Brown, Steve Clemens, Rick Potts, and an anonymous reviewer greatly improved the final manuscript and they are gratefully acknowledged for their input. I'd also like to acknowledge Elisabeth Vrba, Frank Brown, Thure Cerling, Rick Potts, Craig Feibel, Bernard Wood, Kay Behrensmeyer, Clark Howell, and Tim White who have been especially generous and helpful to me over the years as a newcomer to their fascinating field. The marine sediment sequences were made possible through the efforts of the scientists, support staff, and crew of Ocean Drilling Program Legs 108 and 117. Clay mineralogy at Sites 721 and 722 was conducted in collaboration with Drs. Pierre Debrabant and Hervé Chamley (University of Lille). Funding was provided by the Marine Geology and Geophysics and Paleoanthropology divisions of NSF. This is Lamont-Doherty Earth Observatory publication number 6524.*[AH]*

References

[1] A. Gibbons, In search of the first hominids, Science 295 (2002) 1214–1219.

[2] R. Potts, Environmental Hypotheses of Hominin Evolution, Yearb. Phys. Anthropol. 41 (1998) 93–136.

[3] T.E. Cerling, Development of grasslands, savannas in East Africa during the Neogene, Palaeogeogr. Palaeoclimatol. Palaeoecol. 97 (1992) 241–247.

[4] T.E. Cerling, R.L. Hay, An isotopic study of paleosol carbonates from Olduvai Gorge, Quat. Res. 25 (1988) 63–78.

[5] T.E. Cerling, J.M. Harris, B.J. MacFadden, M.G. Leakey, J. Quade, V. Eisenmann, J.R. Erleringer, Global vegetation change through the Miocene/Pliocene boundary, Nature 389 (1997) 153–158.

[6] R. Potts, Evolution and climate variability, Science 273 (1996) 922–923.

[7] R. Potts, Variability selection in hominid evolution, Evol. Anthropol. 7 (1998) 81–96.

[8] R.A. Dart, *Australopithecus africanus*: The man ape of South Africa, Nature 115 (1925) 195–199.

[9] G.A. Bartholomew, J.B. Birdsell, Ecology and the proto-hominids, Am. Anthropol. 55 (1953) 481–498.

[10] Wolpoff, Paleoanthropology, Knopf, New York, 1980.

[11] R. Klein, The Human Career, University of Chicago Press, Chicago, IL, 1989.

[12] B. Senut, M. Pickford, D. Gommery, P. Mein, K. Cheboi, Y. Coppens, First hominid from the Miocene (Lukeino formation, Kenya), C.R. Acad. Sci. Paris IIA 332 (2001) 137–144.

[13] M.G. Leakey, R.L. Hay, Pliocene footprints in the laetoli beds at Laetoli, northern Tanzania, Nature 278 (1979) 317–323.

[14] W.W. Bishop, Pliocene problems relating to human evolution, in: G.L. Isaac, E.R. McCown (Eds.), Human Origins: Perspectives on Human Evolution, vol. 3, WA Benjamin, Menlo Park, CA, 1976, pp. 139–153.

[15] P.B. deMenocal, Plio–Pleistocene African climate, Science 270 (1995) 53–59.

[16] R. Bonnefille, Evidence for a cooler and drier climate in the Ethiopian Uplands towards 2.5 Myr ago, Nature 303 (1983) 487–491.

[17] L.M. Dupont, S.A.G. Leroy, Steps toward drier climatic conditions in northwestern Africa during the Upper Pliocene, in: E. Vrba, G. Denton, L. Burckle, T. Partridge (Eds.), Paleoclimate and Evolution With Emphasis on Human Origins, Yale University Press, New Haven, CT, 1995.

[18] E.S. Vrba, Late Pliocene Climatic Events and Hominid Evolution, The Evolutionary History of the Robust Australopithecines, 1988.

[19] M.L. Prentice, R.K. Matthews, Cenozoic ice-volume history: development of a composite oxygen isotope record, Geology 16 (1988) 963–966.

[20] S.M. Stanley, An ecological theory for the origin of Homo, Paleobiology 18 (1992) 237–257.

[21] E. Vrba, The fossil record of African antelopes (Mammalia, Bovidae) in relation to human evolution and paleoclimate, in: E. Vrba, G. Denton, L. Burckle, T. Partridge (Eds.), Paleoclimate and Evolution With Emphasis on Human Origins, Yale University Press, New Haven, CT, 1995, pp. 385–424.

[22] F.E. Grine, Ecological causality and the pattern of Plio–Pleistocene hominid evolution in Africa, S. Afr. J. Sci. 82 (1986) 87–89.

[23] C.K. Brain, The evolution of man in Africa: was it the result of Cainozoic cooling?, Annex. Transvaal Geol. Soc. S. Afr. J. Sci. 84 (1981) 1–19.

[24] J.D. Clark, Early human occupation of African savanna environments, in: D.R. Harris (Ed.), Human Ecology in Savanna Environments, Academic Press, London, 1980, pp. 41–71.

[25] K.E. Reed, Early hominid evolution and ecological change through the African Plio–Pleistocene, J. Hum. Evol. 32 (1997) 289–322.

[26] R. Bobe, K. Behrensmeyer, R.E. Chapman, Faunal change, environmental variability and late Pliocene hominin evolution, J. Hum. Evol. 42 (2002) 475–497.

[27] E.S. Vrba, Evolution, species, and fossil: how does life evolve?, S. Afr. J. Sci. 76 (1980) 61–84.

[28] E.S. Vrba, G.H. Denton, M.L. Prentice, Climatic influences on early hominid behavior, Ossa 14 (1989) 127–156.

[29] M.L. Prentice, G.H. Denton, The deep-sea oxygen isotope record, the global ice sheet system and hominid evolution, in: F.E. Grine (Ed.), Evolutionary History of the Robust Australopithecines, Aldine de Gruyter, New York, 1988, pp. 383–403.

[30] R. Tiedemann, M. Sarnthein, N.J. Shackleton, Astronomic timescale for the Pliocene Atlantic $\delta^{18}O$ and dust flux records of ODP Site 659, Paleoceanography 9 (1994) 619–638.

[31] S.C. Clemens, D.W. Murray, W.L. Prell, Nonstationary phase of the Plio–Pleistocene Asian monsoon, Science 274 (1996) 943–948.

[32] R. Bobe, G. Eck, Responses of African bovids to Pliocene climatic change, Paleobiology 27 (Suppl. to No. 2) (2001) 1–47.

[33] A.K. Behrensmeyer, N.E. Todd, R. Potts, G.E. McBrinn, Late Pliocene faunal turnover in the Turkana Basin, Kenya and Ethiopia, Science 278 (1997) 1589–1594.

[34] T.D. White, African omnivores: Global climatic change and Plio–Pleistocene hominids and suids, in: E. Vrba, G. Denton, T. Partridge, L. Burckle (Eds.), Paleoclimate and Evolution with Emphasis of Human Origins, Yale University Press, New Haven, CT, 1995, pp. 369–384.

[35] L. Werdelin, M.E. Lewis, Diversity and turnover in eastern African Plio–Pleistocene carnivora, J. Vertebr. Paleontol. 21 (2001) 112–113.

[36] R. Rayner, B. Moon, J. Masters, The Makapansgat australopithecine environment, J. Hum. Evol. 24 (1993) 219–231.

[37] R. Blumenschine, Early Hominid Scavenging Opportunities: Implications of Carcass Availability in the Serengeti and Ngorongoro Ecosystems, British Archeological Reports International Series, Oxford, 1986.

[38] C.S. Feibel, F.H. Brown, I. McDougall, Stratigraphic context of fossil hominids from the Omo Group deposits: Northern Turkana Basin, Kenya and Ethiopia, Am. J. Phys. Anthropol. 78 (1989) 595–622.

[39] F.H. Brown, C.S. Feibel, 'Robust' hominids and the Plio–Pleistocene paleogeography of the Turkana Basin, Kenya and Ethiopia, in: F.E. Grine (Ed.), Evolutionary History of the 'Robsut' Australopithecines, Aldine de Gruyter, New York, 1988, pp. 325–341.

[40] S. Hastenrath, Climate and Circulation of the Tropics, D. Reidel, Boston, MA, 1985.

[41] S.E. Nicholson, An overview of African rainfall fluctuations of the last decade, J. Clim. 6 (1993) 1463–1466.

[42] J.M. Prospero, R.T. Nees, Impact of North African drought and el Niño on mineral dust in the Barbados trade winds, Nature 320 (1986) 735–738.

[43] A. Giannini, R. Saravan, P. Chang, Oceanic forcing of Sahel rainfall on interannual to interdecadal time scales, Science 302 (2003) 1027 1030.

[44] S.C. Clemens, Dust response to seasonal atmospheric forcing: Proxy evaluation and forcing, Paleoceanography 13 (1998) 471–490.

[45] R.R. Nair, V. Ittekot, S. Manganini, V. Ramaswamy, B. Haake, E. Degens, B. Desai, S. Honjo, Increased particle flux to the deep ocean related to monsoons, Nature 338 (1989) 749–751.

[46] K. Pye, Eolian Dust and Dust Deposits, Academic Press, New York, 1987.

[47] G.H. Haug, R. Tiedemann, R. Zahn, A.C. Ravelo, Pole of Panama uplift on oceanic freshwater balance, Geology 29 (2001) 207–210.

[48] N.J. Shackleton, J. Backman, H. Zimmerman, D.V. Kent, M.A. Hall, D.G. Roberts, D. Schnitker, J. Baldauf, A. Despraires, R. Homrighausen, P. Huddlestun, J. Keene, A.J. Kaltenback, K.A.O. Krumsiek, A.C. Morton, J.W. Murray, J. Westberg-Smith, Oxygen isotope calibration of the onset of ice-rafting and history of glaciation in the North Atlantic region, Nature 307 (1984) 620–623.

[49] N.J. Shackleton, A. Berger, W.R. Peltier, An alternative astronomical calibration of the lower Pleistocene time-scale based on ODP Site 677, Trans. R. Soc. Edinburgh Earth Sci. 81 (1990) 251–261.

[50] M.E. Raymo, The initiation of Northern Hemisphere glaciation, Annu. Rev. Earth Planet. Sci. 22 (1994) 353–383.

[51] W.F. Ruddiman, M.E. Raymo, D.G. Martinson, B.M. Clement, J. Backman, Pleistocene evolution: Northern Hemisphere ice sheets and North Atlantic Ocean, Paleoceanography 4 (1989) 353–412.

[52] P.B. deMenocal, J. Bloemendal, J.W. King, A rock-magnetic record of monsoonal dust deposition to the Arabian Sea: Evidence for a shift in the mode of deposition at 2.4 Ma, in: W.L. Prell, N. Niitsuma et al. (Eds.), Proc. ODP Sci. Results 117 (1991) 389–407.

[53] F. Hilgen, W. Krijgsman, C. Langereis, L. Lourens, A. Santarelli, W. Zachariasse, Extending the astronomical (polarity) time scale into the Miocene, Earth Planet. Sci. Lett. 136 (1995) 495–510.

[54] W.F. Ruddiman, T. Janecek, Pliocene–Pleistocene biogenic and terrigenous fluxes at equatorial Atlantic Sites 662, 663, and 664, in: W.F. Ruddiman, M. Sarnthein et al. (Eds.), Proc. ODP Sci. Results 108 (1989) 211–240.

[55] P.B. deMenocal, J. Bloemendal, Plio–Pleistocene subtropical African climate variability and the paleoenvironment of hominid evolution: A combined data-model approach, in: E. Vrba, G. Denton, L. Burckle, T. Partridge (Eds.), Paleoclimate and Evolution With Emphasis on Human Origins, Yale University Press, New Haven, CT, 1995, pp. 262–288.

[56] P.B. deMenocal, W.F. Ruddiman, E.M. Pokras, Influences of high- and low-latitude processes on African climate: Pleistocene eolian records from equatorial Atlantic Ocean Drilling Program Site 663, Paleoceanography 8 (1993) 209–242.

[57] S. Clemens, W.J. Prell, One million year record of summer monsoon winds and continental aridity from the Owen Ridge (Site 722), Northwest Arabian Sea, in: W.J. Prell, N. Niitsuma (Eds.), Proc. ODP 117 (1991) 365–388.

[58] M.E. Raymo, W.F. Ruddiman, N.J. Shackleton, D.W. Oppo, Evolution of Atlantic-Pacific $\delta^{13}C$ gradients over the last 2.5 m.y, Earth Planet. Sci. Lett. 97 (1990) 353–368.

[59] P. Clark, R. Alley, D. Pollard, Climatology – Northern hemisphere ice-sheet influences on global change, Science 286 (1999) 1104–1111.

[60] S. Clemens, W. Prell, D. Murray, G. Shimmield, G. Weedon, Forcing mechanisms of the Indian Ocean monsoon, Nature 353 (1991) 720–725.

[61] P.B. deMenocal, D. Rind, Sensitivity of Asian and African climate to variations in seasonal insolation, glacial ice cover, sea-surface temperature, and Asian orography, J. Geophys. Res. 98 (1993) 7265–7287.

[62] J.E. Kutzbach, P.J. Guetter, The influence of changing orbital parameters and surface boundary conditions on climate simulations for the past 18,000 years, J. Atmos. Sci. 43 (1986) 1726–1759.

[63] W.L. Prell, J.E. Kutzbach, Monsoon variability over the past 150,000 years, J. Geophys. Res. 92 (1987) 8411–8425.

[64] T. Herbert, J. Schuffert, Alkenone unsaturation estimates of late Miocene through late Pliocene sea-surface temperatures at Site 958, in: J.V. Firth et al. (Eds.), Proc. ODP Sci. Results 159T (1998) 17–21.

[65] U. Pflaumann, M. Sarnthein, K. Ficken, K. Grothmann, A.J. Winkler, Variations in eolian and carbonate sedimentation, sea-surface temperature, and productivity over the last 3 m.y. at Site 958 off northwest Africa, in: J.V. Firth et al. (Eds.), Proc. ODP Sci. Results 159T (1998) 1–14.

[66] W.F. Ruddiman, Tropical Atlantic terrigenous fluxes since 25,000 years BP, Mar. Geol. 136 (1997) 189–207.

[67] A. Moreno, J. Targarona, J. Henderiks, M. Canals, T. Freundenthal, H. Meggers, Orbital forcing of dust supply to the North Canary basin over the last 250 kyr, Quat. Sci. Rev. 20 (2001) 1327–1339.

[68] D.W. Parkin, Trade-winds during the glacial cycles, Proc. R. Soc. London A 337 (1974) 73–100.

[69] H. Hooghiemstra, Variations of the NW African trade wind regime during the last 140,000 years: Changes in pollen flux evidenced by marine sediment records, in: M. Sarnthein (Ed.), Paleoclimatology and Paleometeorology: Modern and Past Patterns of Global Atmospheric Transport, Kluwer Academic, Dordrecht, 1989, pp. 733–770.

[70] E.M. Pokras, A.C. Mix, Eolian evidence for spatial variability of Quaternary climates in tropical Africa, Quat. Res. 24 (1985) 137–149.

[71] P.J. Muller, E. Suess, Productivity, sedimentation rate, and sedimentary organic matter in the oceans-I. Organic carbon preservation, Deep-Sea Res. 26A (1979) 1347–1362.

[72] R. Stein, Late Neogene changes of paleoclimate, paleoproductivity off NW Africa (DSDP Site 397), Palaeogeogr. Palaeoclimatol. Palaeoecol. 49 (1985) 47–59.

[73] W.P. Chaisson, A.C. Ravelo, Pliocene development of the east-west hydrographic gradient in the equatorial Pacific, Paleoceanography 15 (2000) 497–505.

[74] W.P. Chaisson, A.C. Ravelo, Changes in upper water-column structure at Site 925, late Miocene–Pleistocene: Planktonic foraminifer assemblage and isotopic evidence, in: N.J. Shackleton, W.B. Curry, C. Richter, T.J. Bralower (Eds.), Proc. ODP Sci. Results 154 (1997) 255–268.

[75] M.A. Cane, P. Molnar, Closing of the Indonesian Seaway as the missing link between Pliocene East African aridification and the Pacific, Nature (2001).

[76] S. Philander, A. Fedorov, Role of tropics in changing the

response to Milankovitch forcing some three million years ago, Paleoceanography 18 (2003) 1045.

[77] G.H. Haug, D.M. Sigman, R. Tiedemann, Onset of permanent stratification in the subarctic Pacific Ocean, Nature 401 (2000) 779–782.

[78] N.W. Driscoll, G.H. Haug, A short circuit in thermohaline circulation: A cause for northern hemisphere glaciation?, Science 282 (1998) 436–438.

[79] M. Rossignol-Strick, Mediterranean quaternary sapropels, an immediate resonse of the African monsoon to variation of insolation, Palaeogeogr. Palaeoclimatol. Palaeoecol. 49 (1985) 237–263.

[80] F.J. Hilgen, Astronomical calibration of Gauss to Matuyama sapropels in the Mediterranean and implication for the geomagnetic polarity timescale, Earth Planet. Sci. Lett. 104 (1991) 226–244.

[81] L.J. Lourens, A. Antonarakou, F. Hilgen, A.A.M. Van Hoof, C. Vergnaud-Grazzini, W.J. Zachariasse, Evaluation of the Plio–Pleistocene astronomical timescale, Paleoceanography 11 (1996) 391–413.

[82] J.P. Sachs, D.J. Repeta, Oligotrophy and nitrogen fixation during eastern Mediterranean sapropel events, Science 286 (1999) 2485–2488.

[83] W. Krijgsman, F. Hilgen, C. Langereis, A. Santarelli, W. Zachariasse, Late Miocene magnetostratigraphy, biostratigraphy and cyclostratigraphy in the Mediterrenean, Earth Planet. Sci. Lett. 136 (1995) 475–494.

[84] M. Zabel, T. Wagner, P.B. deMenocal, Terrigenous signals in sediments of the low-latitude Atlantic – indications to environmental variations during the Late Quaternary: Part II: Lithogenic matter, in: G. Wefer, S. Mulitza, V. Ratmeyer (Eds.), The South Atlantic in the Late Quaternary: Reconstruction of Mass Budget and Current Systems, Springer, Berlin, 2003, pp. 1–23.

[85] J.E. Kutzbach, Model simulations of the climatic patterns during the deglaciation of North America, in: W.F. Ruddiman, H.E. Wright, Jr. (Eds.), North America and Adjacent Oceans during the Last Deglaciation, The Geology of North America K-3, Geological Sciiety of America, Boulder, CO, 1987, pp. 425–446.

[86] J.R. Marlow, C.B. Lange, G. Wefer, A. Rosell-Melé, Upwelling intensification as a part of the Pliocene–Pleistocene climate transition, Science 290 (2000) 2288–2291.

[87] E. Schefuss, S. Schouten, J.H.F. Jansen, J.S. Sinnighe Damsté, African vegetation controlled by tropical sea surface temperatures in the mid-Pleistocene period, Nature 422 (2003) 418–421.

[88] D.L. Griffin, The late Miocene climate of northeastern Africa: unravelling the signals in the sedimentary succession, J. Geol. Soc. London 156 (1999) 817–826.

[89] K. Yemane, R. Bonnefille, H. Faure, Paleoclimatic and tectonic implications of Neogene microflora from the Northwestern Ethiopian Highlands, Nature 318 (1985) 653–656.

[90] G. Wolde Gabriel, Y. Haile-Selassie, P.R. Renne, W.K. Hart, S.H. Ambrose, B. Asfaw, G. Heiken, T. White,

Geology and paleontology of the Late Miocene Middle Awash valley, Afar rift, Ethiopia, Nature 412 (2001) 175–182.

[91] C.S. Feibel, J.M. Harris, F.H. Brown, Paleoenvironmental context for the late Neogene of the Turkana Basin, in: J.M. Harris (Ed.), Koobi Fora Research Project: The Fossil Ungulates: Geology, Fossil Artiodactyls, and Palaeoenvironments, vol. 3, Clarendon Press, Oxford, 1991, pp. 321–370.

[92] F.H. Brown, C.S. Feibel, Stratigraphy, depositional environments, and paleogeography of the Koobi Fora Formation, in: J.M. Harris (Ed.) Koobi Fora Research Project: Stratigraphy, Artiodactyls, and Paleoenvironments, 3, Clarendon, Oxford, 1991, pp. 1–30.

[93] F.H. Brown, The potential of the Turkana Basin for paleoclimatic reconstruction in East Africa, in: E. Vrba, G. Denton, L. Burckle, T. Partridge (Eds.), Paleoclimate and Evolution With Emphasis on Human Origins, Yale University Press, New Haven, CT, 1995, pp. 319–330.

[94] I. McDougall, F.H. Brown, T.E. Cerling, J.W. Hillhouse, A reappraisal of the geomagnetic polarity timescale to 4 Ma using data from the Turkana Basin, East Africa, Geophys. Res. Lett. 19 (1992) 2349–2352.

[95] M.A.J. Williams, G. Assefa, D.A. Adamson, Depositional context of Plio–Pleistocene hominid-bearing formations in the Middle Awash Valley, Southern Afar Rift, Ethiopia, in: L.E. Frostick, R.W. Renaut, J. Reid, J.J. Tiercelin (Eds.), Sedimentation in the African Rifts, Geological Society Special Publication No. 25, Blackwell Scientific, Palo Alto, CA, 1986, pp. 241–251.

[96] D.A. Adamson, M.A.J. Williams, Geological setting of Pliocene rifting and deposition in the Afar Depression of Ethiopia, J. Hum. Evol. 16 (1986) 597–610.

[97] C.S. Feibel, Basin evolution, sedimentary dynamics and hominid habitats in East Africa: an ecosystem approach, in: T. Bromage, F. Schrenk (Eds.), African Biogeography, Climate Change, and Human Evolution, Oxford University Press, Oxford, 1999, pp. 276–281.

[98] T.C. Partridge, B. Wood, P.B. deMenocal, The influence of global climatic change and regional uplift on large mammalian evolution on East and southern Africa, in: E. Vrba, G. Denton, L. Burckle, T. Partridge (Eds.), Paleoclimate and Evolution With Emphasis on Human Origins, Yale University Press, New Haven, CT, 1995, pp. 330–355.

[99] E.S. Vrba, Ecological and adaptive changes associated with early hominid evolution, in: E. Delson (Ed.), Ancestors: The Hard Evidence, A.R. Liss, New York, 1985, pp. 63–71.

[100] J. Kappelman, T. Plummer, L. Bishop, A. Duncan, S. Appleton, Bovids as indicators of Plio–Pleistocene paleoenvironments in East Africa, J. Hum. Evol. 32 (1997) 229–256.

[101] T.W. Plummer, L.C. Bishop, Hominid paleoecology at Olduvai Gorge, Tanzania as indicated by antelope remains, J. Hum. Evol. 27 (1994) 47–75.

[102] L.M. Spencer, Dietary adaptations of Plio–Pleistocene Bovidae: Implications for hominid habitat use, J. Hum. Evol. 32 (1997) 201–228.

[103] H.B. Wesselman, Fossil micromammals as indicators of climatic change about 2.4 Myr ago in the Omo Valley, Ethiopia, S. Afr. J. Sci. 81 (1985) 260–261.

[104] V.M. Sarich, A.C. Wilson, Immunological time scale for hominid evolution, Science 158 (1967) 1200–1203.

[105] S. Kumar, B.A. Hedges, A molecular time scale for vertebrate evolution, Nature 392 (1998) 917–920.

[106] R.L. Stauffer, A. Walker, O.A. Ryder, M. Lyons-Weller, S.B. Hedges, Human and ape molecular clocks and constraints on paleontological hypotheses, J. Hered. 92 (2001) 469–475.

[107] M. Brunet, F. Guy, D. Pilbeam, H.T. Mackaya, A. Likius, D. Ahounta, A. Beauvilain, C. Blondel, H. Bocherens, J.R. Boisserie, L. DeBonis, Y. Coppens, J. Dejax, C. Denys, P. Duringer, V. Eisenmann, G. Fanone, P. Fronty, D. Geraads, T. Lehmann, F. Lihoreau, A. Louchart, A. Mahamat, G. Merceron, G. Mouchelin, O. Otero, P.P. Campomanes, M. PonceDeLeon, J. Rage, M. Sapanet, M. Schuster, J. Sudre, P. Tassy, X. Valentin, P. Vignaud, L. Viriot, A. Zazzo, C. Zollikofer, A new hominid from the Upper Miocene of Chad, Central Africa, Nature 418 (2002) 145–151.

[108] M.D. Leakey, C.S. Feibel, I. McDougall, A. Walker, New four million year old hominid species from Kanapoi and Allia Bay, Kenya, Nature 376 (1995) 565–571.

[109] C.O. Lovejoy, The origin of Man, Science 211 (1981) 341–350.

[110] D. Pilbeam, Human fossil history and evolutionary paradigms, in: M.K. Hecht (Ed.), Evolutionary Biology at the Crossroads, Queens College Press, New York, 1989, pp. 117–138.

[111] T. White, G. Suwa, B. Asfaw, Australopithecus ramidus, a new species of early hominid from Aramis, Ethiopia, Nature 371 (1994) 306–371.

[112] T.D. White et al., New discoveries of Australopithecus at Maka in Ethiopia, Nature 366 (1993) 261–265.

[113] R.A. Foley, Speciation, extinction and climatic change in hominid evolution, J. Hum. Evol. 26 (1994) 275–289.

[114] W. Kimbel, Hominid speciation and Pliocene climatic change, in: E. Vrba, G. Denton, L. Burckle, T. Partridge (Eds.), Paleoclimate and Evolution with Emphasis on Human Origins, Yale University Press, New Haven, CT, 1995, pp. 425–437.

[115] B. Wood, Origin and evolution of the genus Homo, Nature 355 (1992) 783–790.

[116] G. Wolde Gabriel, T.D. White, G. Suwa, P. Renne, J. de Heinzelin, W.K. Hart, G. Heiken, Ecological and temporal placement of early Pliocene hominids at Aramis, Ethiopia, Nature 371 (1994) 330–333.

[117] P. Renne, R. Walter, K. Verosub, M. Sweitzer, J. Aronson, New data from Hadar (Ethiopia) support orbitally-tuned time scale to 3.3 Ma, Geophys. Res. Lett. 20 (1993) 1067–1070.

[118] B. Wood, M. Collard, The human genus, Science 284 (1999) 65–71.

[119] D.E. Lieberman, B. Wood, D. Pilbeam, Homoplasy and early Homo: an analysis of the evolutionary relationships of H. habilis sensu stricto and H. rudolfensis, J. Hum. Evol. 30 (1996) 97–120.

[120] T.D. White, G. Suwa, B. Asfaw, Australopithecus ramidus, a new species of early hominid from Aramis, Ethiopia, Nature 371 (1994) 306–312.

[121] J.G. Wynn, Paleosols, stable carbon isotopes, and paleoenvironmental interpretation of Kanapoi, Northern Kenya, J. Hum. Evol. 39 (2000) 411–432.

[122] M.G. Leakey, C.S. Feibel, I. McDougall, C. Ward, A. Walker, New specimens and confirmation of an early age for Australopithecus anamensis, Nature 393 (1998) 62–66.

[123] M.G. Leakey, F. Spoor, F.H. Brown, P.N. Gathogo, C. Klarie, L.N. Leakey, I. McDougall, New hominin genus from eastern Africa shows diverse middle Pliocene lineages, Nature 410 (2001) 433–440.

[124] D.E. Lieberman, Another face in our family tree, Nature 410 (2001) 419–420.

[125] R.G. Klein, The causes of 'Robust' Australopithecine extinction, in: F.E. Grine (Ed.), Evolutionary History of the 'Robust' Australopithecines, Aldine de Gruyter, New York, 1988.

[126] G. Suwa, B. Asfaw, Y. Beyene, T.D. White, S. Katoh, S. Nagaoka, H. Nakaya, K. Uzawa, P. Renne, G. Wolde Gabriel, The first skull of Australopithecus boisei, Nature 389 (1997) 489–492.

[127] F.E. Grine, The Evolutionary History of the Robust Australopithicines, Aldine de Gruyter, New York, 1988.

[128] F. Schrenk, T.G. Bromage, C.G. Betzler, U. Ring, Y.M. Juwayeyi, Oldest Homo and Pliocene biogeography of the Malawi Rift, Nature 365 (1993) 833–836.

[129] F.C. Howell, P. Haesarts, J. de Heinzelin, Depositional environments, archelogical occurrances, and hominoids from members E and F of the Shungura Formation (Omo Basin, Ethiopia), J. Hum. Evol. 16 (1987) 665–700.

[130] J. de Heinzelin, J.D. Clark, T. White, W. Hart, P. Renne, G. Wolde Gabriel, Y. Beyene, E. Vrba, Environment and behavior of 2.5 million year old Bouri hominids, Science 284 (1999) 625–629.

[131] H. Roche, A. Delagnes, J.P. Brugal, C.S. Feibel, M. Kibunjia, B. Mourre, P.J. Texier, Early hominid stone tool production and technical skill 2.34 Myr ago in West Turkana, Kenya, Nature 399 (1999) 57–60.

[132] S.H. Ambrose, Paleolithic technology and human evolution, Science 291 (2001) 1748–1753.

[133] S. Semaw, P. Renne, J.W.K. Harris, C.S. Feibel, R.L. Bernor, N. Fesseha, K. Mowbray, 2.5-million-year-old stone tools from Gona, Ethiopia, Nature 385 (1997) 333–336.

[134] C.C. Swisher, G.H. Curtis, T. Jacob, A.G. Getty, A. Suprijo, Widiasmoro, Age of earliest known hominids in Java, Indonesia, Science 263 (1994) 1118–1121.

[135] K.E. Reed, Early hominid evolution and ecological

change through the African Plio–Pleistocene, J. Hum. Evol. 32 (1997) 289–322.

[136] T.E. Cerling, J. Quade, Y. Wang, Expansion and emergence of C4 plants, Nature 371 (1994) 112–113.

[137] J.D. Clark, J. de Heinzelin, K.D. Schick, W.K. Hart, T.D. White, G. Wolde Gabriel, R.C. Walter, G. Suwa, B. Asfaw, E. Vrba, Y.H. Selassie, African *Homo erectus*: Old radiometric ages and young Oldowan assemblages in the Middle Awash Valley, Ethiopia, Science 264 (1994) 1907–1910.

[138] E. Abbate et al., A one-million year-old *Homo* cranium from the Danakil (Afar) Depression of Eritrea, Nature 393 (1998) 458–460.

[139] J.W.K. Harris, Cultural beginnings, Plio–Pleistocene archaeological occurrences from the Afar, Ethiopia, in: N. David (Ed.), African Achaeological Review, Cambridge University Press, Cambridge, 1983, pp. 3–31.

[140] C. Ruff, Climate, body size, and body shape in hominid evolution, J. Hum. Evol. 21 (1991) 81–105.

[141] L. Aiello, P. Wheeler, The expensive tissue hypothesis: The brain and the digestive system in human and primate evolution, Curr. Anthropol. 36 (1995) 199–221.

[142] F.H. Brown, A.M. Sarna-Wojcicki, C.E. Meyer, B. Haileab, Correlation of Pliocene and Pleistocene tephra layers between the Turkana Basin of East Africa and the Gulf of Aden, Quat. Int. 13/14 (1992) 55–67.

[143] C.S. Feibel, Tephrostratigraphy and geological context in paleoanthropology, Evol. Anthropol. 8 (1999) 87–100.

[144] A.M. Sarna-Wojcicki, C.E. Meyer, P.H. Roth, F.H. Brown, Ages of tuff beds at East African early hominid sites and sediments in the Gulf of Aden, Nature 313 (1985) 306–308.

[145] P.B. deMenocal, F.H. Brown, Pliocene tephra correlations between East African hominid localities, the Gulf of Aden, and the Arabian Sea, in: J. Agusti, L. Rook, P. Andrews (Eds.), Hominid Evolution and Climatic Change in Europe, vol. 1, Cambridge University Press, Cambridge, 1999, pp. 23–54.

[146] A.C. Mix, N.G. Pisias, W. Rugh, J. Wilson, A. Morey, T.K. Hagelberg, Benthic foraminifer stable isotope record from Site 849 (0–5 Ma): Local and global climate changes, in: N.G. Pisias, L.A. Mayer, T.R. Janecek, A. Palmer-Julson, T.H. Van Andel (Eds.) Proc. ODP Sci. Results 138 (1995).

[147] N.J. Shackleton, An alternative astronomical calibration of the Lower Pleistocene timescale based on ODP Site 677, Trans. R. Soc. Edinburgh Earth Sci. 81 (1990) 251–261.

[148] P. Debrabant, L. Krissek, A. Bouqullon, H. Chamley, Clay mineralogy of Neogene sediments of the western Arabian Sea: Mineral abundances and paleoenvironmental implications, in: W. Prell, N. Niitsuma (Eds.), Proc. ODP Sci. Results 117 (1991) 183–196.

[149] V. Kolla, Clay mineralogy and sedimentation in the western Indian Ocean, Deep-Sea Res. 23 (1976) 949–961.

[150] N. Fagel, P. Debrabant, P. deMenocal, B. Demoulin, Use of sedimentary clay minerals for the reconstruction of periodic paleoclimatic variations in the Arabian Sea, Oceanol. Acta 15 (1992) 125–136.

[151] A.L. Berger, Long-term variations of daily insolation and Quaternary climatic changes, J. Atmos. Sci. 35 (1978) 2362–2367.

[152] M. Rossignol-Strick, African monsoons, an immediate climatic response to orbital insolation forcing, Nature 303 (1983) 46–49.

[153] E. Tuenter, S. Weber, F. Hilgen, L. Lourens, The response of the African summer monsoon to remote and local forcing due to precession and obliquity, Global Planet. Change 36 (2003) 219–235.

Peter B. deMenocal is an Associate Professor at the Lamont-Doherty Earth Observatory of Columbia University. He received a B.S. in Geology from St. Lawrence University, an M.S. in Oceanography from the University of Rhode Island, and a Ph.D. in Geological Sciences from Columbia University in 1991. He is a paleoceanographer who uses sedimentary and geochemical analyses of deep-sea sediments to reconstruct past changes in terrestrial climate and ocean circulation. His interests in African climate and faunal evolution arose from a chance encounter with a roomful of paleoanthropologists at a 1993 meeting in Aerlie, Virginia organized by E. Vrba, L. Burckle, G. Denton, and T. Partridge. He is a resident of New York City.

ELSEVIER

Reprinted from
Earth and Planetary Science Letters 219 (2004) 173–187

www.elsevier.com/locate/epsl

Thermal evolution of the Earth as recorded by komatiites

T.L. Grove*, S.W. Parman

Department of Earth, Atmospheric and Planetary Sciences, Massachusetts Institute of Technology, Cambridge, MA 02139, USA

Received 10 June 2003; received in revised form 18 December 2003; accepted 21 December 2003

Abstract

Komatiites are rare ultramafic lavas that were produced most commonly during the Archean and Early Proterozoic and less frequently in the Phanerozoic. These magmas provide a record of the thermal and chemical characteristics of the upper mantle through time. The most widely cited interpretation is that komatiites were produced in a plume environment and record high mantle temperatures and deep melting pressures. The decline in their abundance from the Archean to the Phanerozoic has been interpreted as primary evidence for secular cooling (up to 500°C) of the mantle. In the last decade new evidence from petrology, geochemistry and field investigations has reopened the question of the conditions of mantle melting preserved by komatiites. An alternative proposal has been rekindled: that komatiites are produced by hydrous melting at shallow mantle depths in a subduction environment. This alternative interpretation predicts that the Archean mantle was only slightly ($\sim 100°C$) hotter than at present and implicates subduction as a process that operated in the Archean. Many thermal evolution and chemical differentiation models of the young Earth use the plume origin of komatiites as a central theme in their model. Therefore, this controversy over the mechanism of komatiite generation has the potential to modify widely accepted views of the Archean Earth and its subsequent evolution. This paper briefly reviews some of the pros and cons of the plume and subduction zone models and recounts other hypotheses that have been proposed for komatiites. We suggest critical tests that will improve our understanding of komatiites and allow us to better integrate the story recorded in komatiites into our view of early Earth evolution.

Keywords: komatiites; mantle plumes; Archean; boninites; subduction zones; mantle melting; experimental petrology; mantle potential temperature; Earth evolution

1. Introduction

Many lines of evidence suggest that the early Earth was hotter than it is today. Increased heating from higher concentrations of radioactive isotopes, heat generated from the segregation of the core and from the initial accretion of the Earth (including a giant moon-forming impact) were likely large enough to have melted much or all of the silicate mantle [1]. While some aspects of the initial thermal state of the mantle seem relatively certain, the subsequent evolution is poorly constrained. An important unknown is the ratio of heat loss to heat production (Urey ratio).

---

* Corresponding author. Tel.: +1-617-253-2878;
Fax: +1-617-253-7102.
 E-mail addresses: tlgrove@mit.edu (T.L. Grove),
parman@mit.edu (S.W. Parman).

0012-821X/04/$ – see front matter © 2004 Elsevier B.V. All rights reserved.
doi:10.1016/S0012-821X(04)00002-0

Higher mantle temperatures would have driven vigorous convection that would have rapidly cooled the mantle. Depending upon the temperature dependence of mantle viscosity chosen, the results of numerical models of secular cooling vary widely, from models that approach current mantle temperatures within the first 500 Myr to models that retain high temperatures into the Proterozoic [2]. The chemical compositions of mafic magmas record the temperature and depth of melt generation, and therefore provide important constraints on mantle conditions throughout Earth history. It is here that komatiites have played an important role, because their compositional characteristics have been used to trace mantle melting depths, temperatures and processes back into the Archean.

As a consequence of erosion, meteorite impacts, early plate tectonic processes or their equivalent there is imperfect preservation of the ancient record of igneous activity on the Earth. Today the basaltic crust formed at mid-ocean ridges and constituting ~70% of the Earth's surface preserves only the latest moments (~170 million years) of its 4.5 billion year magmatic history [3]. Arguably, most of this rock record has been lost and recycled back into the Earth's interior. The rock record that is preserved in the Earth's continental crust extends much further into the geologic past, but preservation deteriorates with increasing age. Detrital zircon crystals found in sandstones at Jack Hills in Australia came from rocks as old as 4.4 Ga [4]. The oldest crustal rocks currently known are the ~4.0 Ga Acasta gneisses [5], a series of granites, granodiorites and tonalites, and the oldest preserved mafic volcanic sequences are ~3.8 Ga [6,7].

Komatiite lavas were first recognized by the Viljoen brothers in the Barberton Mountainland of South Africa in 1965 [8,9]. These rocks represent some of the oldest ultramafic magmatic rocks preserved in the Earth's crust at 3.5 Ga [10]. Shortly after the recognition of these rocks as a record of submarine volcanic activity, komatiites were discovered in the late Archean (2.7 Ga) Abitibi Belt in Munro Township, Ontario, Canada [11]. In 1979 komatiites of Cretaceous age were discovered on Gorgona Island [12]; these repre-

sent the youngest example of komatiite volcanism. Further investigation of Archean terranes led to the recognition of komatiites that range from Archean to Cretaceous in age [13] with most occurring between 3.5 and 1.5 Ga and a peak in komatiite preservation occurring in the late Archean. Recent discoveries of komatiites include a Phanerozoic occurrence in Vietnam [14], and the new compositionally distinct 3.3 Ga komatiites from the Commondale Belt, South Africa [15,16].

Komatiites are distinguished as ultramafic volcanic rocks containing spinifex olivine textures with MgO contents >18 wt% (Fig. 1). The maximum MgO content of komatiites is probably 25–30 wt% MgO, but this is a controversial issue, because all komatiites have been chemically modified by metamorphic processes. When high-MgO komatiites from Barberton were subjected to anhydrous melting experiments in the laboratory, they were found to possess very high liquidus temperatures (>1600°C), a result that was initially interpreted to indicate deep melting conditions (150 km) at high temperatures [17,18] at 3.5 Ga. The progressive decline of komatiites in the Neoproterozoic and their near absence in the Phanerozoic (only three examples) has been used as evidence that conditions in the Earth's mantle had changed in some way through geologic time. The simplest explanation is that this change reflects secular cooling of the mantle.

Typically, komatiites are found interlayered with a variety of igneous lavas. These lower-MgO lavas (variously termed komatiitic basalt, basaltic komatiite or siliceous high-magnesium basalt) along with basalts and olivine cumulates generally form the bulk of the stratigraphy, with komatiites usually representing less than 20% of a sequence [13]. The interpretation of the lower-MgO magmas, especially the basaltic komatiites (Fig. 1), is a key question in the debate over the origin of komatiites. Are they related to komatiites by fractionation or crustal contamination? Were they produced by distinct melting conditions during the komatiite-producing event? Are they totally unrelated?

Archean komatiites are often found in the crust attached to depleted cratonic lithospheric mantle. The mantle beneath Archean cratons is distinct

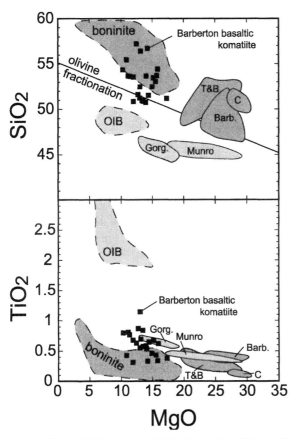

Fig. 1. SiO$_2$ and TiO$_2$ versus MgO in komatiites (fields with solid boundaries), basaltic komatiites (filled squares) and modern mafic magmas (fields with dashed boundaries). Solid line in top panel shows compositional effect that olivine fractionation would have on the most MgO-rich Barberton komatiite. The majority of the basaltic komatiites in Barberton cannot be made by olivine fractionation from a komatiite parent. Boninites (modern subduction-related magmas) show a large compositional overlap with Archean basaltic komatiites. Komatiites show a wide range of major and minor element composition. High-SiO$_2$ komatiites (dark fields) resemble modern boninite magmas that are produced by hydrous melting, while low-SiO$_2$ komatiites (light fields) more closely resemble modern basalts produced by anhydrous decompression melting. The low TiO$_2$ content of komatiite magmas requires their source to have been depleted (i.e. already have had a melt extracted from it). Komatiites: Barb. (Barberton, South Africa, 3.5 Ga [68]), C (Commondale, South Africa, 3.3 Ga [16]), T (Tisdale, Canada, 2.7 Ga [46]), B (Ball, Canada, 2.9 Ga [48]), Munro (Munro, Canada, 2.7 Ga [46]) and Gorg. (Gorgona, South America, 0.088 Ga [93]). Modern magmas: OIB (ocean island basalt, GeoRoc online database), boninite (GeoRoc online database).

from all other mantle reservoirs and one of its distinguishing chemical characteristics is that it is highly depleted. Boyd [19] suggested a genetic link between komatiite and cratonic lithosphere. Age dating using Os isotopic techniques on the mantle xenoliths from the Kaapvaal craton [20] indicates that the melting event that depleted the mantle occurred during the time period that the South African komatiites were being produced (3.0–3.5 Ga). So there is permissive evidence that komatiites and cratonic mantle may somehow be linked in their origin. If cratonic mantle (or some part of it) is the residue of komatiite extraction, the pressure–temperature conditions recorded by cratonic xenoliths place additional constraints on komatiite melting conditions [21].

2. Historical overview

The first discussions concerning the existence of ultramafic magmas occurred in the first half of the 20th century, before there was any consensus that the chemical composition of the Earth's upper mantle was also ultramafic; specifically that it consisted of peridotite. One early interpretation of alpine peridotites was that they represented igneous intrusions of ultramafic magmas [22,23]. We know now that these rocks do not represent mantle melts, but are fragments of the Earth's upper mantle that have been tectonically emplaced into the crust. However, in the 1960s the question of the significance and/or existence of alpine peridotites as magmatic rocks was still very much an open topic [24]. Hess [25] was one of the proponents of the peridotitic magma hypothesis. At the same time Engel et al. were searching for a single primary magma [26] that had been produced throughout geologic time by mantle melting. In this context it is interesting to note that both Hess [27] and Engel et al. [28] were among the early visitors to see the Viljoens' exciting discovery of komatiites, and both thought that these new igneous rocks held promise for increasing our understanding of early Earth igneous processes.

The time interval of the discovery of the komatiites overlapped with the plate tectonic revolution, the detailed exploration of the Earth's ocean floor and the return of the first lunar samples. Thoughts on the significance of komatiite magmas rapidly evolved in response to the context provided by these new discoveries. The original interpretation of the Viljoens [8] was that komatiites represented the first crust formed on Earth as a result of a catastrophic melting event triggered by convective overturn during core formation. A meteorite impact origin for komatiites [29] was also proposed. Fyfe [30] suggested that komatiites formed in an Archean plume-dominated environment. This mode of origin has become the most widely accepted. However, discussion of the processes and/or tectonic setting of komatiites did not end with the proposal put forth by Fyfe. Using chemical information and/or field setting, komatiites were proposed to form in mid-ocean ridge environments [31,32]. Others pointed out the uniqueness of komatiites and suggested that they could not be specifically associated with a tectonic environment [33]. Green [18] suggested that melting occurred as a result of the ascent of hot mantle diapirs from depths as great as 150 km. His later re-evaluation of this melting model led to the suggestion that melting began at this depth, but extended to very shallow depths [34,35]. The origin of komatiite in a wet melting environment was proposed by Allègre [36]. The similarity between modern mafic arc lavas and komatiitic lavas was noted by Brooks and Hart [33] and Cameron et al. [37] and implied komatiite generation in a subduction zone. The possibility of a planet-wide melting event was revitalized by Nisbet and Walker [38]. Others [39] have suggested that plate tectonics did not operate in the Archean, and that vertical tectonics drove magma generation processes. Of these many hypotheses, the plume model is currently the most popular, and proponents of it have recently reasserted their arguments for producing komatiites by this process [40].

3. Present status of the field: a re-evaluation of komatiite melting conditions

Several recent discoveries have led to a recon-

sideration of the significance of komatiite magmatism. New petrologic evidence from the komatiites in the Barberton Mountainland resulted in: (1) a revision of the MgO contents of the igneous protolith downward, lowering the anhydrous liquidus temperature to 1430°C, and (2) new estimates of magmatic water contents, indicating at least 3 wt% dissolved H_2O in the komatiite magmas [41]. Petrologic evidence preserved in igneous minerals and melt inclusions that have escaped metamorphic alteration also shows dissolved H_2O abundance levels from 0.3 to >2% in komatiite magmas from Munro Township, Canada, Belingwe, Zimbabwe and Kola Peninsula, Russia [42–45]. New occurrences of komatiites have been discovered where the komatiites are interlayered with Archean lavas that resemble modern ultramafic arc lavas (boninites). These include komatiites in the late Archean Tisdale volcanic belt [46,47], 3.0 Ga komatiites from the Uchi subprovince in the Superior Craton of Canada [48] and komatiite–boninite associations in the Nondweni, South Africa [49]. Finally, high-MgO, high-SiO_2 lavas, the 3.3 Ga Commondale ultramafic suite, have been discovered in the southern Kaapvaal Craton, South Africa [15,16] and represent a new variety of orthopyroxene-bearing komatiites. Wilson et al. [16] propose a hydrous melting origin in a subduction environment to explain this distinctive type of komatiite.

An increasing appreciation of the complex chemical diversity of komatiites is a major driving force for the re-evaluation of the existing melting models. In Fig. 1 we show the chemical compositional variations in MgO, SiO_2 and TiO_2 for komatiites from a variety of localities as well as modern plume and arc magmas. SiO_2 contents vary from 45 to 53 wt%, MgO contents from 18 (by definition) to ~30 wt% and TiO_2 varies by a factor of 5. Producing such chemical diversity with a single melting process is a challenge for all models of komatiite generation. Plume advocates tend to focus on the low-SiO_2 komatiites such as those from Munro [11,46] and Gorgona [12,50] and point to their compositional similarity to modern ocean island basalts. Subduction advocates point to the similarity of high-SiO_2 komatiites such as those in Barberton [51] and Commondale [15,16] with the most MgO-rich modern arc magmas (boninites), as well as the compositional overlap of boninites with basaltic komatiites interlayered with the komatiites (Fig. 1). The plume advocates dismiss the high-SiO$_2$ compositions as formed by either fractionation or crustal contamination of komatiite magmas, while subduction advocates question whether Gorgona magmas should be considered analogs of Archean komatiites. Neither model has successfully explained the complete spectrum of komatiite compositions.

3.1. Evidence from experimental petrology on the depth and temperature of komatiite melting

The first experimental study on komatiites was carried out by Green et al. [17] who concluded that komatiites were produced at high temperatures ($\sim 1650°C$) and estimated a depth of 150 km or greater for melt segregation from the source. This initial conclusion of a deep melting origin was revisited in light of new evidence from experimental studies of the best available estimate of the primary komatiite magmas from Munro Township and Western Australia. Green reasoned that any process should be consistent with the existence of a thick, cold lithosphere of > 150 km thick while other parts of the Earth yielded hot komatiite. He proposed that komatiite melting began at depths of > 150 km, and a high-temperature mantle melting column extended to shallow depths and left a mantle residue that had olivine as its only residual phase [34,35]. Takahashi and Scarfe [52] carried out experiments at higher pressures and showed that a depth of 150 km and a temperature > 1700°C were required to produce the most Mg-rich komatiite magmas. In the Takahashi model the depth of segregation was initially tied to the inferred liquidus temperature and melting was assumed to stop deep in the Earth's mantle, where the komatiite magma separated from its source, and the melt was delivered to the surface (batch melting). In more recent manifestations of the plume hypothesis, adiabatic decompression melting occurs and the initial depth of melting is controlled by the intersection of the adiabat with the peridotite solidus (Fig. 2). Compositional evidence (CaO/Al$_2$O$_3$ and Gd/Yb

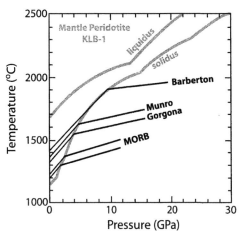

Fig. 2. Depths and temperatures of melting of komatiites in a plume setting inferred by Herzberg [86]. Black lines show the pressure–temperature path experienced by a body of mantle as it ascends adiabatically towards the surface. The gray lines are the solidus and liquidus of KLB-1, a fertile mantle composition. The solidus is the first point at which the mantle will begin melting. The liquidus is the point at which the mantle would be completely molten. Below the solidus, the slope of the lines reflects slight cooling due to decompression. The potential temperature is the temperature at which this path intersects the surface. The greater slope above the solidus reflects both the heat loss required to produce melting and the greater compressibility of melts relative to solid. In the plume model, the oldest komatiites (Barberton at 3.5 Ga) record the highest mantle potential temperatures, while late Archean (2.7 Ga) Munro komatiites and Cretaceous (88 Ma) Gorgona komatiites are formed in progressively cooler mantle plumes. Note that if melting occurs by adiabatic decompression, then higher melting temperatures require that melting begins at greater depths. The range of melting temperatures for present-day mid-ocean ridge basalts (MORB) is also shown.

ratios in the komatiite) is interpreted to indicate garnet was present along with olivine and orthopyroxene in the source region of komatiites. The latest experimental investigation of this melting process has been that of Walter [53] who melted peridotite and tried to produce a melt composition that was identical to komatiite. Walter matched some of the compositional characteristics at pressures of 6 GPa, but not all (e.g. the CaO/Al$_2$O$_3$ of low-Al komatiites). Some recent estimates of komatiite melting conditions suggest even higher pressures (> 10 GPa) and temperatures > 2000°C [54,55]. However, these investiga-

tors use indirect evidence to arrive at these estimates.

In the plume hypothesis the depth and temperature of komatiite generation decrease with time.
Archean komatiites (like Barberton at 3.5 Ga) are
the hottest, and come from the greatest depth,
while late Archean komatiites (like Munro at 2.7
Ga) come from shallower depths and the youngest komatiites from Gorgona (at 0.086 Ga) come
from the shallowest depths (Fig. 2). However, this
model has been called into question by the recent
discovery of high-SiO_2 komatiites from Commondale. One indication of depth of melting is the
SiO_2 content of the magma. As depth of melting
increases, the effect of increasing pressure is to
move the phase boundaries for melting to lower
SiO_2 content and higher MgO and FeO contents.
As Fig. 1 shows, the oldest komatiites (e.g. Barberton and Commondale) have higher SiO_2 than
the younger ones (Gorgona and Munro). Thus,
the age of komatiites and their depth of melting
do not correlate.

If komatiites were produced in a mid-ocean
ridge setting, the melt generation process would
be adiabatic decompression fractional melting.
The distinctive signature of this style of melting
is that it begins when mantle ascending along an
adiabat crosses its solidus and continues to shallow depths until it encounters lithospheric mantle
(Fig. 3). This melting style characterizes the major
mode of melt production in the Earth's present-
day mid-ocean ridge system. The compositional
systematics of modern mid-ocean ridge basalts
[56–58] are interpreted to indicate variations in
the temperature of 150–200°C for mantle upwelling beneath mid-ocean ridges [56]. Hotter mantle
starts melting deeper, continues melting until shallow depths and records the maximum amount of
melting in the ridge system through formation of
the thickest crustal basaltic section. The cooler
upwelling mantle does not reach its solidus temperature until a shallow depth, melting takes place
over a smaller temperature–depth interval and less
basaltic crust is formed. Presumably, the extrapolation of this melting process back into Archean
mid-ocean ridges would result in a melting process that began at much greater depths, extended
over a larger pressure–depth range to shallow

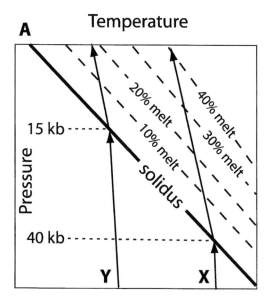

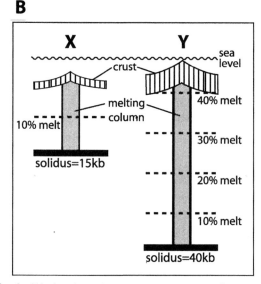

Fig. 3. (A) A schematic pressure–temperature diagram for
melting of mantle peridotite by adiabatic decompression.
Path X represents hotter mantle potential temperature that
crosses the solidus at greater depth and higher temperature
and undergoes a higher extent of melting. Path Y crosses the
solidus at lower temperature and undergoes a lesser extent of
melting. See caption of Fig. 2 for further explanation. (B)
Cross-section of the oceanic crust and mantle corresponding
to the melting histories shown by paths X and Y. Hotter
mantle begins melting at greater depths and melts to a greater extent than cooler mantle [56]. This model was developed
for mid-ocean ridge magmas, but is applicable to any decompression melting regime, such as mantle plumes.

mantle depths and resulted in larger degrees of melting and mantle depletion. Komatiites require at least 30–40% melting of the mantle. Models for this melting process predict melting rates of 0.15–0.3% per km of upwelling. Thus, to produce komatiites melting would have to begin at ~ 200 km depth and extend to beneath the ridge.

The experiments discussed above were all done under anhydrous conditions, in the absence of H_2O. Green et al. [17] also performed hydrous experiments, but ruled out high magmatic H_2O contents for komatiites on the grounds that such magmas would devolatilize and vesiculate as they reached the Earth's surface. Other proponents of dry komatiites have put forth similar arguments [59]. Nevertheless, Green's experiments suggest that Barberton komatiite with 30 wt% MgO could be generated by hydrous melting at 1500°C. Recent experiments on a less altered, lower-MgO (25 wt%) composition indicates melting temperatures of 1400–1450°C and depths of ~ 70 km [60] (Fig. 4).

The pre-eruptive H_2O contents of Barberton komatiites were estimated using petrologic evidence by Parman et al. [41], who found that at least 3 wt% dissolved H_2O in the komatiite magma was required to reproduce the compositions of the igneous minerals preserved in the rock samples. They reasoned that such high H_2O contents might be preserved if the komatiites were shallow intrusives, as suggested by de Wit et al. [32]. Field observations have led Dann [61] to the interpretation that the Barberton komatiites are lava flows, and this evidence has been used to dismiss the petrologic estimates of 3–6 wt% H_2O in the Barberton komatiites [40,59]. However, there are abundant examples of modern high-MgO submarine lavas that have erupted and failed to vesiculate and degas. High-MgO (22 wt%) boninites have glassy chill margins that contain 2 wt% H_2O that is primary in origin [62] and the interiors of some of these pillow lavas contain up to 7 wt% H_2O in the interstitial glass. Thus, the petrologic evidence of hydrous komatiite magmas does not rely on whether they are extrusive or intrusive.

If some komatiites do contain high H_2O, then the next issue is: where did the H_2O come from?

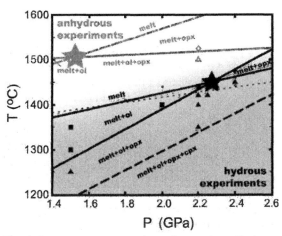

Fig. 4. Pressure–temperature phase diagram for a Barberton komatiite composition under hydrous (black lines) and anhydrous (gray lines) conditions [60]. Mafic melts are in equilibrium with various solid phases depending upon the pressure, temperature and H_2O content of the system. The melt is inferred to have been produced at the conditions where it is saturated with the minerals that it left as a residue in the mantle, in this case olivine and orthopyroxene. Under anhydrous conditions, the Barberton komatiite is multiply saturated at ~1.5 GPa and ~1510°C (gray star). With 6 wt% H_2O added, the melt saturates at a lower temperature (~1450°C) but a higher pressure (~2.3 GPa, black star).

Modern plume magmas contain H_2O but in much lower amounts (< 0.5 wt%) [63,64] than the amount estimated for the Barberton komatiites. Kawamoto et al. [65] have suggested that komatiites were produced by hydrous Archean plumes, thus reconciling the evidence for high H_2O contents in some komatiite magmas with the plume model. They suggest that the Archean mantle contained higher volatile contents, left over from the accretion of the Earth. In this model the initial few percent of melting in a plume would carry away all of the H_2O in the mantle. Subsequent high-degree melting would be anhydrous. Alternatively, proponents of the plume hypothesis have suggested that the H_2O was introduced by crustal contamination [59].

Modern hydrous melts are produced primarily by volatile fluxing in subduction zones [60,66]. This process leads to the generation of H_2O contents as high as 10 wt% in some mafic arc magmas [67]. Hydrous melting in subduction zones only reaches high extents because the mantle is continuously fluxed by volatiles released from the sub-ducting slab (i.e. the water–rock ratio is high) and because the inverted thermal gradient in the mantle wedge can drive the melting process. Melts ascend from cooler to hotter mantle, and continuously re-equilibrate with their enclosing mantle by increasing their melt fraction.

In many occurrences of komatiites, it has been recognized that they have compositional similarities to or are interlayered with lavas similar to boninites [16,47,49,68–71]. Boninites are generated by high extents of hydrous melting in subduction zones, and are characterized by high SiO_2 and high MgO (Fig. 1). Boninites are produced by melting conditions associated with the inception of subduction [72]. In the early stages of subduction zone development, subduction is initiated along an active fault in the oceanic crust (Fig. 5) and the subducted plate rapidly sinks into the mantle, drawing hot, buoyant mantle asthenosphere into the forearc [73]. This early magmatism involves melting of hydrated, depleted mantle. Addition of water and asthenospheric upwelling leads to catastrophic, high-extent melting. Eventually as the subduction zone matures, this extreme form of melting ends and more normal island arc magmas predominate (Fig. 5).

The close spatial and temporal association of Archean komatiites with boninite-like magmas has led to the suggestion that the two magma types were produced by similar melting processes (i.e. hydrous melting in subduction zones) [51,68]. In modern settings this melting process occurs over a very broad zone in the forearc, is characterized by high magma production and eruption rates, an extensional environment and progressive migration of volcanism away from the trench [72]. In the Archean a similar scenario is envisioned where early boninite–komatiite activity is succeeded by island arc tholeiitic magmatism (Fig. 5). The absence of komatiites from modern boninite settings presumably reflects the thermal evolution of the Earth's mantle [68]. When komatiites were first recognized boninites were also in the process of being discovered [74–76], and their distinctive compositional characteristics and tectonic setting had not yet been recognized. This is perhaps responsible for the delayed apprecia-

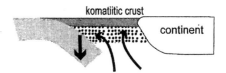

1. subduction initiation

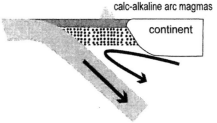

2. mature subduction zone

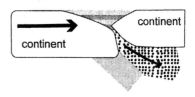

3. subduction termination - continent collision

highly depleted mantle

Fig. 5. Subduction model for the origin of komatiites in subduction zones [51], modeled after melting process inferred for boninites [72]. (1) Boninite-like melting event produces a range of high-MgO melts: komatiite, komatiitic basaltic magmas and low-Ti tholeiites (dark shading) and a corresponding heavily depleted mantle residue (stippled region). (2) The continued subduction of the lithosphere cools the mantle and establishes a mature subduction zone. Subsequent lower-temperature hydrous magmas (calc-alkaline andesites) are emplaced on top of early, ultramafic crust. The cold, buoyant residual mantle remains beneath the forearc. (3) The komatiitic crust is obducted onto continents during continent collision at the end of subduction. The depleted mantle is thickened and incorporated into the continental lithospheric mantle.

tion of the compositional similarities between boninites and Archean komatiites.

4. Constraints on early Earth thermal history

In the modern Earth a range of mantle temperatures is recorded in plume, mid-ocean ridge and subduction zone environments (Fig. 6). The best-characterized melt generation process in a plume

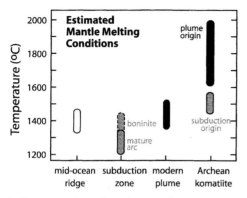

Fig. 6. The range of mantle melt generation temperatures estimated for various modern tectonic settings compared to temperatures inferred for komatiite melt generation by plume model (black filled field) and subduction model (gray filled field). MORB estimates are from Kinzler and Grove [58]. The Hawaiian plume temperature is based on experimental work of Eggins [77] and Wagner et al. [78]. Subduction zone estimates are separated in boninite melting (dashed boundary [83]) and melting in steady-state subduction environments (solid boundary [79–82]). Considering the maximum estimated temperatures for each category, the plume model requires ~ 500°C cooling between 3.5 Ga and now, while the subduction model requires ~ 100°C cooling.

environment is Hawaii. Primitive Hawaiian picrites provide evidence for melt segregation temperature and depth at the top of the Hawaiian plume of 1420°C at 66 km [77,78]. The Hawaiian plume erodes thick lithosphere beneath Hawaii, altering the major and trace element compositions of melts; depending on the model parameters, the mean mantle potential temperature is 1530°C. For mid-ocean ridges, estimates of mantle potential temperatures range from 1320 to 1475°C [58]. In subduction zones, primitive magmas from mature arcs provide evidence of mantle wedge melting temperatures that range from 1200–1250°C [79,80] to 1300–1350°C [81,82]. At the initiation of subduction, boninite magmas record melting conditions of 1400–1450°C at shallow depths (15–45 km) [83].

Geodynamic models of mantle convection have used the estimates of modern mantle temperatures as boundary conditions to solve for the thermal evolution of the mantle [2,84]. The models have considered the effects of variable rheology, variable contributions of plumes to cooling and the effects of phase transitions in the mantle, among

other parameters. The temperature evolution curves produced have been diverse, from near-linear cooling, to rapid early cooling, to complex, episodic cooling. The plume model is most consistent with models that argue for slow cooling of the Earth, so as to retain high temperatures in the mantle 1–2 billion years after the formation of the Earth (Fig. 7). The subduction model argues for only moderately higher mantle temperatures in the Archean, and therefore for convection models in which the mantle cools rapidly in the first 500 million years.

In either case, linking komatiite melting temperatures to the temperature structure of the Archean mantle is not straightforward. If komatiites are the product of plumes, the geodynamic models indicate they likely originated at the core–mantle boundary, and reflect unusually hot lower mantle temperatures (or they would not form thermal plumes). If komatiites are subduction-related, then they record the temperature of shallow upper mantle that was drawn towards the surface and hydrated by the movements of lithospheric plates. Modern boninites are only produced in a minority of subduction zones, presumably where subduction has occurred over relatively hot mantle. If the boninite analogy holds for komatiites, then they record only the high end of the spectrum of Archean upper mantle temperatures. In

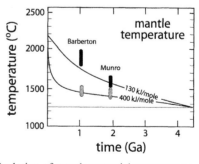

Fig. 7. Evolution of mantle potential temperatures with time in an Earth that is cooled by both plumes and plates (from Davies [84]). Curves show thermal evolution for a highly temperature-dependent mantle viscosity (activation energy = 400 kJ/mol) and for a weakly temperature-dependent viscosity (130 kJ/mol). Melting temperature estimates for hydrous (gray fields) and dry (black fields) komatiite melting of Barberton komatiites at 3.5 and 2.7 Ga (Barberton [40,54,55,60].

both cases, the inferred mantle temperatures are likely to be upper bounds and not representative of the entire temperature range present in the convecting mantle.

5. Future challenges and directions

5.1. Constraining temperatures of melt production

The extremely high temperature estimates for komatiite generation are based upon the assumption that komatiite samples with the highest MgO contents are representative of a preserved igneous rock composition. These high-MgO compositions lead to liquidus temperature estimates that range from 1800°C to 1675°C [53,85,86]. In contrast, Parman et al. [41] conclude that these samples have been altered during greenschist metamorphism to enrich the MgO content over that of the original igneous rock. When the least altered samples are considered, the liquidus temperatures are significantly lower (Figs. 6 and 7), from 1450 to 1550°C. Estimating the melting conditions of these complex metamorphosed igneous rocks is problematic. It should only be attempted when evidence can be found in the rock itself that it is representative of an igneous magma composition. The composition of preserved magmatic olivine provides the best mineralogical evidence to determine that a rock represents a liquid composition, but such olivines are rarely preserved. Even when olivine is preserved, its composition can be modified by metamorphic processes. In the case of the more fluid-mobile major and trace elements (e.g. Na and Sr), there is little hope that these will be directly preserved in the metamorphosed igneous rock, but preserved igneous minerals (e.g. pyroxene, [87]) can provide an estimate. Temperature constraints obtained from natural komatiite bulk rock compositions in the absence of evidence that they represent liquid compositions should be considered questionable. All komatiites are altered metamorphic rocks and the use of all samples for temperature estimation purposes is not justified. A few well-constrained temperature estimates are probably better than many unconstrained estimates.

Experimental studies of the well-constrained komatiite bulk compositions can provide further evidence of melt segregation conditions. There are only a few recent experimental studies of the liquidus phase relations of natural komatiites [34,41,60]. The first studies on komatiites [88,89] were done in the early days of experimental melting and some suffered from shortcomings that have been eliminated by advances and improvements in technique. Direct determinations of komatiite liquidus phase relations provide complementary information to the magmatic compositions obtained from peridotite melting experiments. In the latter, a peridotite starting composition is partly melted over a range of pressures and temperatures to define a set of conditions that produce a komatiite liquid. Melting experiments have successfully produced some of the major element characteristics of komatiites, but have not matched other key major element characteristics, notably the high Ca/Al, that characterize some Archean komatiites [53]. The major element, trace element and isotopic compositions of komatiites require that the source was strongly depleted, while the experimental studies of mantle melting [53,90,91] have almost exclusively used mantle compositions that have never been melted (i.e. primitive or fertile). There are many potential depleted mantle compositions whose melting behavior should be experimentally investigated. Furthermore, at the high temperatures required by these melting studies it is often difficult to preserve and analyze the liquid composition [54]. Comparing experiments on natural komatiite compositions and peridotites should yield better constraints on pressure and temperature of melting, possibly eliminating some of the variability of liquidus temperature estimates that currently exist.

5.2. Field evidence for the tectonic setting of komatiites

Field evidence can provide valuable constraints on tectonic processes related to komatiite production. A weakness of many komatiite studies, particularly experimental ones, is that they focus just on the komatiites and ignore the majority of the magmatic sequence in which the komatiites are

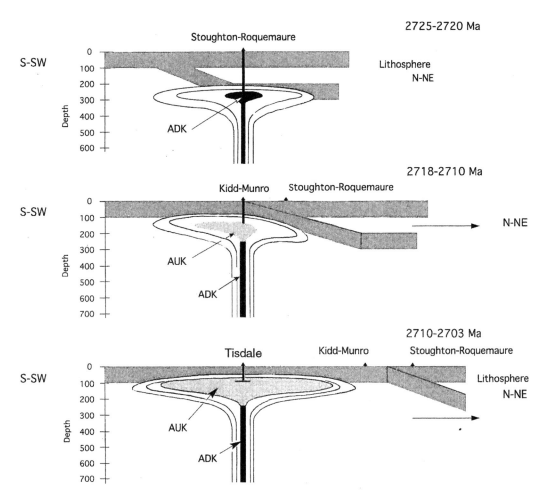

Fig. 8. Model that explains the interlayering of komatiites with subduction-related magmas by the interaction of a plume and a subduction zone. Alumina-undepleted komatiites (AUK) and alumina-depleted komatiites (ADK) are considered to come from different parts of the ascending plume [94].

found, particularly the basaltic komatiites. Currently there is no consensus on whether these magmas are just fractionated and/or crustally contaminated komatiites or are truly primitive magmas that record mantle melting conditions distinct from the komatiites. If they are primary and their arc-like geochemical signatures were not produced by crustal contamination, they pose a significant challenge to the plume hypothesis, requiring unappealing models of plumes repeatedly impinging upon arcs (Fig. 8).

Likewise, the interpretation of the trace element geochemistry of the komatiites themselves is

clouded. Like the basaltic komatiites, many komatiites have chemical features that are typical of modern subduction-related magmas (e.g. [48]). These characteristics include variable enrichments in large ion lithophile and light rare earth elements along with depletions in high field strength elements. Both models agree that these indicate input of trace element-enriched material into the komatiites. The difference is whether it was added as a hydrous fluid into the komatiite source in a subduction zone (subduction model), or was introduced by the addition of granitic material as the komatiite traversed the crust (plume model).

6. Conclusions

The recent discovery of an entirely new class of komatiites in Commondale demonstrates that there is still much to learn from fundamental field observations. Experimental studies should be undertaken to constrain the range of possible melting conditions implied by both models, though the data are particularly scarce for hydrous mantle melting. The challenge for trace element geochemistry is to distinguish the chemical signatures of crustal contamination from the addition of a subduction-derived hydrous fluid component. This can be a difficult task in unaltered modern magmas, let alone in altered and metamorphosed samples. Perhaps advances in microsampling techniques such as laser-ablation inductively coupled mass spectrometry and ion microprobe will allow the preserved igneous minerals (olivine, pyroxene and spinel) to be used more effectively.

In sum, the origin and implications of komatiites is an open and active debate that has far-reaching implications for the thermal and chemical evolution of the Earth. The number of competing hypotheses seems to have multiplied in the recent literature with plumes, hydrous plumes, plumes interacting with subduction zones and boninite-like melting all being advocated. It is possible that komatiites were generated by more than one process in the Archean, in the same way that basalts are produced in a range of tectonic settings today. This may explain why advocates of the plume and subduction models tend to focus on different komatiite localities and would challenge us to discover reliable means to distinguish between them. Kuhn [92] argues that paradigm shifts are preceded by times of confusion, heated debate and a multiplicity of hypotheses that explain differing subsets of the existing data. This description seems to provide an accurate account of the current state of komatiite research and promises an exciting future.

Acknowledgements

The authors thank J. Bedard, R. Stern and an anonymous reviewer for comments on this paper. We also thank our collaborators, Maarten de Wit and Jesse Dann, for stimulating discussions on the origins of komatiites. Support for work in the Barberton Mountainland of South Africa has been provide by NSF Continental Dynamics Grant EAR-9526702 and by the FRD of South Africa.*[AH]*

References

[1] V.S. Solomatov, Fluid dynamics of a terrestrial magma ocean, in: R.M. Canup, K. Righter (Eds.), Origin of the Earth and Moon, The University of Arizona Press, Tuscon, AZ, 2000, pp. 323–338.

[2] G.F. Davies, Cooling the core and mantle by plume and plate flows, Geophys. J. Int. 115 (1993) 132–146.

[3] A. Bartolini, R.L. Larson, Pacific microplate and the Pangea supercontinent in the Early to Middle Jurassic, Geology 29 (2001) 735–738.

[4] S.A. Wilde, J.W. Valley, W.H. Peck, C.M. Graham, Evidence from detrital zircons for the existence of continental crust and oceans on the Earth 4.4 Gyr ago, Nature 409 (2001) 175–178.

[5] S.A. Bowring, I.S. Williams, Priscoan (4.00–4.03 Ga) orthogneisses from northwestern Canada, Contrib. Mineral. Petrol. 134 (1999) 3–16.

[6] A. Polat, A.W. Hofmann, M.T. Rosing, Boninite-like volcanic rocks in the 3.7–3.8 Ga Isua greenstone belt, West Greenland: geochemical evidence for intra-oceanic subduction zone processes in the early Earth, Chem. Geol. 184 (2002) 231–254.

[7] J. David, M. Parent, R. Stevenson, P. Nadeau, L. Godin, La séquence supracrustale de Porpoise Cove, région d'Inukjuak: un exemple unique de croute paléo-Archéene (ca. 3.8 Ga) dans la Province du Supérieur, Ministère de Ressources Naturelles, Quebec DV2002-10, 2002, 34 pp.

[8] M.J. Viljoen, R.P. Viljoen, The geology and geochemistry of the Lower Ultramafic Unit of the Onverwacht Group and a proposed new class of igneous rocks, Upper Mantle Project, Spec. Publ. Geol. Soc. S. Afr. 2 (1969) 55–85.

[9] M.J. Viljoen, R.P. Viljoen, Evidence for the existence of a mobile extrusive peridotitic magma from the Komati Formation of the Onverwacht Group, Upper Mantle Project, Spec. Publ. Geol. Soc. S. Afr. 2 (1969) 87–112.

[10] A. Kröner, G.R. Byerly, D.R. Lowe, Chronology of early Archaean granite-greenstone evolution in the Barberton Mountain Land, South Africa, based on precise dating by single zircon evaporation, Earth Planet. Sci. Lett. 103 (1991) 41–54.

[11] D.R. Pyke, A.J. Naldrett, O.R. Eckstran, Archean ultramafic flows in Munro Township, Ontario, Geol. Soc. Am. Bull. 84 (1973) 955–977.

[12] L.M. Echeverria, Tertiary or Mesozoic komatiites from

Gorgona Island, Colombia – Field relations and geochemistry, Contrib. Mineral. Petrol. 73 (1980) 253–266.

[13] M.J. de Wit, L. D.Ashwal, Convergence towards divergent models of greenstone belts, in: M.J. de Wit, L.D. Ashwal (Eds.), Greenstone Belts, Oxford University Press, New York, 1997, pp. ix–xix.

[14] A.I. Glotov, G.V. Polyakov, T.T. Hoa, P.A. Balykin, V.A. Akimtsev, A.P. Krivenko, N.D. Tolstykh, N.T. Phuong, H.H. Thanh, T.Q. Hung, T.E. Petrova, The Ban Phuc Ni-Cu-PGE deposit related to the phanerozoic komatiite-basalt association in the Song Da Rift, northwestern Vietnam, Can. Mineral. 39 (2001) 573–589.

[15] A.H. Wilson, S.B. Shirey, R.W. Carlson, Archaean ultradepleted komatiites formed by hydrous melting of cratonic mantle, Nature 423 (2003) 858–861.

[16] A.H. Wilson, A new class of silica enriched, highly depleted komatiites in the southern Kaapvaal Craton, South Africa, Precambrian Res. 127 (2003) 125–141.

[17] D.H. Green, I.A. Nicholls, M. Viljoen, R. Viljoen, Experimental demonstration of the existence of peridotitic liquids in earliest Archean magmatism, Geology 3 (1975) 11–14.

[18] D.H. Green, Genesis of Archean peridotitic magmas and constraints on Archean geothermal gradients and tectonics, Geology 3 (1975) 15–18.

[19] F.R. Boyd, Compositional distinction between oceanic and cratonic lithosphere, Earth Planet. Sci. Lett. 96 (1989) 15–26.

[20] D.G. Pearson, R.W. Carlson, S.B. Shirey, F.R. Boyd, P.H. Nixon, Stabilisation of Archaean lithospheric mantle; a Re-Os isotope study of peridotite xenoliths from the Kaapvaal Craton, Earth Planet. Sci. Lett. 134 (1995) 341–357.

[21] T.H. Jordan, Continents as a chemical boundary layer: The origin and evolution of the Earth's continental crust, Phil. Trans. R. Soc. London A 301 (1981) 359–373.

[22] N.L. Bowen, The origin of ultrabasic and related rocks, Am. J. Sci. 14 (1927) 89–108.

[23] J.M.L. Vogt, Magmas and igneous ore deposits, Econ. Geol. 21 (1926) 207–233.

[24] J.S. Turner, J. Verhoogen, Igneous and Metamorphic Petrology, McGraw-Hill, New York, 1960, 694 pp.

[25] H.H. Hess, A primary peridotitic magma, Am. J. Sci. 209 (1938) 321–344.

[26] A.E.J. Engel, C.G. Engel, R.G. Havens, Chemical characteristics of the oceanic mantle and the upper mantle, Geol. Soc. Am. Bull. 76 (1965) 719–734.

[27] J.S. Dickey Jr., A primary peridotitic magma revisited: Olivine quench crystals in a peridotitic lava, Mem. Geol. Soc. Am. 132 (1972) 289–297.

[28] A.E.J. Engel, The Barberton Mountainland, in: P.J. Cloud (Ed.), Adventures in Earth History, W.H. Freeman, San Francisco, CA, 1970, pp. 431–445.

[29] D.H. Green, Archaean greenstone belts may include terrestrial equivalents of lunar maria?, Earth Planet. Sci. Lett. 15 (1972) 263–270.

[30] W.S. Fyfe, The evolution of the Earth's crust: Modern plate tectonics to ancient hot spot tectonics, Chem. Geol. 23 (1978) 89–114.

[31] A.Y. Glikson, Primitive Archean element distribution patterns: Chemical evidence and geotectonic significance, Earth Planet. Sci. Lett. 12 (1971) 309–320.

[32] M.J. de Wit, R.A. Hart, R.J. Hart, The Jamestown Ophiolite Complex, Barberton Mountain Belt – a section through 3.5 Ga oceanic-crust, J. Afr. Earth Sci. 6 (1987) 681–730.

[33] C. Brooks, S.R. Hart, On the significance of komatiite, Geology 2 (1974) 107–110.

[34] D.H. Green, Petrogenesis of Archean ultramafic magmas and implications for Archean tectonics, in: A. Kröner (Ed.), Precambrian Plate Tectonics, Elsevier, New York, 1981, pp. 469–489.

[35] D.H. Green, The Earth's lithosphere and asthenosphere – Concepts and constraints derived from petrology and high pressure experiments, Geol. Soc. Austr. Spec. Publ. 17 (1987) 1–22.

[36] C.J. Allègre, Genesis of Archaean komatiites in a wet ultramafic subducted plate, in: N.T. Arndt, E.G. Nisbet (Eds.), Komatiites, Springer-Verlag, Berlin, 1982, pp. 495–500.

[37] W.E. Cameron, E.G. Nisbet, V.J. Dietrich, Boninites, komatiites and ophiolitic basalts, Nature 280 (1979) 550–553.

[38] E.G. Nisbet, D. Walker, Komatiites and the structure of the Archaean mantle, Earth Planet. Sci. Lett. 60 (1982) 105–113.

[39] J. Bédard, P. Brouillette, L. Madore, A. Berclaz, Archean cratonization and deformation in the northern Superior Province, Canada: an evaluation of plate tectonic versus vertical tectonic models, Precambrian Res. 127 (2003) 61–87.

[40] N.T. Arndt, Komatiites, kimberlites and boninites, J. Geophys. Res. 108 (2003) 2293.

[41] S.W. Parman, J.C. Dann, T.L. Grove, M.J. de Wit, Emplacement conditions of komatiite magmas from the 3.49 Ga Komati Formation, Barberton Greenstone Belt, South Africa, Earth Planet. Sci. Lett. 150 (1997) 303–323.

[42] W.F. McDonough, T.R. Ireland, Intraplate origin of komatiites inferred from trace-elements in glass inclusions, Nature 365 (1993) 432–434.

[43] K. Shimizu, T. Komiya, K. Hirose, N. Shimizu, S. Maruyama, Cr-spinel, an excellent micro-container for retaining primitive melts – implications for a hydrous plume origin for komatiites, Earth Planet. Sci. Lett. 189 (2001) 177–188.

[44] W.E. Stone, E. Deloule, M.S. Larson, C.M. Lesher, Evidence for hydrous high-MgO melts in the Precambrian, Geology 25 (1997) 143–146.

[45] E.J. Hanski, Petrology of the Pechenga ferropicrites and cogenetic, Ni-bearing gabbro-wehrlite intrusions, Kola Peninsula, Russia, Geol. Surv. Finland Bull. 367 (1992) 11 192.

[46] J. Fan, R. Kerrich, Geochemical characteristics of aluminum depleted and undepleted komatiites and HREE-en-

riched low-Ti tholeiites, western Abitibi greenstone belt; a heterogeneous mantle plume-convergent margin environment, Geochim. Cosmochim. Acta 61 (1997) 4723–4744.

[47] R. Kerrich, D. Wyman, J. Fan, W. Bleeker, Boninite series; low Ti-tholeiite associations from the 2.7 Ga Abitibi greenstone belt, Earth Planet. Sci. Lett. 164 (1998) 303–316.

[48] P. Hollings, D. Wyman, R. Kerrich, Komatiite-basalt-rhyolite volcanic associations in northern Superior Province greenstone belts; significance of plume-arc interaction in the generation of the proto continental Superior Province, Lithos 46 (1999) 137–161.

[49] A.H. Wilson, J.A. Versfeld, The early Archaean Nondweni greenstone belt, southern Kaapvaal Craton, South Africa; Part II, Characteristics of the volcanic rocks and constraints on magma genesis, Precambrian Res. 67 (1994) 277–320.

[50] L.M. Echeverria, B.G. Aitken, Pyroclastic rocks – Another manifestation of ultramafic volcanism on Gorgona Island, Colombia, Contrib. Mineral. Petrol. 92 (1986) 428–436.

[51] S.W. Parman, T.L. Grove, J.C. Dann, M.J. de Wit, A subduction origin for komatiite and cratonic lithospheric mantle, S. Afr. J. Geol. 107 (2004) in press.

[52] E. Takahashi, C.M. Scarfe, Melting of peridotite to 14 GPa and the genesis of komatiite, Nature 315 (1985) 566–568.

[53] M.J. Walter, Melting of garnet peridotite and the origin of komatiite and depleted lithosphere, J. Petrol. 39 (1998) 29–60.

[54] C. Herzberg, Generation of plume magmas through time – an experimental perspective, Chem. Geol. 126 (1995) 1–16.

[55] M.J. Cheadle, D. Sparks, Can komatiites be 'dry' plume-type magmas?, Geochim. Cosmochim. Acta 66 (2002) A133.

[56] E.M. Klein, C.H. Langmuir, Global correlations of ocean ridge basalt chemistry with axial depth and crustal thickness, J. Geophys. Res. 92 (1987) 8089–8115.

[57] E.M. Klein, C.H. Langmuir, Local versus global variations in ocean ridge basalt composition – a reply, J. Geophys. Res. 94 (1989) 4241–4252.

[58] R.J. Kinzler, T.L. Grove, Primary magmas of midocean ridge basalts. 2. Applications, J. Geophys. Res. 97 (1992) 6907–6926.

[59] N. Arndt, C. Ginibre, C. Chauvel, F. Albarede, M. Cheadle, C. Herzberg, G. Jenner, Y. Lahaye, Were komatiites wet?, Geology 26 (1998) 739–742.

[60] T.L. Grove, S.W. Parman, J.C. Dann, Conditions of magma generation for Archean komatiites from the Barberton Mountainland, South Africa, in: Y. Fei, C.M. Bertka, B.O. Mysen (Eds.), Mantle Petrology; Field Observations and High-pressure Experimentation; A Tribute to Francis R. (Joe) Boyd, Geochem. Soc. Spec. Publ. 6 (1999) 155–167.

[61] J.C. Dann, The 3.5 Ga Komati Formation, Barberton

Greenstone Belt, South Africa, Part I: New maps and magmatic architecture, S. Afr. J. Geol. 103 (2000) 47–68.

[62] D. Ohnenstetter, W.L. Brown, Compositional variation and primary water contents of differentiated interstitial and included glasses in boninites, Contrib. Mineral. Petrol. 123 (1996) 117–137.

[63] J.E. Dixon, L. Leist, C. Langmuir, J.G. Schilling, Recycled dehydrated lithosphere observed in plume-influenced mid-ocean-ridge basalt, Nature 420 (2002) 385–389.

[64] R.J. Stern, Subduction zones, Rev. Geophys. 40 (4) (2002) 1012.

[65] T. Kawamoto, R.L. Hervig, J.R. Holloway, Experimental evidence for a hydrous transition zone in the early Earth's mantle, Earth Planet. Sci. Lett. 142 (1996) 587–592.

[66] T.L. Grove, S.W. Parman, S.A. Bowring, R.C. Price, M.B. Baker, The role of an H_2O-rich fluid component in the generation of primitive basaltic andesites and andesites from the Mt. Shasta region, N California, Contrib. Mineral. Petrol. 142 (2002) 375–396.

[67] I.S.E. Carmichael, The andesite aqueduct: perspectives on the evolution of intermediate magmatism in west-central (105–99 degrees W) Mexico, Contrib. Mineral. Petrol. 143 (2002) 641–663.

[68] S.W. Parman, T.L. Grove, J.C. Dann, The production of Barberton komatiites in an Archean subduction zone, Geophys. Res. Lett. 28 (2001) 2513–2516.

[69] P. Hollings, R. Kerrich, Trace element systematics of ultramafic and mafic volcanic rocks from the 3 Ga North Caribou greenstone belt, northwestern Superior Province, Precambrian Res. 93 (1999) 257–279.

[70] J.-L. Poidevin, Boninite-like rocks from the Palaeoproterozoic greenstone belt of Bogoin, Central African Republic; geochemistry and petrogenesis, Precambrian Res. 68 (1994) 97–113.

[71] S.-S. Sun, R.W. Nesbitt, M.T. McCulloch, Geochemistry and petrogenesis of Archaean and early Proterozoic siliceous high-magnesian basalts, in: A.J. Crawford (Ed.), Boninites and Related Rocks, Unwin and Hyman, London, 1989, pp. 148–173.

[72] R.J. Stern, S.H. Bloomer, Subduction zone infancy; examples from the Eocene Izu-Bonin-Mariana and Jurassic California arcs, Geol. Soc. Am. Bull. 104 (1992) 1621–1636.

[73] C.E. Hall, M. Gurnis, M. Sdrolias, L.L. Lavier, R.D. Muller, Catastrophic initiation of subduction following forced convergence across fracture zones, Earth Planet. Sci. Lett. 212 (2003) 15–30.

[74] W.B. Dallwitz, D.H. Green, J.E. Thompson, Clinoenstatite in a volcanic rock from Cape Vogel Area Papua, J. Petrol. 7 (1966) 375–402.

[75] R.G. Coleman, Plate tectonic emplacement of upper mantle peridotites along continental edges, J. Geophys. Res. 76 (1971) 1212–1222.

[76] V. Dietrich, R. Emmermann, R. Oberhansli, H. Puchelt, Geochemistry of basaltic and gabbroic rocks from West Mariana Basin and Mariana Trench, Earth Planet. Sci. Lett. 39 (1978) 127–144.

[77] S.M. Eggins, Petrogenesis of Hawaiian tholeiites – 1. Phase-equilibria constraints, Contrib. Mineral. Petrol. 110 (1992) 387–397.

[78] T.P. Wagner, T.L. Grove, Melt/harzburgite reaction in the petrogenesis of tholeiitic magma from Kilauea volcano, Hawaii, Contrib. Mineral. Petrol. 131 (1998) 1–12.

[79] Y. Tatsumi, Origin of high-magnesian andesites in the Setouchi Volcanic Belt, Southwest Japan 2. Melting phase-relations at high-pressures, Earth Planet. Sci. Lett. 60 (1982) 305–317.

[80] M.B. Baker, T.L. Grove, R. Price, Primitive basalts and andesites from the Mt Shasta region, N California – Products of varying melt fraction and water-content, Contrib. Mineral. Petrol. 118 (1994) 111–129.

[81] Y. Tatsumi, M. Sakuyama, H. Fukuyama, I. Kushiro, Generation of arc basalt magmas and thermal structure of the mantle wedge in subduction zones, J. Geophys. Res. 88 (1983) 5815–5825.

[82] K.S. Bartels, R.J. Kinzler, T.L. Grove, High-pressure phase-relations of primitive high-alumina basalts from Medicine Lake Volcano, Northern California, Contrib. Mineral. Petrol. 108 (1991) 253–270.

[83] T.J. Falloon, L.V. Danyushevsky, Melting of refractory mantle at 1.5, 2 and 2.5 GPa under, anhydrous and H_2O-undersaturated conditions: Implications for the petrogenesis of high-Ca boninites and the influence of subduction components on mantle melting, J. Petrol. 41 (2000) 257–283.

[84] G.F. Davies, Punctuated tectonic evolution of the Earth, Earth Planet. Sci. Lett. 136 (1995) 363–379.

[85] E.G. Nisbet, M.J. Cheadle, N.T. Arndt, M.J. Bickle, Constraining the potential temperature of the Archean mantle – a review of the evidence from komatiites, Lithos 30 (1993) 291–307.

[86] C. Herzberg, Depth and degree of melting of komatiites, J. Geophys. Res. 97 (1992) 4521–4540.

[87] S.W. Parman, N. Shimizu, T.L. Grove, J.C. Dann, Constraints on the pre-metamorphic trace element composition of Barberton komatiites from ion probe analyses of preserved clinopyroxene, Contrib. Mineral. Petrol. 144 (2003) 383–396.

[88] M.J. Bickle, C.E. Ford, E.G. Nisbet, Petrogenesis of peridotitic komatiites – evidence from high-pressure melting experiments, Earth Planet. Sci. Lett. 37 (1977) 97–106.

[89] N.T. Arndt, Melting of ultramafic lavas (komatiites) at 1 atm and high pressures, Carnegie Inst. Washington Yearb. 75 (1976) 555–557.

[90] C. Herzberg, J.Z. Zhang, Melting experiments on komatiite analog compositions at 5 GPa, Am. Mineral. 82 (1997) 354–367.

[91] E. Takahashi, T. Shimazaki, Y. Tsuzaki, H. Yoshida, Melting study of a peridotite KLB-1 to 6.5 GPa, and the origin of basaltic magmas, Phil. Trans. R. Soc. London A 342 (1993) 105–120.

[92] T.S. Kuhn, The Structure of Scientific Revolutions, University of Chicago Press, Chicago, IL, 1962, 212 pp.

[93] B.G. Aitken, L.M. Echeverria, Petrology and geochemistry of komatiites and tholeiites from Gorgona-Island, Colombia, Contrib. Mineral. Petrol. 86 (1984) 94–105.

[94] R.A. Sproule, C.M. Lesher, J.A. Ayer, P.C. Thurston, C.T. Herzberg, Spatial and temporal variations in the geochemistry of komatiites and komatiitic basalts in the Abitibi greenstone belt, Precambrian Res. 115 (2002) 153–186.

Tim Grove is a Professor of Geology at the Massachusetts Institute of Technology. He received his Bachelor's degree from the University of Colorado in 1971 and his Ph.D. degree from Harvard University in 1976. His primary research area is experimental petrology and geochemistry. A sabbatical stay with Maarten de Wit at the University of Cape Town, South Africa in 1993–194 brought him to the Barberton Mountainland, and sparked an interest in the origin of komatiites.

Stephen Parman is currently a postdoctoral researcher at the Massachusetts Institute of Technology. In 2004, he will move to Harvard University to work with Charles Langmuir, combining experimental and geochemical approaches to understanding igneous processes. He received his Bachelor's degree from Harvard University in 1994 and his Ph.D. degree from the Massachusetts Institute of Technology in 2001. Steve's interest in komatiites goes back to his senior thesis, when he was introduced to the worlds of both experimental research and Archean geochemistry by Carl Agee.

Index

Printed and bound by CPI Group (UK) Ltd, Croydon, CR0 4YY

03/10/2024

01040332-0017